Communications and Control Engineering

Communications and Control Engineering is a high-level academic monograph series publishing research in control and systems theory, control engineering and communications. It has worldwide distribution to engineers, researchers, educators (several of the titles in this series find use as advanced textbooks although that is not their primary purpose), and libraries.

The series reflects the major technological and mathematical advances that have a great impact in the fields of communication and control. The range of areas to which control and systems theory is applied is broadening rapidly with particular growth being noticeable in the fields of finance and biologically-inspired control. Books in this series generally pull together many related research threads in more mature areas of the subject than the highly-specialised volumes of *Lecture Notes in Control and Information Sciences*. This series's mathematical and control-theoretic emphasis is complemented by *Advances in Industrial Control* which provides a much more applied, engineering-oriented outlook.

Indexed by SCOPUS and Engineering Index.

Publishing Ethics: Researchers should conduct their research from research proposal to publication in line with best practices and codes of conduct of relevant professional bodies and/or national and international regulatory bodies. For more details on individual ethics matters please see:
https://www.springer.com/gp/authors-editors/journal-author/journal-author-helpdesk/publishing-ethics/14214

More information about this series at http://www.springer.com/series/61

Yury Orlov

Nonsmooth Lyapunov Analysis in Finite and Infinite Dimensions

Springer

Yury Orlov
Electronics and Telecommunication
Department
CICESE Research Center
Ensenada, Baja California, Mexico

ISSN 0178-5354 ISSN 2197-7119 (electronic)
Communications and Control Engineering
ISBN 978-3-030-37624-6 ISBN 978-3-030-37625-3 (eBook)
https://doi.org/10.1007/978-3-030-37625-3

Mathematics Subject Classification (2010): 93C10, 93C15, 93C20, 93C23, 93C30

This Springer imprint is published by the registered company Springer Nature Switzerland AG
The registered company address is: Gewerbestrasse 11, 6330 Cham, Switzerland

To loving memory of my parents
Vladimir and Anna

To my wife Marina and my daughters
Liubove and Nadezhda

And to my brother Valery

Foreword

Professor Orlov is a well-known scientist, one of the leading experts in the areas of discontinuous dynamic systems and infinite-dimensional systems. He joined my laboratory in 1979 after graduation from Math. Department of Moscow State University, the best university in the former USSR. Sliding mode control was the research area of the laboratory and the main theoretical results had been published by the end of the 70s. All of them were oriented toward finite-dimensional systems and all attempts of generalization for infinite-dimensional systems, in particular for a system governed by partial differential equations, failed. Yury demonstrated that all main concepts of the existing theory were to be revised and developed original mathematical methods for deriving sliding motion equations, analysis, and design feedback control systems. His results were published in leading international and Soviet journals.

The tracks of the research interests at the laboratory can be found in the further part of the scientific career of Prof. Orlov in Mexico oriented toward dynamic systems, which are beyond the conventional theory of differential equations and conventional stability theory.

The present monograph is primarily a research monograph on constructive Lyapunov stability tools, applicable to a broad class of discontinuous and hybrid dynamic systems of finite and infinite dimensions.

It is worth noticing that no special background is required since the proposed methodologies are introduced at an appropriate conceptual level. Reader's attention is particularly drawn to first-order and second-order sliding modes as well as to Zeno modes in variable structure systems, and to Andronov–Hopf bifurcation of the hybrid Van der Pol oscillator, impacting a unilateral constraint.

The stability theory based on Lyapunov functions may not be applicable for differential equations with non-Lipschitz continuous or discontinuous right-hand sides. The monograph does reflect this modern mainstream of capturing nonsmooth Lyapunov functions, analyzed in terms of directional Dini (super- and sub-) derivatives and contingent (hypo- and epi-) derivatives. Various kinds of (multiple, semiglobal, finite time stable, homogenous, strict and non-strict) Lyapunov functions and functionals are introduced to analyze the stability of dynamic systems, respectively of finite and infinite dimensions.

The combination of the conventional Lyapunov stability theory with the concept homogeneity is the key idea to design feedback control systems with finite-time convergence, which is not accessible for Lipschitzian systems. Both stability analysis and control design methods based on Lyapunov functions as a quadratic form plus fractional power functions of state can be found in the monograph. Robust stability and disturbance attenuation are addressed in terms of viscosity and proximal Clarke solutions of Hamilton–Jacobi–Isaacs nonlinear PDE, using the special form of Lyapunov functions.

A very interesting new result should be mentioned. Well-known Lyapunov–Razumikhin and Lyapunov–Krasovskii methods widely used for stability analysis of finite-dimensional systems were generalized for infinite-dimensional systems by Prof. Orlov and then used for nontrivial system applications (adaptive control of linear diffusion–reaction process, time-delay systems, and periodic tracking of bipedal gait).

The practical value of theoretical results of the monograph was confirmed by their use in the set of application-oriented projects: "Control of Mechanical Systems under Energy and Magnitude Constraints", "Finite Time Orbitally Stabilizing Synthesis of Complex Dynamic Systems with Bifurcations with Application to Biological Systems", "Stability Analysis and Control Synthesis of Electromechanical Systems with Friction".

Professor Orlov is recognized by the colleagues as one of the leading experts in the area of nonlinear control theory. His research achievements are always of interest that explains numerous invitations of him as a visiting professor in Belgium, France, Italy, UK, and Sweden. The prestigious journals in control are looking for his expertise as a reviewer, Prof. Orlov is a member of Editorial Boards of IEEE Transactions on Control Systems Technology, International Journal of Robust and Nonlinear Control, and IMA Journal of Mathematical Control and Information. The list can be complemented by plenary presentations at IEE/IFAC conferences, invited lectures, and short courses in many countries.

I am sure that the monograph will cause great interest of colleagues, as all previously published books by Prof. Orlov.

Ohio, USA
October 2019

Vadim Utkin
Ohio State University

Preface

The Lyapunov approach is a powerful tool for the stability analysis and feedback design of dynamic systems. Converse Lyapunov function theory guarantees the existence of Lyapunov functions for the most asymptotically stable nonlinear systems. A particular Lyapunov function, being useful in the input-to-state stabilizing feedback design, is not, however, provided by the converse theory explicitly. Although differentiable Lyapunov functions are often identified, using standard constructions, e.g., the Hamiltonian of Euler–Lagrange systems, their general construction is, in general, a challenging research topic on modern (relay, variable structure, impulsive, switched, hybrid, event-triggered) dynamic systems, for which the need of nonsmooth Lyapunov functions has well been recognized from Sontag and Sussman (1995).

A systematic construction of possibly non-differentiable Lyapunov functions forms a core of the present research monograph, focusing on variable structure systems with resets and coming with elements of a textbook on such systems. The extension to other discontinuous systems (e.g., to those mentioned above) is possible and the development is rather involved.

The monograph consists of three parts. Part I outlines relevant fundamentals for the investigation to be conducted. Chapter 1 previews discontinuous benchmark models of variable structure systems (VSS) with resets in continuous- and discrete-time perspectives. Background material and mathematical tools are presented for these models in Chap. 2 in the finite-dimensional setting and in Chap. 3 in the infinite-dimensional setting. Special attention is given to peculiar motions such as sliding modes (SMs), Zeno modes, and bifurcations under unilateral constraints that will arise in the sequel.

Part II constructs Lyapunov functions for discontinuous systems of interest. For such systems, Chap. 4 updates modern Lyapunov tools, capturing strict and nonstrict semiglobal Lyapunov functions as well as so-called input-to-state stable (ISS), finite time stable (FTS), and homogeneous Lyapunov functions. The construction of Lyapunov–Krasovskii functionals is additionally presented in a Hilbert space. Constructive Lyapunov functions are then introduced in Chap. 5 as specific (proximal Filippov) solutions of properly interpreted linear gradient Lyapunov and quadratic Hamilton–Jacobi partial differential equation/inequality (PDE/PDI), derived for discontinuous systems. Generalized quadratic Lyapunov functions are particularly introduced in terms of the augmented state vector which is composed not only of the original state components, but also of their fractional degrees. Such a generalized Lyapunov function is guaranteed to possess a negative definite time derivative, computed along the system trajectories, provided that a certain linear matrix inequality (LMI) is feasible. The feasibility of the resulting LMI yields an explicit sufficient condition of the underlying discontinuous system to be asymptotically stable. Once the system is verified to be homogeneous of a negative degree, its finite time stability is additionally established.

Attractive features of the proposed Lyapunov function constructions are illustrated in Part III. Based on the Hamilton–Jacobi PDE approach, the gains of first- and second-order sliding mode algorithms are properly tuned in Chap. 6 not only for rejecting matched external disturbances but also for attenuating mismatched ones. A remarkable extension to the boundary control of a heat process is additionally presented here. In Chap. 7, the Lyapunov redesign is applied in the infinite-dimensional setting for adaptive identification of linear distributed parameter systems (DPS) and time-delay systems (TDS). Finally, Chap. 8 advertises engineering control applications such as robust stabilization of a bipedal gait under unilateral constraints as well as the sine-Gordon energy control, constituting an academic example of controlling continuum of oscillators.

The monograph complements a recent one (Malisoff and Mazenc 2009) by Malisoff and Mazenc, which was confined to smooth Lyapunov function constructions in the finite-dimensional setting. Being emphasized on building nonsmooth Lyapunov functions for a broader class of discontinuous systems, involving distributed parameter and time-delay systems, the present work is distinct from the existing books in the area.

The presentation is given at an advanced level to be a complement to standard textbooks on nonlinear and sliding mode control, e.g., by Khalil (2002), Isidori (Isidori 1995 and 1999), and Utkin (1992). The general theoretical framework of the monograph is flavored with engineering control applications such as orbital stabilization of biped robots and academic examples of controlling distributed parameter and time-delay systems. All this makes the book reasonably complete and attractive for advanced graduate students, researchers, and practitioners, interested in the analysis and synthesis of discontinuous systems.

This work summarizes the author's experience on the Lyapunov function design during his long-term staying with the CICESE control group and it is benefitted from numerous innovative discussions the author had with leading specialists in the area. The author wishes to thank his mentor Vadim Utkin and his coauthors Joseph Bentsman, Jean-Pierre Richard, Michele Dambrine and Lotfi Belkoura, Alessandro Pisano and Elio Usai, Sarah Spurgeon, Harshal Oza, Bernard Brogliato, Alexander Poznyak and Andrey Polyakov, Laurent Autrique, Christine Chevallereau and Yannick Aoustin, Luis Aguilar and Oscar Montano, Alexandr Fradkov, and Boris Andrievsky for their contribution to forming the scope of the present monograph.

The author also appreciates the wonderful research environment at Lund University where he was on the sabbatical leave in the 2017–2018 academic year and where much of this monograph was written. The supporting collaboration and hospitality of Anders Rantzer, who kindly invited the author for spending the sabbatical year, are also greatly acknowledged.

The work was partially supported by the CONACYT grant A1-S-9270.

Lund, Sweden–Ensenada, Mexico
September 2017–October 2019

Yury Orlov

References

Isidori A (1995) Nonlinear control systems, 3rd edn. Springer, London
Isidori A (1999) Nonlinear control systems II. Springer, London
Khalil H (2002) Nonlinear systems, 3rd edn. Prentice Hall, Upper Saddle River
Malisoff M, Mazenc F (2009) Constructions of strict Lyapunov functions. Springer, London
Sontag ED, Sussman HJ (1995) Nonsmooth control-Lyapunov functions. Proceedings of the 34th IEEE Decision Control Conference 2799–2805
Utkin VI (1992) Sliding modes in control optimization. Springer, Berlin

Contents

Part I Introduction

1 Benchmark Models 3
1.1 Variable Structure Systems 3
1.1.1 First- and Higher-Order Sliding Modes 4
1.1.2 Chattering Phenomenon and Discrete-Time Sliding Modes 7
1.1.3 Infinite-Dimensional Sliding Modes 8
1.2 Dynamics Under Unilateral Constraints 9
1.2.1 Zhuravlev–Ivanov Transformation 10
1.2.2 Bouncing Ball: State Resets and Zeno Behavior 11
1.2.3 Constrained Van der Pol Oscillator: Limit Cycles and Hopf Bifurcation 12
1.3 Concluding Remarks 14
References 14

2 Mathematical Background 17
2.1 Comparison Principle and Barbalat's Lemma 17
2.2 Discontinuous and Multi-valued Vector Fields 18
2.2.1 Filippov Solutions 18
2.2.2 Equivalent Control Method and Other Solution Concepts 20
2.2.3 Ambiguous Sliding Modes 22
2.2.4 Uniqueness of Sliding Modes in Affine Systems 23
2.2.5 Regularization of Discontinuous Systems in Hilbert Space 24
2.3 Complementarity Formulation of Constrained Lagrange Dynamics 28
2.3.1 Implicit Euler Integration of Sliding Modes 29
2.4 Hopf Bifurcation of Discontinuous Limit Cycles: Case Study 32

2.4.1 Constrained Van der Pol Oscillator 34
2.4.2 Existence of a Constrained Limit Cycle 34
2.4.3 Numerical Analysis of Phenomenological Behaviors 38
2.4.4 Hopf Bifurcation Analysis via Poincaré Method 39
2.4.5 Constrained Van der Pol Oscillator with Manipulated Parameters 43
2.5 Concluding Remarks 44
References 44

3 Mathematical Tools of Dynamic Systems in Hilbert Spaces 45
3.1 Sobolev Spaces and Instrumental Inequalities 45
3.2 Linear Partial Differential Equations 48
3.2.1 Linear Differential Operators 48
3.2.2 Parabolic, Elliptic, and Hyperbolic Operators 49
3.2.3 Green Function and Mild Solutions 50
3.2.4 Weak Solutions 51
3.3 Sturm–Liouville Operators and Their Properties 52
3.3.1 Eigenvalue Estimates 54
3.3.2 Uniform Boundedness of the Eigenfunctions 58
3.4 Separation of Variables 61
3.4.1 Parabolic Case Study 61
3.4.2 Hyperbolic Case Study 66
3.5 Nonlinear First-Order Partial Differential Equations 68
3.5.1 Viscosity Solutions of First-Order PDEs 68
3.5.2 Discontinuous Strict Hamilton–Jacobi Inequality and Its Proximal Solutions 71
3.6 Stability in Euclidean and Hilbert Spaces 74
3.6.1 Abstract Dynamic Systems and Relevant Stability Concepts 74
3.6.2 Robust Stability of Uncertain Dynamic Systems: Basic Definitions 76
3.6.3 Sliding Mode Dynamics in Hilbert Space 79
3.6.4 Hilbert Space-Valued Dynamics with Delay 81
3.6.5 Homogeneous Differential Inclusions and Their Finite Time Stability 83
3.7 Concluding Remarks 89
References 90

Part II Construction of Nonsmooth Lyapunov Functions

4 Modern Lyapunov Tools 95
4.1 Strict Lyapunov Functionals 95
4.1.1 Multiple Lyapunov Functionals 98
4.1.2 Semi-global Lyapunov Functionals 99

4.1.3 Finite Time Stable Lyapunov Functionals 100
4.1.4 Homogeneous Lyapunov Functions 102
4.1.5 Input-to-State Stable Lyapunov Functions 103
4.2 Non-strict Lyapunov Functionals 104
4.2.1 Invariance Principle 105
4.2.2 Invariance Principle Extension 107
4.3 Lyapunov Functionals Under Unilateral Constraints 108
4.4 Concluding Remarks 109
References 110

5 Control Lyapunov Functions 113
5.1 Lyapunov Algebraic Equation and Quadratic Forms 113
5.2 Generalized Forms 115
5.2.1 Semiglobal Strict Lyapunov Functions of Twisting VSS 115
5.2.2 Strict Lyapunov Functions of Supertwisting VSS 128
5.2.3 GF Lyapunov Functions of Homogeneous Systems and Their LMI-Based Construction 132
5.3 Construction of Multiple FTS Lyapunov Functions via Solving Lyapunov Gradient Equation 135
5.4 Lyapunov Minmax Approach and Speed Gradient Method 136
5.5 Construction of Lyapunov Functions Using Proximal Solutions of Hamilton–Jacobi PDI 140
5.6 Concluding Remarks 145
References 145

Part III Lyapunov Redesign

6 Lyapunov-Based Tuning 149
6.1 $\mathscr{L}_2$-Gain Tuning of First-Order Sliding Modes 149
6.1.1 Tuning Under Full-State Information 151
6.1.2 Tuning of SM Estimator Gains 154
6.1.3 Tuning Under Incomplete State Information 161
6.2 $\mathscr{L}_2$-Gain Tuning of Second-Order Sliding Modes 162
6.2.1 Tuning of Twisting Controller 162
6.2.2 Tuning of Supertwisting Estimator 165
6.2.3 Output Feedback Tuning 167
6.3 Settling Time Tuning of Enforced Double Integrator 170
6.3.1 Switched Control Synthesis 171
6.3.2 Reaching Time Estimate of Linear Feedback 172
6.3.3 Settling Time Estimate of Twisting Controller 178
6.3.4 Settling Time Tuning 185

6.4 ISS Point-Wise Feedback Synthesis of Parabolic Systems 189
6.4.1 Control Synthesis 191
6.4.2 Existence of Closed-Loop Solutions 192
6.4.3 ISS Analysis and Tuning 196
6.4.4 Supporting Simulation 200
6.5 Concluding Remarks 203
References 203

7 Lyapunov Approach to Adaptive Identification and Control in Infinite-Dimensional Setting 205
7.1 Lyapunov–Razumikhin Redesign and Identification of Linear Time-Delay Systems 206
7.1.1 State-Space Representation and Weak Controllability 207
7.1.2 Identifiability Analysis 210
7.1.3 Razumikhin-Based Adaptive Identifier Design 219
7.1.4 SISO Case Study 225
7.1.5 Application to Engine Transient Fuel Identification 237
7.2 Lyapunov–Krasovskii Redesign and Identification of Linear Parabolic PDE 246
7.2.1 Identification over In-Domain Sensing 247
7.2.2 Identification over Boundary Sensing 261
7.2.3 Simulation Results 278
7.3 Concluding Remarks 279
References 280

8 Control Applications 283
8.1 Synthesis of Mechanical Systems Under Unilateral Constraints 283
8.1.1 Robust Tracking Problem and Hybrid Error Dynamics 284
8.1.2 Pre-feedback Design and $\mathscr{H}_\infty$ Synthesis 287
8.1.3 $\mathscr{H}_\infty$-Control of Mass–Spring–Barrier System 294
8.1.4 $\mathscr{H}_\infty$ Tracking of a Periodical Bipedal Gait 307
8.2 Energy Control of Continuum of Oscillators 311
8.2.1 Sine-Gordon Nonlinear PDE Model and Problem Statement 312
8.2.2 Control Objective 314
8.2.3 Energy Control Synthesis Using State Feedack 315
8.2.4 Luenberger Observer Design 320
8.2.5 Energy Control Synthesis Using Output Feedback 324
8.2.6 Numerical Study 327

8.3 Concluding Remarks 331
Appendix A: Proof of Theorem 8.4 331
Appendix B: Proof of Theorem 8.6 335
References 335

Index 339

Abbreviations

CLF	Control Lyapunov functions
DOF	Degree of freedom
DPS	Distributed parameter system
FTS	Finite time stable/stability
FOSM	First-order sliding mode
GF	Generalized form
HOSM	Higher-order sliding mode
ISS	Input-to-state stable/stability
iISS	integral ISS
LMI	Linear matrix inequality
MRAC	Model reference adaptive control
MIMO	Multi-input-multi-output
ODE	Ordinary differential equation
PDE	Partial differential equation
PDI	Partial differential inequality
PD	Proportional derivative
SISO	Single-input-single-output
SM	Sliding Mode
SOSM	Second-order sliding mode
TDS	Time-delay system
VSS	Variable structure system
ZOH	Zero-order holder

Part I
Introduction

The primary concern of the monograph is the construction of Lyapunov functions for discontinuous (possibly, distributed parameter and time delay) dynamic systems. Various kinds of discontinuous systems are well-recognized from the modern control theory such as relay, impulsive, variable structure, switched, hybrid, and event-triggered systems to name a few.

The present investigation focuses on VSS with potential SMs and state resets, capturing essential features of the general discontinuous systems. Throughout, such systems are regarded as the ones composed of a family of continuous vector fields which are separated by switching manifolds and which are equipped with a state restitution rule applied at time instants of hitting a switching manifold. After hitting a switching manifold, the system dynamics undergo discontinuities to either cross it or to evolve along the manifold for a while. A motion that occurs on a switching manifold is typically referred to as the SM.

In Part I, basic models and complex phenomena of discontinuous systems of interest are previewed, and the mathematical background of these systems is revised.

Chapter 1
Benchmark Models

Nowadays, different paradigms are used to deal with systems, exhibiting discontinuities due to, e.g., changes in their environment or collisions with physical constraints. The present study focuses on VSS with SMs and resets. To begin with, first- and higher-order sliding modes (FOSM and HOSM) are conceptually introduced in continuous- and discrete-time perspectives. After that, an undesired chattering phenomenon is discussed. SM dynamics in the infinite-dimensional setting are exemplified next. Finally, specific modes such as state resets, Zeno behavior, and Hopf bifurcation are featured to occur in the presence of the constraints. While dealing with certain impact dynamics, the Zhuravlev–Ivanov transformation is invoked for avoiding the state resets of the transformed vector field which turns out to be discontinuous (rather than impulsive), even if the original field is continuous.

1.1 Variable Structure Systems

The state-space representation is the standard framework of VSS whose operation domain G, belonging to a Euclidean space $\mathbb{R}^n$, is normally partitioned into a finite number of connected subdomains $G_j \subset \mathbb{R}^n$, $j = 1, \ldots, N$ with disjoint interiors and boundaries ∂G_j of measure zero. Within each of these subdomains, an autonomous VSS is governed by an ordinary differential equation (ODE)

$$\dot{x} = f_j(x), \; x \in G_j, \; j = 1, \ldots, N \tag{1.1}$$

with a continuous vector field $f_j(x)$, possessing a finite limit $f_j(x)$ as the state argument $x \in G_j$ approaches a boundary point $x^* \in \partial G_j$.

The overall vector field is thus represented by a discontinuous state function $f(x)$ on the entire operation domain G because the different vector fields are generally speaking distinct of each other on their mutual boundaries. A *nonautonomous VSS model*

Y. Orlov, *Nonsmooth Lyapunov Analysis in Finite and Infinite Dimensions*,
Communications and Control Engineering,
https://doi.org/10.1007/978-3-030-37625-3_1

$$\dot{x} = f_j(x,t),\ x \in G_j,\ j = 1,\ldots,N \tag{1.2}$$

with time-varying vector fields $f_j(x,t) : G_j \times \mathbb{R} \to \mathbb{R}^n,\ j = 1,\ldots,N$ is readily reproduced from the autonomous one by viewing its time variable as a state component with trivial dynamics $\dot{t} = 1$.

Due to the potential presence of sliding modes along the boundaries ∂G_j, $j = 1,\ldots,N$ of the operation subdomains, a VSS may or may not inherit specific features of its individual subsystems. For instance (Utkin 1992), being orchestrated by unstable subsystems, the resulting VSS may be capable of generating asymptotically stable dynamics, and vice versa. Thus, the overall VSS analysis is far from being a trivial extension of that of the individual subsystems, and it should certainly involve sliding mode dynamics. The subsequent overview of VSS and SM features is extracted from Utkin et al. (2020).

1.1.1 First- and Higher-Order Sliding Modes

Starting from the original use, the term of sliding mode was associated with discontinuities in motion equations and with trajectories, belonging to the corresponding discontinuity manifold. However, an attempt to apply such a sliding mode concept to discretized systems turned out to be unsuccessful because it resulted in chattering oscillations around (rather than sliding along) the discontinuity manifold. An alternative approach is in discretizing the underlying system in such a manner that a similar behavior occurs at an integral manifold in finite time (Acary and Brogliato 2010; Drakunov and Utkin 1989). It is the latter feature adopted in this book to determine the SM concept.

Postponing a discussion on the meaning of the ODE (1.1) for a while, let us assume that in a domain $G \in \mathbb{R}^n$, the underlying VSS (1.1) possesses absolutely continuous solutions, which are differentiable almost everywhere. Let $x(x_0,t)$ denote such a solution, initialized with $x(0) = x_0$. The formal SM definition is then as follows (Drakunov and Utkin 1992).

Definition 1.1 A nonempty subset $D \subset S^p$ of p-dimensional manifold $S^p \subset \mathbb{R}^n$, $0 \le p < n$ is said to be a *sliding domain* of (1.1) if it is locally finite-time attractive, i.e., if any $y \in D$ possesses $\varepsilon > 0$ such that once $x_0 \in B_\varepsilon(y) = \{x \in \mathbb{R}^n : \|x_0 - y\| < \varepsilon\}$, there exists $T = T(x_0) \ge 0$ for which $x(x_0,T) \in D$ and for which $T(x_0) \to 0$ as $x_0 \to y$. Plots of the trajectories of (1.1), evolving within the sliding domain D, are then referred to as *sliding modes of codimension* $r = n - p$ provided that there is no manifold $S^q \subset \mathbb{R}^n$ of lower dimension $q < p$ such that $D \subset S^q$.

Apart from VSS (1.1), the above definition appears to be applicable to both discrete-time systems and to non-Lipschitzian continuous-time systems whose integral manifolds are reachable in finite time. It is also worth noticing that for the well-known FOSMs and HOSMs, the aforementioned SM codimension concept appears to be relevant to that of the SM order which is roughly speaking defined (see Shtessel et al. 2014 for the rigorous definition) as the relative degree of the sliding variable with respect to the singular input. These and other aspects of Definition 1.1 are further illustrated by simple examples.

Example 1.1 The simplest FOSM example is given by the first-order ODE

$$\dot{x}(t) = -\text{sign}\, x(t) + d(t), \quad x(0) = x_0 \tag{1.3}$$

with a discontinuous right-hand side where $t, x(t), d(t) \in \mathbb{R}$ are, respectively, the time variable, instant state, and unknown disturbance such that

$$|d(t)| \leq d_0 < 1 \quad \text{for all } t \geq 0. \tag{1.4}$$

For $x(t) \neq 0$ one has

$$\frac{d}{dt}|x(t)| \leq -(1 - d_0) < 0, \tag{1.5}$$

thereby concluding that any solution $x(t)$ reaches the discontinuity surface $x = 0$ of codimension 1 in finite time $T(x_0) \leq \frac{|x_0|}{1-d_0}$ and $\lim_{x_0 \to 0} T(x_0) = 0$. By virtue of (1.5), the state trajectory cannot escape from the surface $x = 0$ after the time instant $T(x_0)$. Thus, due to the infinite frequency of switching of the signum function the trivial FOSM occurs in the origin $x = 0$ regardless of whichever disturbance (1.4) affects the system.

Example 1.2 The first-order ODE

$$\dot{y} = -\sqrt{|y|}\text{sign}\, y \tag{1.6}$$

with a continuous right-hand side is topologically equivalent to the disturbance-free ($d \equiv 0$) ODE (1.3) of Example 1.1. Indeed, the continuously invertible state transformation $x = 2\sqrt{|y|}\text{sign}\, y$, being differentiated on solutions of (1.6), results in the disturbance-free version $\dot{x} = -\text{sign}\, x$ of (1.3) which is finite-time stable. Thus, (1.6) represents a *continuous system with FOSM* on the SM manifold $y = 0$ of co-dimension 1. Clearly, the continuous system (1.6) is no longer capable of producing the sliding mode $y = 0$ under uniformly bounded, but nonvanishing disturbances $d(t)$ because its right-hand side is nullified at $y = 0$ and the time derivative $\frac{d}{dt}|y(t)| = -\sqrt{y} + d(t)\text{sign}\, y$ of the norm $|y(\cdot)|$, computed on the trajectories of the perturbed system $\dot{y} = -\sqrt{y}\text{sign}\, y + d(t)$, is not negative definite anymore.

Example 1.3 Figure 1.1 illustrates the FOSM that occurs in the second-order ODE

$$\ddot{x}(t) = -\text{sign}\, s(t), \;\; s = \dot{x} + cx, \;\; c \neq 0 \tag{1.7}$$

within the sliding domain $(m, n) \subset S^1 = \{(x, \dot{x}) \in \mathbb{R}^2 : \dot{x} + cx = 0\}$ which is specified with $m = -n = (-c^{-2}, c^{-1})$ to ensure that $\frac{d}{dt}|s(t)| = -1 + c\dot{x}\, \text{sign}\, s < 0$. The

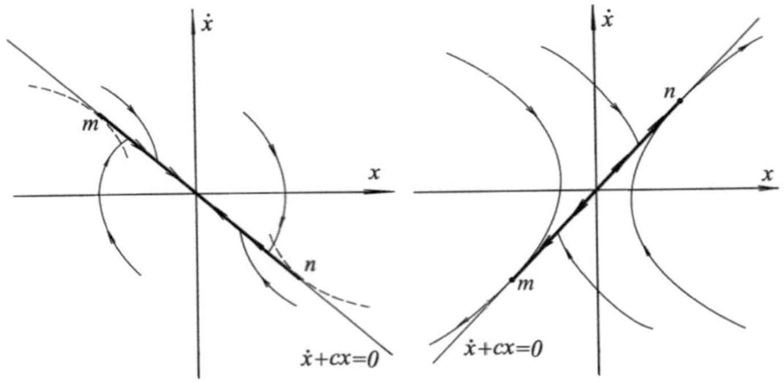

(a) Stable FOSM ($c > 0$). (b) Unstable FOSM ($c < 0$).

Fig. 1.1 First-order sliding modes

latter inequality stands to prevent the system dynamics of crossing the sliding domain as it happens for sufficiently large state values, located on the switching surface $s = 0$ of codimension 1 outside the interval (m, n). Actually, the SM is of the first order because system (1.7) is of the relative degree 1 with respect to the sliding variable s and the sliding motion $\dot{x} + cx = 0$ is stable for $c > 0$ (Fig. 1.1a) and it is unstable for $c < 0$ (Fig. 1.1b).

Example 1.4 (*Fuller phenomenon*) The optimal control law for the second-order system

$$\ddot{s}(t) = u(s(t), \dot{s}(t)), \quad t > 0, \quad s(t), u(t) \in \mathbb{R}, \quad |u| \le 1 \tag{1.8}$$

with the criterion $\int_0^\infty s^2(t)dt$ is well-known (Zelikin and Borisov 1994) to be the switching state function

$$u(s, \dot{s}) = -\text{sign } S(s, \dot{s}), \quad S(s, \dot{s}) = s + c\dot{s}^2 \text{sign } \dot{s}. \tag{1.9}$$

The optimal constant parameter $c > 0$ is such that the closed-loop dynamics, schematically depicted in Fig. 1.2, cross the switching line $S = 0$ at a countably infinite number of points, located within a finite time interval. The time intervals between successive switches decrease following a geometric progression and as a result produce a finite accumulation point, known as *Zeno behavior*. The existence of a finite accumulation point of the optimal switching time instants, discovered by Fuller (1961), means the finite-time state convergence to the origin where there appears a SM of codimension 2. Since system (1.8) is of relative degree 2 with respect to the position s, it generates a second-order SM (SOSM). Remarkably, Fuller effect disappears for high (nonoptimal) values of c because just in the case, the state trajectory slides along the parabolic switching curve $S = 0$ of codimension 1 similar to that of Example 1.3, cf. Fig. 1.1a.

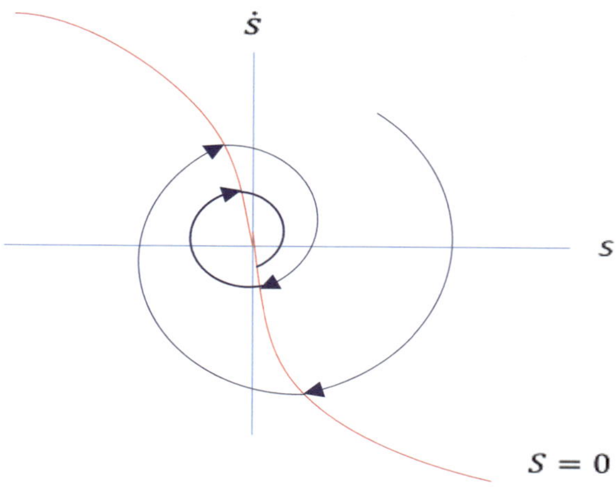

Fig. 1.2 Fuller phenomenon with SOSM in the origin

1.1.2 Chattering Phenomenon and Discrete-Time Sliding Modes

It is well-known that SMs are deformed if before enforcing the underlying system, a discontinuous input (e.g., the relay signal $u = -\text{sign}\, x(t)$ in Example 1.1) passes through a dynamic (low-pass) filter. Just in the case, SMs are transformed into high-frequency oscillations in the vicinity of the discontinuity surface. Such a motion is extremely undesired in practice and it is typically referred to as a *chattering* mode.

The digital implementation of SMs with a finite sampled frequency is another source of the generation of chattering in a vicinity of the discontinuity surface. Chattering-free SMs can, however, arise in discrete-time approximations of continuous-time systems provided that an implicit Euler approximation is applied. For instance, SMs occur in the nonlinear multi-input-multi-output (MIMO) system

$$x_{k+1} = F(x_k) + B(x_k)u_k, \quad x_k \in \mathbb{R}^n, \quad u_k \in \mathbb{R}^m \tag{1.10}$$

on the linear surface $s(x) = Cx = 0$ with a nonsingular matrix $CB(x_k)$ if the input signal u_k is set to

$$u_k = -\left(CB(x_k)\right)^{-1} CF(k, x_k) \tag{1.11}$$

to ensure that $s(x_{k+1}) = 0$ for any x_k.

It is straightforward to verify that the input signal (1.11) makes $s(x)$ equal to zero after one step. It is not realistic, however, to implement (1.11) for small values of the sampling period δ, if the discrete system (1.10) is an approximation of its continuous-time counterpart

$$\dot{x}(t) = f(x(t)) + b(x(t))u(t), t > 0. \tag{1.12}$$

Just in the case, $F(x) = x + \delta f(x)$, $B(x) = \delta b(x)$, $s(x) = Cx$, and

$$u = -\frac{1}{\delta}[Cb(x)]^{-1} s(x) - [Cb(x)]^{-1} Cf(x). \tag{1.13}$$

Given $s(x) \neq 0$, the value of (1.13) becomes infinitely large with δ decaying to zero. It means that the proposed law (1.13) should be modified to obey realistic constraints of any real-life system. Under the norm bound $\|u\| \leq M = const$ the constraint is satisfied with

$$u = \begin{cases} u_k & \text{if } \|u_k\| \leq M \\ M\frac{u_k}{\|u_k\|} & \text{if } \|u_k\| > M \end{cases} \tag{1.14}$$

where u_k is given by (1.11). The saturated input (1.14) is collinear to (1.11) and its norm does not exceed the admissible value M.

The formal definition of SMs in the discrete-time system

$$x_{k+1} = F(x_k), \quad F : \mathbb{R}^n \to \mathbb{R}^n \tag{1.15}$$

is inspired from Definition 1.1 of continuous-time SMs and it is summarized as follows (Drakunov and Utkin 1989).

Definition 1.2 In the discrete-time system (1.15), it is said that a SM takes place on a subset Σ of a manifold $s(x) = 0$, $s : \mathbb{R}^n \to \mathbb{R}^m$, $m < n$, if there exists an open neighborhood U of this subset such that $s(F(x)) \in \Sigma$ for any $x \in U$.

As opposed to continuous-time SMs, which are admitted by non-Lipschitz vector fields only, discrete-time SMs can arise even in linear systems (1.10), (1.11) with $F(x) = Ax$ and $B(x) = B$ where A, B, C are matrices of appropriate dimensions and CB is nonsingular.

1.1.3 Infinite-Dimensional Sliding Modes

A simple infinite-dimensional SM example

$$\dot{x} = -\frac{x}{\|x\|} + h(x, t) \tag{1.16}$$

is given in terms of the state evolution $x(t)$ in a real Hilbert space H. Potential solutions of (1.16) are affected by an additive dynamic disturbance $h \in H$, which is assumed to be uniformly bounded $\|h(x, t)\| \leq \varepsilon$ with a norm bound $\varepsilon < 1$ and of class C^1 in its arguments. The vector field $-x/\|x\|$ of the disturbance-free system (1.16) is discontinuous in the origin and it is smooth outside so that classical solutions of the Hilbert space-valued equation (1.16) are well locally posed beyond the origin.

Since the norm $\|x\| = \sqrt{\langle x, x \rangle}$ in the Hilbert space is defined via the inner product $\langle \cdot, \cdot \rangle$, differentiating the functional

$$V(x(t)) = \|x(t)\|^2 \tag{1.17}$$

on the solutions of (1.16), initialized beyond the origin, yields

$$\dot{V}(x(t)) = \frac{d\|x(t)\|^2}{dt} = 2\langle x(t), \dot{x}(t)\rangle \le -2(1-\varepsilon)\|(x(t)\| = -2(1-\varepsilon)\sqrt{V(t)}. \tag{1.18}$$

By the well-known comparison principle (Khalil 2002, p. 102) (see Lemma 2.1 of the next chapter), an arbitrary solution of the scalar differential inequality (1.18) is dominated

$$V(t) \le V_0(t) \; for \; all \; t \ge 0 \tag{1.19}$$

by a solution of its equality counterpart

$$\dot{V}_0(t) = -2(1-\varepsilon)\sqrt{V_0(t)}, \tag{1.20}$$

provided that both are equally initialized with

$$V_0(0) = V(0) = \sqrt{\|x(0)\|}. \tag{1.21}$$

Solving the differential equation (1.18) analytically, it follows that $V_0(t) = 0$ for all $t \ge (1-\varepsilon)^{-1}\|x^0\|$, and by virtue of (1.19), $V(t)$ vanishes after the same time instant $T(x^0) = \|x^0\|$, independent of an admissible disturbance h and such that $T(x^0) \to 0$ as $x^0 \to 0$. Thus, all solutions of (1.16), initiated beyond the origin, steer to the origin in finite time and due to (1.18) they never leave the origin. Hence, in the infinite-dimensional system (1.16), there appears a sliding mode on the trivial surface $x = 0$ after the time instant $T(x_0) = \|x^0\|$. Clearly, this sliding mode is unambiguously set by the surface equation $x = 0$ regardless of whichever additive dynamic disturbance $\|h(x,t)\| \le \varepsilon < 1$ affects the system.

1.2 Dynamics Under Unilateral Constraints

While operating under a unilateral time-varying constraint

$$F(x,t) \ge 0, \tag{1.22}$$

both autonomous VSS dynamics (1.1) and nonautonomous ones (1.2) are to be enforced to instantaneous state resets

$$x(t+) = \sigma(x(t-), t) \tag{1.23}$$

at impact time instants of hitting the constraint. Hereinafter, $F(x,t) : \mathbb{R}^{n+1} \to \mathbb{R}$ and $\sigma(x,t) : \mathbb{R}^{n+1} \to \mathbb{R}^n$ are continuous functions, $x(t-) = \lim_{s\to t-} x(s), x(t+) = \lim_{s\to t+} x(s)$, and for certainty, $x(t) = x(t+)$. The solutions of such VSS with impacts can be defined, based on the Filippov concept and using existing methods (see Brogliato 2016; Monteiro-Marques 1993; Moreau 1988; Stewart 2000 for the solution concept definition via differential inclusions with both friction and collisions terms on the right-hand side). The resulting discrete behavior at impact instants is subsequently exemplified by specifying the restitution rule (1.23).

1.2.1 Zhuravlev–Ivanov Transformation

The *Zhuravlev–Ivanov transformation* (Ivanov 1993; Zhuravlev 1978) constitutes a powerful analysis tool of unilaterally constrained systems because it results in VSS which are well-posed, an attribute absent for the original impact systems (Stewart 2000). The transformation is applicable to n degree-of-freedom (n-DOF) mechanical systems with unilateral (generally speaking, inelastic) constraints of codimension one (Brogliato 2016, Sect. 1.4.3). For transparency, it is illustrated here for a simple constrained double integrator (1-DOF system)

$$\dot{x}_1 = x_2, \tag{1.24a}$$

$$\dot{x}_2 = \omega(x_1, x_2, t), \tag{1.24b}$$

$$x_1 \geq 0, \tag{1.24c}$$

$$x_2(t_k+) = -e\, x_2(t_k-), \quad \text{if} \quad x_2(t_k-) < 0, x_1(t_k) = 0, \quad i = 0, 1, \ldots \tag{1.24d}$$

where the scalar state components x_1, x_2 are the position and the velocity, respectively, the external force $\omega(x_1, x_2, t)$ is piecewise continuous, t_k is the kth jump time instant where the velocity undergoes a reset or jump, $e \in (0, 1)$ is the (elasticity) parameter, responsible for the loss of energy. The equalities (1.24a) and (1.24b) represent the continuous dynamics without jumps in velocity. The inequality (1.24c) represents the unilateral constraint on the position x_1 which evolves in a domain with a boundary. It is assumed that the jump event occurs instantaneously within an infinitesimally small time and hence (Brogliato 2016) mathematically can be represented by *Newton's restitution rule*, given by (1.24d).

For the constrained double integrator (1.24), the basic idea behind the nonsmooth Zhuravlev–Ivanov coordinate transformation

$$x_1 = |s|, x_2 = R\, v \,\mathrm{sign}\,(s) \tag{1.25}$$

is to mimic the actual position $x_1(t)$ on each segment $[t_{2k}, t_{2k+1}]$, $k = 0, 1, \ldots$, symmetrically with respect to the origin, and to specify the complement coordinate v with the multiplier R in such a manner

$$R = 1 - \frac{1-e}{1+e} \,\mathrm{sign}\,(s\, v) \tag{1.26}$$

that along with the "mirror" position $s(t)$, its complement $v(t)$ remains everywhere continuous. In terms of the new coordinates, the transformed dynamics are thus jump free, and being straightforwardly computed from (1.24a) and (1.24b), they are represented in the VSS form

$$\begin{aligned} \dot{s} &= R\, v \\ \dot{v} &= R^{-1}\text{sign}\,(s)\omega(|s|, Rv\text{sign}\,(s), t). \end{aligned} \tag{1.27}$$

It should be noted that by virtue of the state transformation (1.25), (1.26), the origin $s = v = 0$ of the transformed system (1.27) corresponds to the origin $x_1 = x_2 = 0$ of the to-be-transformed system (1.24). Although the Zhuravlev–Ivanov transformation (1.25) is not invertible, the original dynamics are nevertheless recovered via (1.25). Moreover, there is a one-to-one correspondence between the two coordinate systems at the origin and the stability analysis can, therefore, be performed in the transformed coordinates. Another benefit of the Zhuravlev–Ivanov transformation (1.27) is that both sliding modes (to possibly occur in the original system due to, e.g., dry friction) and collision phenomena (caused by the unilateral constraint) are adsorbed into a unique framework of the resulting VSS. The VSS reformulation (1.27) is also capable of capturing the so-called *Zeno behavior*, coming with infinite rebounds, similar to those generated by a bouncing ball.

1.2.2 Bouncing Ball: State Resets and Zeno Behavior

The dimensionless model

$$\ddot{x} = -1 \tag{1.28}$$

of the free motion of a unit-mass bouncing ball is deduced from the Newton mechanics for the height x and velocity $\dot{x}$ of the ball by normalizing the gravitational constant (Liberzon 2003, Sect. 1.1.2). Because of the ground constraint $x \geq 0$, the ball changes its velocity when it hits the constraint $x(t) = 0$. The Newton restitution rule

$$\dot{x}(t+) = -r\dot{x}(t-) \tag{1.29}$$

with an elasticity parameter $r \in (0, 1)$ is typically involved to adequately describe the instantaneous velocity change.

Initializing the ball with $x(0) = 0$ and $\dot{x}(0) = 1$, its free motion (1.28) is integrated to

$$\dot{x}(t) = -t + 1, \quad x(t) = -\frac{1}{2}t^2 + t. \tag{1.30}$$

It follows that $x(2) = 0$ at the time instant $t = 2$, when the ball hits the constraint for the first time with the pre-impact velocity $\dot{x}(2-) = -1$. The post-impact velocity $\dot{x}(2+) = r$ is then computed in accordance with (1.29).

With the state reinitialization, thus determined, the free motion (1.28) after the first impact-time instant $t = 2$ is then integrated to

$$\dot{x}(t) = -t + 2 + r, \quad x(t) = -\frac{1}{2}(t-2)^2 + (t-2)r, \quad t \geq 2. \tag{1.31}$$

The latter integration figures out that the next impact-time instant occurs at $t = 2 + 2r$ when $x(2 + 2r) = 0$ and the post-impact velocity takes the value $\dot{x}(2 + 2r) = r^2$.

By iterating on these successive integrations, it is concluded that the impact time instants form the sequence

$$t_k = 2\Sigma_{i=1}^{k} r^{i-1}, \; k = 1, 2, \dots \tag{1.32}$$

possessing a finite accumulation point

$$T = 2\Sigma_{i=1}^{\infty} r^{i-1} = \frac{2}{r-1}, \tag{1.33}$$

which is the double sum of the geometric progression $\{r^i\}_{i=1}^{\infty}$. Thus, the ball makes infinitely many bounces before the time instant T, while its state approaches the equilibrium $x(t_k) = 0, \; \dot{x}(t_k) = r^k \to 0$ as $t_k \to T$. Similar to the Fuller phenomenon of Example 1.4, the dynamics of the bouncing ball exhibit the Zeno behavior.

1.2.3 *Constrained Van der Pol Oscillator: Limit Cycles and Hopf Bifurcation*

The Van der Pol oscillator, governed by the second-order nonlinear differential equation

$$\ddot{x} + \varepsilon[x^2 - \rho^2]\dot{x} + \mu^2 x = 0 \tag{1.34}$$

with positive parameters ε, ρ, μ is of fundamental value in nonlinear oscillation theory. It possesses (see, e.g., Khalil 2002) a stable limit cycle that attracts all other solutions except a unique equilibrium point, being the origin $(x, \dot{x}) = (0, 0)$. With the value ε, escaping to zero, the nonlinear Van der Pol oscillator degenerates to the linear one, which possesses the center (non-asymptotically stable equilibrium) and harmonic orbits of amplitude and frequency dependent on the initial conditions. Such a phenomenon of degenerating a stable limit cycle under small parameter variations to a (possibly, asymptotically) stable equilibrium is referred to as Poincaré–Andronov–Hopf bifurcation, or simply Hopf bifurcation (Hale and Koçak 1991, p. 208).

To reshape the generated limit cycle geometry to an appropriate form of an ellipse or of a circle, in particular (see Fig. 1.3 for a typical phase trajectory of the Van der Poll oscillator vs. that of its reshaped modification), the following Van der Pol modification

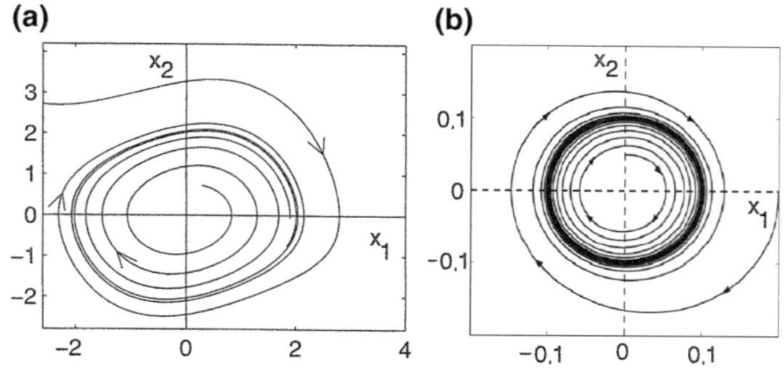

Fig. 1.3 **a** Phase trajectories of the Van der Pol Oscillator (1.34) with $\rho = 1$, $\mu = 1$, $\varepsilon = 0.2$. **b** Phase trajectories of the modified Van der Pol Oscillator (1.35) with $\rho = 0.01$, $\mu = 1$, $\varepsilon = 1000$ (Reprinted from Herrera et al. 2017, Copyright 2017, with permission from Elsevier)

$$
\begin{aligned}
\dot{x}_1 &= x_2, \\
\dot{x}_2 &= -\varepsilon\left[\left(x_1^2 + \frac{x_2^2}{\mu^2}\right) - \rho^2\right]x_2 - \mu^2 x_1
\end{aligned}
\tag{1.35}
$$

is due where the state vector $x = (x_1, x_2)^T$ consists of the position x_1 of the oscillator and its velocity x_2. It is worth noticing that by rescaling the time variable $t = \mu^{-1}\tau$ and the state variables $x_1 = \rho y_1$, $x_2 = \rho\mu y_2$, the modified Van der Pol oscillator (1.35) is re-parameterized with the only parameter $\alpha = \varepsilon\rho^2\mu^{-1}$. Clearly, once represented in terms of the new state variables y_1 and $y_2 = dy_1/d\tau$ where the differentiation is made with respect to the artificial time variable τ, the dynamics of the re-scaled state vector $y = (y_1, y_2)^T$ are governed by

$$
\begin{aligned}
\frac{dy_1}{d\tau} &= y_2, \\
\frac{dy_2}{d\tau} &= -\alpha[(y_1^2 + y_2^2) - 1]y_2 - y_1, \quad \alpha = \frac{\varepsilon\rho^2}{\mu} > 0.
\end{aligned}
\tag{1.36}
$$

As shown in Orlov et al. (2004), the proposed modification (1.35) still belongs to a class of dumped systems, its limit cycle, which inherits the stability property from its original version, is explicitly governed by the ellipse equation

$$
x_1^2 + \frac{x_2^2}{\mu^2} = \rho^2, \tag{1.37}
$$

and it is remarkably generated by harmonic oscillations

$$
\dot{x}_1 = x_2, \; \dot{x}_2 = -\mu^2 x_1 \tag{1.38}
$$

initialized on ellipse (1.37). Thus, while being initialized outside the origin, the modified Van der Pol oscillator (1.35) produces stable harmonic oscillations of magnitude ρ and frequency μ, with the transient limit cycle speed (damping) ε. Due to the above features, the Van der Pol modification (1.35) has become extremely suited for its use in the model reference adaptive control (Orlov et al. 2008; Roup and Bernstein 2001) where the desired magnitude and frequency of the resulting oscillation are on-line manipulatable.

Similar to the modified Van der Pol oscillator (1.35), its hybrid version, operating under the unilateral constraint

$$x_1 \geq 0 \tag{1.39}$$

subject to the elastic velocity restitution rule

$$x_2(t^+) = -ex_2(t^-) \text{ iff } x(t) = 0 \tag{1.40}$$

with the restitution parameter $e \in (0, 1)$, possesses an equilibrium in the origin that is straightforwardly verified by inspection. As shown in Herrera et al. (2017), such a constrained oscillator (1.35), (1.39), (1.40) is capable of generating a stable limit cycle as well and it exhibits the Hopf bifurcation under certain variations of the oscillator parameters. Thus, in contrast to the unconstrained Van der Pol oscillator, exhibiting the Hopf bifurcation phenomenon only when the transient speed (damping) parameter ε is degenerated to $\varepsilon = 0$, its constrained counterpart bifurcates to a stable limit cycle under yet positive damping $\varepsilon > 0$; details are postponed for Sect. 2.4.

1.3 Concluding Remarks

Peculiar VSS behaviors with potential state resets such as HOSMs, chattering, Zeno modes, and bifurcations have been illustrated. Special mathematical tools, addressing such peculiarities, are presented next to compose the background material of the subsequent construction of nonsmooth Lyapunov functions of uncertain VSS with resets.

References

Acary V, Brogliato B (2010) Implicit Euler numerical scheme and chattering-free implementation of sliding mode systems. Syst Control Lett 59:284–293

Brogliato B (2016) Nonsmooth mechanics, 3rd edn. Springer, London

Drakunov S, Utkin V (1989) On discrete-time sliding modes. In: Proceedings of the IFAC conference on nonlinear control system design, pp 273–278

Drakunov S, Utkin V (1992) Sliding mode control in dynamic systems. Int J Control 55(4):1029–1037

Fuller A (1961) Relay control systems optimized for various performance criteria. In: Proceedings of the 1st IFAC triennial world congress, vol 1, pp 510–519
Hale JK, Koçak H (1991) Dynamics and bifurcations. Springer, New York
Herrera L, Montano O, Orlov Y (2017) Hopf bifurcation of hybrid Van der Pol oscillators. Nonlienar Analysis: Hybrid Systems 26:225–238
Ivanov AP (1993) Analytical methods in the theory of vibro-impact systems. J Appl Math Mech 57:221–236
Khalil H (2002) Nonlinear systems, 3rd edn. Prentice Hall, Upper Saddle River
Liberzon D (2003) Switching in systems and control. Birkhäuser, Boston
Monteiro-Marques M (1993) Differential inclusions in nonsmooth mechanical problems: shocks and dry friction. Birkhäuser, Boston
Moreau JJ (1988) Unilateral contact and dry friction in finite freedom dynamics. In: Moreau JJ, Panagiotopoulos PD (eds) Nonsmooth mechanics and applications. CISM courses and lectures. Springer, Wien
Orlov Y, Acho L, Aguilar L (2004) Quasihomogeneity approach to the pendubot stabilization around periodic orbits. In: Proceedings of the 2nd IFAC symposium on systems, structure and control, pp 448–453
Orlov Y, Aguilar L, Acho L (2008) Asymptotic harmonic generator and its application to finite time orbital stabilization of a friction pendulum with experimental verification. Int J Control 81:227–234
Roup AV, Bernstein DS (2001) Adaptive stabilization of a class of nonlinear systems with nonparametric uncertainty. IEEE Trans Auto Control 46:1821–1825
Shtessel Y, Edwards C, Fridman L, Levant A (2014) Sliding mode control and observation. Birkhäuser, New York
Stewart DE (2000) Rigid-body dynamics with friction and impact. SIAM Rev 42:3–39
Utkin VI (1992) Sliding modes in control and optimization. Springer, Berlin
Utkin V, Poznyak A, Orlov Y, Polyakov A (2020) Road map for sliding mode design (to be published)
Zelikin M, Borisov V (1994) Theory of chattering control with applications to cosmonautics, robotics, economics, and engineering. Birkhäuser, Boston
Zhuravlev VF (1978) Equations of motion of mechanical systems with ideal onesided links. J Appl Math Mech 42:839–847

Chapter 2
Mathematical Background

Differential equations (inclusions) with discontinuous right-hand side and complementarity formulation of impact dynamics are adopted as mathematical models of VSS with resets. For later use, mathematical tools of these systems are revisited in ODE settings. First, useful instrumental tools of Barbalat's lemma and comparison principle are reproduced. After that, various solution concepts such as Filippov solutions and Utkin equivalent control method are introduced for finite-dimensional discontinuous (multi-valued) vector fields, and physical sense behind them is then revealed. Implicit Euler integration is additionally involved in numerical simulations of such vector fields to preserve sliding motions for their discrete approximations. Apart from this, the Poincaré method is invoked to illustrate the Poincaré–Andronov–Hopf bifurcation of impact dynamics, generating stable limit cycles.

2.1 Comparison Principle and Barbalat's Lemma

To begin with, popular instrumental tools, which are particularly invoked in the stability analysis, are recalled for later use.

Lemma 2.1 (Comparison Principle) *Consider the scalar differential equation*

$$\dot{v} = f(v, t), \quad v(t_0) = v_0, \tag{2.1}$$

where $f(v, t)$ is continuous in t and locally Lipschitz in v for all $t \in \mathbb{R}$ and all $v \in J \subset \mathbb{R}$. Let t_0, T ($T > t_0$ is finite or infinite) be the maximal interval of existence of the solution $v(t)$, and let $v(t) \in J$ for all $t \in [t_0, T)$. Along with this, suppose that $V(t)$ is a continuous function, whose upper right-hand derivative $D^+V(t)$ satisfies the differential inequality

$$D^+V(t) \leq f(V(t), t), \quad V(t_0) \leq v_0 \tag{2.2}$$

Y. Orlov, *Nonsmooth Lyapunov Analysis in Finite and Infinite Dimensions*, Communications and Control Engineering,
https://doi.org/10.1007/978-3-030-37625-3_2

with $V(t) \in J$ *for all* $t \in [t_0, T)$. *Then,* $V(t) \leq v(t)$ *for all* $t \in [t_0, T)$.

Lemma 2.2 (Barbalat's Lemma) *Let* $\phi(t) : \mathbb{R} \to \mathbb{R}$ *be a uniformly continuous function on* $[0, \infty)$. *Suppose that* $\lim_{t\to\infty} \int_0^t \phi(\tau)d\tau$ *exists and is finite. Then,* $\phi(t) \to 0$ *as* $t \to \infty$.

For the proofs of the above results, the interested reader may respectively refer to Khalil (2002, Lemmas 3.4 and 8.2).

2.2 Discontinuous and Multi-valued Vector Fields

The underlying time-varying VSS is determined by a piecewise continuous vector field $f(x, t) : \mathbb{R}^{n+1} \to \mathbb{R}^n$, and it is governed by

$$\dot{x} = f(x, t), \ x \in \mathbb{R}^n, \ t \in \mathbb{R}. \tag{2.3}$$

A function $f : \mathbb{R}^{n+1} \to \mathbb{R}^n$ is said to be piecewise continuous iff $\mathbb{R}^{n+1}$ consists of a finite number of open connected subdomains $G_j \subset \mathbb{R}^{n+1}$, $j = 1, 2, \ldots, N$ with disjoint interiors and the boundary set $\partial G = \bigcup_{i=1}^{N} \partial G_j$ of measure zero such that $f(x, t)$ is continuous in each G_j up to the boundary.

The meaning of the ODE (2.3) is well-posed only within any subdomain $G_j \subset \mathbb{R}^{n+1}$, $j = 1, 2, \ldots, N$ where $f(x, t)$ remains continuous, whereas it should additionally be specified on the boundary set ∂G where the vector field undergoes discontinuities and therefore it cannot uniquely be determined.

2.2.1 *Filippov Solutions*

A standard approach to overcome the mathematical obstructions of the discontinuous ODE (2.3) is due to Filippov (1988). The main idea of the Filippov approach is to embed the ill-posed ODE (2.3) into a well-posed differential inclusion, which coincides with the original ODE in any continuity subdomain.

The Filippov procedure involves the differential inclusion

$$\dot{x} \in F(x(t), t), \ x \in \mathbb{R}^n, \ t \in \mathbb{R}, \tag{2.4}$$

generated by a set-valued function $F : \mathbb{R}^{n+1} \to 2^{\mathbb{R}^n}$, which is specified with

$$F(x, t) = \bigcap_{\varepsilon>0} \bigcap_{\mu(N)=0} \operatorname{co} f\left(x + \varepsilon B \backslash N, t\right), \tag{2.5}$$

where $B = \{x \in \mathbb{R}^n : \|x\| \leq 1\}$ is the unit ball in $\mathbb{R}^n$, $\mu(N)$ stands for the Lebesgue measure of a set N, and $\text{co}(M)$ is for the *convex* hull of a set M.

Definition 2.1 (*Filippov solution*) An absolutely continuous function $x : \mathbb{R} \to \mathbb{R}^n$, defined on some interval I, is said to be a *Filippov solution* of (2.3) if it satisfies the differential inclusion (2.4), (2.5) almost everywhere on I.

It is worth noticing that the Filippov solutions are defined in the conventional Carethéodory sense whenever they are within a continuity subdomain G_j, $j = 1, 2, \ldots, N$ where the Filippov multiple set (2.5) is nothing else than the singleton $F(x, t) = f(x, t)$.

It is also reasonable to comment on why a set of zero measure should be ruled out in the solution continuation procedure (2.5). The velocity vector on the discontinuity set ∂G can be found in the system context only. Attempts to rule it beyond this context would be contradictive as seen from the 1-DOF mechanical system $m\ddot{x} = -F_0 \text{sign}\,\dot{x}(t) - p(t)$. If such a system is affected by an external force p of magnitude $p_0 = \sup_{t\geq} |p(t)|$ less than the Coulomb level $F_0 > p_0$ of the dry friction $-F_0\text{sign}\,\dot{x}(t)$ than a trivial SM occurs on the discontinuity surface $\dot{x} = 0$ where the velocity is nullified. Once $\dot{x}(t) \equiv 0$ it is clear that $F_0\text{sign}\,\dot{x}(t) = p(t)$ so that the dry friction value is far from being arbitrary at an equilibrium.

It is well-known (Filippov 1988) that if a set-valued function F is locally convex and compact, and upper semicontinuous around an initial state

$$x(t_0) = x_0 \tag{2.6}$$

the Cauchy problem (2.4), (2.6) possesses at least one local solution. It is particularly the case if such a function is generated by the Filippov procedure (2.5) with a piecewise continuous function $f(x, t)$. The resulting multi-valued function $F(x, t)$ is then a convex polyhedron (2.5) (particularly, a segment or a point).

The Filippov procedure is subsequently illustrated for a function $f(x, t)$ which is piecewise continuous in x and continuous in t. Let $f(x, t)$ undergoes discontinuities on a smooth surface $S = \{x \in \mathbb{R}^n : s(x) = 0\}$, separating the state domain $\mathbb{R}^n$ on two subdomains $G_1 = \{x \in \mathbb{R}^n : s(x) > 0\}$ and $G_2 = \{x \in \mathbb{R}^n : s(x) < 0\}$, and the boundary ∂G.

Then the Filippov set (2.5) is specified to a segment, connecting the vectors

$$f^+(x, t) = \lim_{\tilde{x} \to x\,:\,\tilde{x} \in G_1} f(\tilde{x}, t), \;\; f^-(x, t) = \lim_{\tilde{x} \to x\,:\,\tilde{x} \in G_2} (\tilde{x}, t).$$

Provided that this segment crosses the plane $P(x)$, which is tangential to the surface S at a point $x \in S$, the Filippov solutions of (2.3) slide along the surface S. The *sliding mode equation* $\dot{x} = f^0(x, t)$ on this surface is then derived from (2.4) with the function

$$f^0(x, t) = \frac{\langle grad\, s(x), f^-(x, t)\rangle f^+(x, t) + \langle grad\, s(x), f^+(x, t)\rangle f^-(x, t)}{\langle grad\, s(x), f^+(x, t) - f^-(x, t)\rangle}$$

being such a Filippov velocity vector, which is determined by a cross-point of the segment and the plane $P(x)$. Thus, $f^0(x, t) = \mu(x, t) f^+(x, t) + [1 - \mu(x, t)]f^-(x, t)$ where $\mu(x, t) \in [0, 1]$ is such that the gradient vector $grad\ s(x)$ is orthogonal to $f^0(x, t)$, i.e.,

$$\langle grad\, s(x), \mu(x, t) f^+(x, t) + [1 - \mu(x, t)] f^-(x, t)\rangle = 0.$$

Otherwise, $grad\ s(x)$ is not orthogonal to $\mu f^-(x, t) + (1 - \mu) f^+(x, t)$ for any $\mu \in [0, 1]$ and the Filippov solutions of (2.4) cross the surface, thus yielding an isolated "switching" of the right-hand side of (2.3).

In general, the Filippov velocity vector on a sliding manifold is constructed in the form

$$f^0(x, t) = \sum_{j=1}^{N} \mu_j(x, t) f^j(x, t),$$

where nonnegative functions $\mu_j(x, t) : \mathbb{R}^{n+1} \to \mathbb{R}$ should be determined according to the Carethéodory convex hull theorem in such a way that

$$\sum_{j=1}^{N} \mu_j(x, t) = 1$$

and the vector $f^0(x, t)$ belongs to the tangential space to a sliding manifold at the point $(x, t) \in \mathbb{R}^{n+1}$.

2.2.2 Equivalent Control Method and Other Solution Concepts

For controlled systems of the form

$$\dot{x} = f(x, u(x, t), t) \tag{2.7}$$

with a continuous (in all arguments) right-hand side $f = (f_1, \ldots, f_n)^T$ and a piecewise continuous input function $u = (u_1, \ldots, u_m)^T$, an alternative approach of defining the velocity vector on a discontinuity set is based on the *equivalent control method*, proposed by Utkin (1992). Let the components $u_i(x, t)$, $i = 1, \ldots, m$ of the control input $u(x, t)$ undergo discontinuities on possibly intersecting smooth surfaces $S_i = \{x \in \mathbb{R}^n : s_i(x) = 0\}$ and vary within the segments $U_i(x, t) = [u_i^-(x, t), u_i^+(x, t)]$ where

$$u_i^-(x, t) = \lim_{(\tilde{x}, t) \in S_i^-,\ \tilde{x} \to x} u_i(\tilde{x}, t), \quad S_i^- = \{(x, t) \in \mathbb{R}^{n+1} : s_i(x) < 0\}$$

and

$$u_i^+(x,t) = \lim_{(\tilde{x},t)\in S^+,\ \tilde{x}\to x} u_i(\tilde{x},t),\ \ S_i^+ = \{(x,t) \in \mathbb{R}^{n+1} : s_i(x) > 0\}.$$

According to the equivalent control method, SMs of (2.7) are defined by means of the following multi-valued function

$$U(x,t) = \bigcap_{\varepsilon>0} \bigcap_{\mu(N)=0} \operatorname{co} u\left(x + \varepsilon B \backslash N, t\right), \tag{2.8}$$

resulting in the right-hand side

$$F_U(x,t) = f(x, U(x,t), t), \tag{2.9}$$

which is clearly distinct from the multi-valued function (2.5) of Definition 2.1 to be specified in the present case to

$$F(x,t) = \bigcap_{\varepsilon>0} \bigcap_{\mu(N)=0} \operatorname{co} f(x, u\left(x + \varepsilon B \backslash N, t\right), t). \tag{2.10}$$

Definition 2.2 (*Utkin solution*) An absolutely continuous function $x : \mathbb{R} \to \mathbb{R}^n$, defined on some interval I, is said to be an *Utkin solution* of (2.7) if there exists a locally measurable function $u^{eq} : \mathbb{R} \to \mathbb{R}^m$ such that $u^{eq}(t) \in U(t, x(t))$ and

$$\dot{x}(t) = f(x(t), u^{eq}(t), t) \tag{2.11}$$

almost everywhere on I.

It is clear that $u^{eq}(t) = u(t, x(t))$ for $x \notin \cup_{i=1}^m S_i$ and it remains to define $u^{eq}(t)$ on the discontinuity manifold only. Definition 2.2 yields the following procedure:

$$\begin{array}{lll} \langle grad\, s_i(x), f(x, u^{eq}, t)\rangle = 0 & \text{if} & x \in S_i, \\ u_i^{eq} = u_i(x,t) & \text{if}\ x \notin S_i, & i = 1, 2, \ldots, m \end{array} \tag{2.12}$$

of determining $u^{eq} \in \mathbb{R}^m$ as a solution of nonlinear algebraic equations. Such a solution $u^{eq}(t, x)$ is called *equivalent control* (Utkin 1992) if it belongs to $U(x, t)$.

The existence of Utkin solutions is reduced to analysis of the corresponding differential inclusion in accordance with the lemma, extracted from Filippov (1962, page 78).

Lemma 2.3 *Consider a continuous function $f : \mathbb{R}^{n+m+1} \to \mathbb{R}^n$ and its set-valued counterpart $U(x,t) : \mathbb{R}^{n+1} \to 2^{\mathbb{R}^m}$, governed by* (2.8)*. Let $U(x,t)$ be upper semi-continuous on an open set $\Omega \times I$, where $\Omega \subseteq \mathbb{R}^n$, and let $F_U(x,t)$, given by* (2.9)*, be non-empty, compact and convex for every $(x,t) \in \Omega \times I$. Suppose that an absolutely continuous function $x(t) \in \Omega$ for $t \in \mathscr{I}$ satisfies the differential inclusion*

$$\dot{x}(t) \in F_U(x(t), t) \tag{2.13}$$

for almost all $t \in I$. *Then there exists a measurable function* $u^{eq}(t) \in U(x(t), t)$ *such that* (2.11) *holds almost everywhere on* I.

In general, the differential inclusion (2.13) may be non-convex. Since the existence of solutions for such a non-convex differential inclusion is far from being trivial (Bartolini and Zolezzi 1985), the algebraic approach, resulting from the equivalent control method, is the most effective way of finding Utkin solutions.

To avoid dealing with non-convex differential inclusions Aizerman and Pyatnitskii involved (Aizerman and Pyatnitskii 1974) the following convex multi-valued function

$$F_{AP}(x,t) = \mathrm{co} f(x, U(x,t), t), \quad t \in \mathbb{R} \tag{2.14}$$

obtained by applying the Filippov convex hull construction (2.5) to the Utkin multi-valued function (2.9).

Definition 2.3 (*Aizerman–Pyatnitskii solution*) An absolutely continuous function $x : \mathbb{R} \to \mathbb{R}^n$, defined on some interval I, is said to be an *Aizerman–Pyatnitskii solution* of (2.7) if almost everywhere on I, it satisfies the differential inclusion

$$\dot{x}(t) \in F_{AP}(x,t) \tag{2.15}$$

specified with (2.14).

Clearly, both Filippov and Utkin solutions represent particular cases of Aizerman–Pyatnitskii solutions.

2.2.3 Ambiguous Sliding Modes

The Filippov and Utkin solution concepts, generally speaking, determine distinct sliding motions whereas Aizerman–Pyatnitskii approach may lead to an ambiguous behavior in itself. To better realize which sliding motion occurs in a real-life system one should view a detailed system description at a microscopic level, admitting the uniqueness of the system dynamics. This claim is additionally exemplified with a non-affine VSS (Utkin 1992, p. 35), generating both Filippov and Utkin dynamics, dependent on which way the nonlinear relay is implemented.

Example 2.1 Consider the VSS

$$\dot{x}_1 = 0.3x_2 + x_1 u, \; \dot{x}_2 = -0.7x_1 + 4x_1 u^3 \tag{2.16}$$

with the relay input

$$u = \begin{cases} 1 & \text{if } (x_1 + x_2)x_1 < 0 \\ -1 & \text{if } (x_1 + x_2)x_1 > 0 \end{cases}. \tag{2.17}$$

The switching line $x_1 = 0$ does not contain any system trajectories, which are straightforwardly verified to cross this line. Contrariwise, SMs occur in system (2.16), (2.17) on the switching line $x_1 + x_2 = 0$, which attracts the system dynamics, pointing towards each other from both sides of the line.

The equivalent control value is then obtained from the first relation of (2.12), which is now specified to $(-1 + u^{eq} + 4(u^{eq})^3) = 0$. Substituting the resulting equivalent control value $u^{eq} = 0.5$ into (2.16) for u yields the unstable equation $\dot{x}_1 = 0.2x_1$, governing Utkin solutions on the switching line $x_1 + x_2 = 0$.

The Filippov vector field $x_1 = -0.1x_1$, obtained by specifying the Filippov velocity (2.10) for the present system, proves to be asymptotically stable.

It is worth noticing that both Filippov SMs and Utkin SMs are described by reduced-order equations because they are confined to the one-dimensional state subspace $x_1 + x_2 = 0$. Clearly, the Aizerman–Pyatnitskii vector field (2.14) in the present example is also confined to the one-dimensional state subspace $x_1 + x_2 = 0$. The Aizerman–Pyatnitskii SMs on the switching line $x_1 + x_2 = 0$ are thus governed by the differential inclusion $\dot{x}_1 \in [-0.1, 0.2]x_1$, representing both stable and unstable dynamics.

The resulting SM dynamics have long been recognized to match to the Filippov solutions for the relay input with hysteresis and to the Utkin solutions for the saturated high gain amplifier. In Utkin (1992), this claim is theoretically validated for general VSS.

Apart from the hysteresis-based and saturated high gain approximations of the relay input, other kinds of VSS *regularization*, e.g., involving unmodeled dynamics and switching with delay, can be considered for defining SMs. In general, such a regularization replaces the underlying discontinuous input $u(x, t)$ by a new one $u^\delta(x, t)$ such that it is rather close

$$\|u^\delta(x,t) - u^\delta(x,t)\| \le \delta \quad \delta > 0 \tag{2.18}$$

to $u(x, t)$ outside the δ-layer $s(x)\| \le \delta$ of the discontinuity manifold $s(x) = 0$ for all $t \in \mathbb{R}$ and all $x \in \mathbb{R}^n$, and such that the conventional solution uniqueness–existence theorems are applicable to the regularized system. Then the set of all possible SM equations (2.4), taking into account the regularization, can be found in terms of the limiting solutions $u^\delta(x, t)$ as the positive parameter δ escapes to zero.

2.2.4 Uniqueness of Sliding Modes in Affine Systems

The Aizerman–Pyatnitskii multi-valued vector field (2.14) generalizes the Filippov procedure (2.5) towards the controlled VSS (2.7) and it relies on the Filippov observation that if in a vicinity of a discontinuity surface, the velocity vector of (2.7) switches between several values with possibly an infinite frequency then it belongs to the minimal convex hull of these values and no other motions can be generated.

Once the Utkin and Filippov procedures (2.9) and (2.10) generate a unique velocity vector, the Aizerman–Pyatnitskii multi-valued vector field (2.14) is degenerated to the same velocity singleton, thereby ensuring the well-posedness of SMs. This is indeed the case of a class of affine VSS (2.7) of the form

$$f(x, u(x,t), t) = g(x,t) + b(x,t)u(x,t) \tag{2.19}$$

provided that

$$\det\left[\frac{\partial s(x)}{\partial x} b(x,t)\right] \neq 0 \quad \forall x \in S, \ \forall t \in \mathbb{R}, \tag{2.20}$$

where $s(x) = (s_1(x), s_2(x), \ldots, s_m(x))^T$ and $S = \{x \in \mathbb{R} : s(x) = 0\}$ is the discontinuity set of $u(x,t)$. The following result is extracted from Utkin (1992).

Theorem 2.1 *Consider an affine VSS* (2.7), (2.19), *specified with continuous functions* $g : \mathbb{R}^{n+1} \to \mathbb{R}^n$ *and* $b : \mathbb{R}^{n+1} \to \mathbb{R}^{n\times m}$, *and piecewise continuous function* $u : \mathbb{R}^{n+1} \to \mathbb{R}^m$ *such that each component* $u_i(x,t)$, $i = 1, \ldots, m$ *of* $u(x,t)$ *undergo discontinuities on the corresponding time-invariant switching surface* $s_i(x) = 0$ *of class* C^1. *Then under condition* (2.20), *any Filippov solution constitutes an Utkin solution and vise versa, any Utkin solution constitutes a Filippov solution.*

Due to Theorem 2.1, the sliding motion equation (2.11), governing both Utkin and Filippov (and hence, Aizerman–Pyatnitskii) solutions of the affine system (2.7), (2.19), is uniquely determined by substituting the equivalent control function

$$u^{eq}(x,t) = -\left[\frac{\partial s(x)}{\partial x} b(x,t)\right]^{-1}\left[\frac{\partial s(x)}{\partial x} g(x,t)\right] \tag{2.21}$$

analytically extracted from (2.12), into (2.11) for $u(x,t)$. The resulting ODE

$$\dot{x} = g(x,t) - \left[\frac{\partial s(x)}{\partial x} b(x,t)\right]^{-1}\left[\frac{\partial s(x)}{\partial x} g(x,t)\right] \tag{2.22}$$

governs any (Utkin, Filippov, Aizerman–Pyatnitskii) SMs, occurring in the system in question, and it proves to be well-posed under the conditions of Theorem 2.1. The SMs well-posedness is thus established for the affine VSS (2.7), (2.19) provided that condition (2.20) is satisfied.

2.2.5 Regularization of Discontinuous Systems in Hilbert Space

In order to address SM dynamics in the infinite-dimensional setting the boundary-layer regularization procedure of Sect. 2.2.3 is further developed for a differential equation

$$\dot{x} = Ax + f(x,t) + bu(x,t), \ t > 0, \ x(0) = x^0 \in \mathscr{D}(A), \tag{2.23}$$

where the state variable $x(t)$ and the input signal $u(x,t)$ evolve in Hilbert spaces H and U, respectively. Hereinafter, the infinitesimal operator A with domain $\mathscr{D}(A)$ generates a strongly continuous semigroup $T_A(t)$ on H, the operator function $f(x,t)$ with values in H is of class C^1 in all arguments, and b is a linear bounded operator, acting from U to H.

It is said (Curtain and Zwart 1995) that a family $\{T(t)\}_{t\geq 0}$ of linear bounded operators $T(t), \ t \geq 0$ forms a strongly continuous semigroup on a Hilbert space H if the identity $T(t+\tau) = T(t)T(\tau)$ is satisfied for all $t, \tau \geq 0$, and the functions $T(t)x$ are continuous with respect to $t \geq 0$ for all $x \in H$. Just in the case, the induced operator norm $\|T(t)\|$ of the semigroup satisfies the inequality $\|T(t)\| \leq \omega e^{\beta t}$, $t \geq 0$ with some growth bound β and some $\omega > 0$.

The domain of an operator A, generating a strongly continuous semigroup, is well-known (Curtain and Zwart 1995) to form the Hilbert space $\mathscr{D}(A)$ with the graph inner product

$$\langle x, y \rangle_{\mathscr{D}(A)} = \langle x, y \rangle_H + \langle Ax, Ay \rangle_H, \ x, y \in \mathscr{D}(A)$$

defined by means of the inner product $\langle \cdot, \cdot \rangle_H$ of the underlying Hilbert space H. It is worth noticing that $\mathscr{D}(A) \hookrightarrow H$ ($\mathscr{D}(A)$ is continuously embedded into H), i.e., it is dense in H and the inequality $\|x\|_H \leq \omega_0 \|x\|_{\mathscr{D}(A)}$ holds for all $x \in \mathscr{D}(A)$ and some positive constant ω_0. If β is a growth bound of the semigroup, then given $\lambda > \beta$, there holds $(A - \lambda I)^{-1} H = \mathscr{D}(A)$ where I is the identity operator, and the norm of $x \in \mathscr{D}(A)$ given by $\|(A - \lambda I)x\|_H$ is equivalent to the graph norm $\|x\|_{\mathscr{D}(A)}$ of $\mathscr{D}(A)$. Particularly, $\|x\|_{\mathscr{D}(A)} = \|Ax\|_H$ if A possesses a growth bound $\beta < 0$.

If the input function u is as smooth as the system nonlinearity f, the above equation locally possesses a unique strong solution $x(t)$, whose standard definition is as follows.

Definition 2.4 A continuous function $x(t)$, defined on $[0, T)$, is said to be a strong solution of the initial-value problem (2.23) with a continuously differentiable input $u(x,t)$ iff $\lim_{t \downarrow 0} \|x(t) - x^0\|_H = 0$, and $x(t)$ is continuously differentiable and satisfies the equation for $t \in (0, T)$.

The precise meaning of the solutions of (2.23), corresponding to piecewise continuously differentiable inputs, is defined as a limiting result obtained through the boundary-layer regularization procedure, similar to that proposed in Sect. 2.2.3 for the finite-dimensional VSS.

Let the input $u(x,t)$ be continuously differentiable beyond a linear manifold

$$cx = 0 \tag{2.24}$$

with $c \in \mathcal{L}(H, S)$ being a linear bounded operator from H to some Hilbert space S, and let $u(x, t)$ undergo discontinuities on this manifold. Then the strong solutions of (2.23) are only considered whenever they are beyond the discontinuity manifold (2.24) whereas in a vicinity of this manifold, the original system is replaced by a related system, which takes into account all possible imperfections in the new input function $u^\delta(x, t)$ (e.g., delay, hysteresis, and saturation) and for which there exists a strong solution. A generalized solution of the Hilbert space-valued system (2.23) is then obtained by making the characteristics of the new system approach those of the original one.

Although beyond the discontinuity manifold the meaning of the initial value problem is confined to strong solutions of (2.23), the treatment of the case where such a solution is admitted to be in a mild sense as a solution to a corresponding integral equation, is rather involved. Similar to the finite-dimensional case, a motion along the discontinuity manifold is referred to as a SM, and it is rigorously introduced by invoking the *generalized solution* concept from Orlov (2009).

Let the subspace

$$H_1 = ker\ c = \{x_1 \in H : cx_1 = 0\} \subseteq H$$

complement a subspace $H_2 \subseteq H$ in the sense that

$$H = H_1 \oplus H_2.$$

Then the discontinuity manifold (2.24) takes the form $x_2 = 0$, if written in terms of the new coordinates $x_1(t) = P_1 x(t) \in H_1$ and $x_2(t) = P_2 x(t) \in H_2$ where $P_i,\ i = 1, 2$ is the projector on the subspace H_i and $A_i = A|_{H_i}$ the operator restriction on H_i. SMs on $x_2 = 0$ are the generalized solutions of (2.23), which are defined as follows (Orlov 2009).

Definition 2.5 An absolutely continuous function $x^\delta(t)$, defined on some interval $[0, \tau)$, is said to be an *approximate δ-solution* of system (2.23) if it is a strong solution of

$$\dot{x}^\delta = Ax^\delta + f(x^\delta, t) + bu^\delta(x^\delta, t) \tag{2.25}$$

with some $u^\delta(x, t)$ such that

$$\|u^\delta(x, t) - u(x, t)\| \le \delta \tag{2.26}$$

for all $t \ge 0$ and for all $x = (x_1, x_2) \in H = H_1 \oplus H_2$ subject to $\|x_2\|_{\mathcal{D}(A_2)} \ge \delta$.

Definition 2.6 An absolutely continuous function $x(t)$, defined on some interval $[0, \tau)$, is said to be a *generalized solution* of system (2.23) if there exists a family of approximate δ-solutions $x^\delta(t)$ of the system such that

$$\lim_{\delta \to 0} \|x^\delta(t) - x(t)\|_{\mathcal{D}(A)} = 0\ uniformly\ in\ t \in [0, \tau).$$

By definition, SMs are in general nonunique and involve all possible approximations of the discontinuous input within a boundary-layer of the discontinuity manifold. As in the finite-dimensional setting, an equivalent control value $u^{eq}(x,t)$, maintaining the system motion on this manifold, is imposed by the original system itself.

For describing SMs in the infinite-dimensional system, their dynamics (2.23) are represented in terms of the variables $x_1(t) \in H_1$ and $x_2(t) \in H_2$:

$$\dot{x_1} = A_{11}x_1 + A_{12}x_2 + f_1(x_1, x_2, t) + b_1 u(x_1, x_2, t), \; x_1(0) = x_1^0, \tag{2.27}$$

$$\dot{x_2} = A_{21}x_1 + A_{22}x_2 + f_2(x_1, x_2, t) + b_2 u(x_1, x_2, t), \; x_2(0) = x_2^0, \tag{2.28}$$

where $A_{ij} = P_i A_j, \; i, j = 1, 2$ are the operators from H_j to H_i, and $f_i = P_i f$, $b_i = P_i b$. Provided that the operator b_2 is non-singular and the inverse operator b_2^{-1} is bounded there exists a unique solution of the algebraic equation $A_{21}x_1 + f_2(x_1, 0, t) + b_2 u(x, t) = 0$ with respect to u. This solution

$$u^{eq}(x_1) = -b_2^{-1}[A_{21}x_1 + f_2(x_1, 0, t)] \tag{2.29}$$

is accepted as the equivalent control value because (2.29) is the only input that ensures the identity $\dot{x}_2 = 0$ on the discontinuity manifold $x_2 = 0$, and under appropriate initial conditions, it maintains system (2.23) on the manifold $x_2 = 0$.

Setting $\tilde{A} = A_{11} - b_1 b_2^{-1} A_{21}$ and $f_0(x_1, x_2, t) = f_1(x_1, x_2, t) - b_1 b_2^{-1} f_2(x_1, x_2, t)$, the SM equation

$$\dot{x} = \tilde{A}x_1 + f_0(x_1, 0, t) \tag{2.30}$$

governing the system motion on the discontinuity manifold $cx = 0$, is obtained by substituting the equivalent control value (2.29) into (2.28) for u. The equivalent control method is thus straightforwardly extended to the Hilbert space-valued dynamics (2.23) under the following assumptions:

(A1) The linear operator b is bounded and its projection b_2 on the subspace H_2 is continuously invertible, i.e., the operator b_2^{-1} from H_2 to U is bounded, too;

(A2) The infinitesimal operators A and $\tilde{A} = A_{11} - b_1 b_2^{-1} A_{21}$ generate strongly continuous semigroups $T_A(t), t \geq 0$ and $T_{\tilde{A}}(t), t \geq 0$ on the Hilbert spaces H and H_1, respectively;

(A3) The system nonlinearity $f(x, t)$ is everywhere continuously differentiable in (x, t) and it satisfies the linear growth condition in x;

(A4) The input function $u(x, t)$ is everywhere continuously differentiable beyond the discontinuity manifold $x_2 = 0$;

(A5) The operator $G_0 = \tilde{A} b_1 b_2^{-1}$ from H_2 to H_1 is governed by A_{12} in the sense that $\mathscr{D}(G_0) \subseteq \mathscr{D}(A_{12})$ and $\|G_0 y\| \leq K\|A_{12}y\|$ for all $y \in \mathscr{D}(G_0)$ and some $K > 0$.

Assumption (A1) is inherited from the finite-dimensional case to ensure the SM uniqueness. Assumptions (A2)–(A4), coupled to Assumption (A1), ensure the existence and uniqueness of local strong solutions of (2.23) beyond the discontinuity manifold $x_2 = 0$. Moreover, these assumptions ensure that the equivalent control value (2.30), relying on the operator function $f_0(x_1, x_2, t) = f_1(x_1, x_2, t) - b_1 b_2^{-1} f_2(x_1, x_2, t)$, reads Assumption (A4), too, and the existence and uniqueness of local strong solutions of the sliding mode equation (2.30) are then also guaranteed. Assumption (A5) is intrinsic for infinite-dimensional systems and if it would fail to hold other generalized solutions, not governed by (2.30), could appear. The following result (Orlov 2009, Theorem 2.4) is in force.

Theorem 2.2 *Consider the dynamic system* (2.23) *under Assumptions (A1)–(A5). Let the system, being initialized in the discontinuity manifold* $x_2 = 0$ *with* $x^0 = (x_1^0, 0) \in H$, *start evolving in this manifold on some time interval* $[0, \tau)$. *Then the initial-value problem* (2.23) *possesses a unique generalized solution* $x(t)$, *and on the time interval* $[0, \tau)$ *this solution, starting from* $x_1(0) = x_1^0$ *is governed by the SM equation* (2.30).

The interested reader may refer to Orlov (2009) for the detailed proof of Theorem 2.2. An alternative Filippov-flavored approach to the meaning of discontinuous systems in the infinite-dimensional setting may be found in Levaggi (2002a, b).

2.3 Complementarity Formulation of Constrained Lagrange Dynamics

Constrained Lagrange dynamics

$$M(q)\ddot{q} + G(q, \dot{q}, t, \lambda) = 0 \tag{2.31}$$

$$F(q(t), t) \geq 0, \lambda(t) \geq 0, \; \lambda^T(t)(t) F(q(t), t) = 0 \tag{2.32}$$

of the generalized coordinates $q(\cdot) \in \mathbb{R}^n$ form an important subclass of VSS with resets of Sect. 1.2. The dynamic description (2.31) relies on the inertia matrix $M(\cdot) : \mathbb{R}^{n+1} \to \mathbb{R}^{n \times n}$, which is typically everywhere differentiable, symmetric, and nonnegative definite. Along with this, it relies on the term $G(\cdot) : \mathbb{R}^{2n+l+1} \to \mathbb{R}^n$, which accounts for the external forces, including friction and contact reaction, where the contact reaction is represented through a time-varying Lagrange multiplier $\lambda(t) \in \mathbb{R}^l$. During the impact phase (2.32), the unilateral constraint $F(q(t), t)$, and multiplier $\lambda(t)$ are to simultaneously be nonnegative and orthogonal to each other, thereby imposing the so-called *complementarity constraint* on the system in question (Brogliato 2016).

A particular feature of the constrained Lagrange dynamics (2.31), (2.32) is that Coulombs friction or another tangent action can enforce sticking/sliding modes on the constraint where the Lagrange multiplier method is relevant to the Filippov vector field definition. The implicit Euler integration yields the correct choice of the

Lagrange multiplier and it is therefore (Acary and Brogliato 2008; Acary et al. 2012) an appropriate simulation tool of constrained Lagrange dynamics with sliding modes.

2.3.1 Implicit Euler Integration of Sliding Modes

The *implicit Euler method* does not modify the sliding surface (Acary et al. 2012) and it is therefore well-suited for the numerical simulation of continuous-time SM systems.

In what follows, the method is illustrated for a first-order dynamic system

$$\begin{cases} \dot{x}(t) = -M\tau(t) + \varphi(t) \\ \tau(t) \in \operatorname{sgn} x(t), \end{cases} \tag{2.33}$$

where $\operatorname{sgn} x = \begin{cases} +1 & \text{if } x > 0 \\ -1 & \text{if } x > 0 \\ [-1, 1] & \text{if } x = 0 \end{cases}$ is a multi-valued sign function, $\varphi(\cdot)$ is a Lebesgue measurable function such that

$$\|\varphi\|_\infty < \rho < M. \tag{2.34}$$

Extensions to higher order dynamic systems, including twisting and supertwisting algorithms, are rather technical and are available in Acary et al. (2012), Huber et al. (2016).

As shown in Example 1.1, system (2.33) is globally asymptotically stable around its equilibrium $x = 0$, which is attained in finite time, regardless of whichever input $\varphi(\cdot)$ affects the system.

Choosing correctly the Filippov velocity value is the object of the following discretization:

$$\begin{cases} \tilde{x}_{k+1} = x_k - Mh\tau_{k+1} \\ \tau_{k+1} \in \operatorname{sgn}(\tilde{x}_{k+1}) \\ x_{k+1} = x_k - Mh\tau_{k+1} + h\varphi_{k+1}. \end{cases} \tag{2.35}$$

The first two lines of (2.35) are relevant for the *nominal* system, which is not affected by $\varphi(\cdot)$ and from which the multi-valued sign function is unambiguously computed at the time instants $t_k,\ k = 0, 1, \ldots$. The proposed numerical integration is said to be implicit because it involves $\tilde{x}_{k+1}$ in the sign multifunction. However, $\tilde{x}_{k+1}$ is just an intermediate variable which is computed in the implicit integration, thereby yielding a causal sign function implementation (in the sense that it is not explicitly dependent on future state variables). The third line of (2.35) is the Euler approximation of the perturbed system (2.33), affected by $\varphi(\cdot)$ and prespecified with $\tau(t) = \tau_{k+1}$ on the time interval $[t_k, t_{k+1})$. The following result is in force (Acary et al. 2012).

Proposition 2.1 *Consider the discrete system* (2.33) *with the assumptions above and with initial condition* $x_0 \in \mathbb{R}$. *Then the nominal state variable* $\tilde{x}_k$ *is nullified after a finite number of steps* k_0 *whereas the perturbed variable integrates to* $x_k = h\varphi_k$ *for all* $k > k_0$, *thus presenting the external input attenuation by a factor* h. *Moreover, their dirty derivatives satisfy* $\frac{\tilde{x}_{k+1}-\tilde{x}_k}{h} = 0$ *for all* $k > k_0$ *and* $\frac{x_{k+1}-x_k}{h} = \varphi_{k+1} - \varphi_k$ *for all* $k > k_0 + 1$.

Proof For later use, recall that for $K \subset \mathbb{R}^n$ being a closed non-empty convex set, the normal cone to K at $x \in K \subset \mathbb{R}^n$ is $N_K(x) = \{z \in \mathbb{R}^n \mid z^T(y-x) \leq 0 \text{ for all } y \in K\}$. Then for any $x \in \mathbb{R}^n$ and $y \in \mathbb{R}^n$, one has

$$x \in \operatorname{sgn}(y) \Leftrightarrow y \in N_{[-1,1]}(x) \tag{2.36}$$

and

$$\begin{aligned} -x + y \in L^{-1}N_K(x) &\Leftrightarrow x = \operatorname{proj}_L(K; y) \\ &\Leftrightarrow x = \operatorname{argmin}_{z\in K} \tfrac{1}{2}(z-y)^T L(z-y), \end{aligned} \tag{2.37}$$

where $L = L^T > 0$ is an $n \times n$ positive definite matrix and $\operatorname{proj}_L(K; y)$ denotes the orthogonal projection of y on K in the metric defined by L.

Provided that $|x_0| > Mh > 0$, the generalized equation $\tilde{x}_{k+1} = x_k - Mh\tau_{k+1}$, being coupled to $\tau_{k+1} \in \operatorname{sgn}(\tilde{x}_{k+1})$, is found by using (2.36) and (2.37) to be equivalent to the inclusion $\tau_{k+1} - \frac{x_k}{Mh} \in -N_{[-1,1]}(\tau_{k+1})$. The latter inclusion is in turn equivalent to $\tau_{k+1} = \operatorname{proj}([-1,1]; \frac{x_k}{Mh})$. Thus one arrives at the conclusions

- If $x_k > Mh$ then $\tilde{x}_{k+1} = x_k - Mh$ and $\operatorname{sgn}(\tilde{x}_{k+1}) = 1$,
- If $x_k < -Mh$ then $\tilde{x}_{k+1} = x_k + Mh$ and $\operatorname{sgn}(\tilde{x}_{k+1}) = -1$,
- If $x_k \in [-Mh, Mh]$ then $\tilde{x}_{k+1} = 0$,

thereby inferring that

- If $x_k > Mh$ then $x_{k+1} = x_k + h\varphi_{k+1} - Mh = x_k + h(\varphi_{k+1} - M) < x_k + h(\rho - M)$. Since $\rho - M < 0$ the state is strictly decreased from step k to step $k+1$.
- If $x_k < -Mh$ then $x_{k+1} = x_k + h\varphi_{k+1} + Mh = x_k + h(\varphi_{k+1} + M) > x_k + h(M - \rho)$. Since $M - \rho > 0$ the state is strictly increased from step k to step $k+1$.

It follows that for sufficiently large initial values $|x_0| > Mh$, the state escapes to zero $\tilde{x}_{k_0} = 0$ after $k_0 = \lceil \frac{x_0}{h|M-\rho|} \rceil$ steps where $\lceil x \rceil$ is the integer part of $x \in \mathbb{R}$. Indeed, at k_0 the state x_k reaches the interval $(-Mh, Mh)$ and then the unique solution for $\tilde{x}_k$ is zero. From $\tilde{x}_{k_0} = 0$ one deduces that $|x_{k_0}| < Mh$. Actually, in the case of sufficiently small initial conditions $|x_0| \leq Mh$, it is straightforward to verify that $k_0 = 1$.

To compute the next value of $\tilde{x}_k$ one has to solve the generalized equation

$$\begin{cases} \tilde{x}_{k_0+1} = x_{k_0} - Mh\tau_{k_0+1} \\ \tau_{k_0+1} \in \operatorname{sgn}(\tilde{x}_{k_0+1}), \end{cases} \tag{2.38}$$

whose unique solution $\tilde{x}_{k_0+1} = 0$ is detected by inspection (it is the maximal monotonicity of the sign multifunction that makes this hold). The reasoning can be repeated to conclude that $\tilde{x}_k = 0$ for all $k \geq k_0$. Therefore $\frac{\tilde{x}_{k+1}-\tilde{x}_k}{h} = 0$ for all $k > k_0$. To complete the proof it remains to note that if

$$\tilde{x}_{k+1} = x_k - Mh\tau_{k+1} = 0, \quad k \geq k_0, \tag{2.39}$$

then

$$\tau_{k+1} = \frac{x_k}{hM}, \tag{2.40}$$

and the state x_{k+1} is given by

$$x_{k+1} = h\varphi_{k+1}. \tag{2.41}$$

Hence,

$$x_k = h\varphi_k,\ \tau_{k+1} = \frac{\varphi_k}{M} \text{ for all } k \geq k_0 + 1, \tag{2.42}$$

so that $\frac{x_{k+1}-x_k}{h} = \varphi_{k+1} - \varphi_k$ for all $k > k_0 + 1$. ■

For the purpose of the disturbance compensation, the auxiliary compensator variable $\hat{x}$ is introduced through the dynamic equation $\dot{\hat{x}}(t) = -M\tau_1(t),\ \tau_1(t) \in \mathrm{sgn}(x(t))$, the error variable $e = x - \hat{x}$, the augmented input $u = -M\mathrm{sgn}(x(t)) - M_e\mathrm{sgn}(e(t))$, and the ordered gains $M_e > M > 0$. The resulting system is thus given by

$$\begin{cases} \dot{x}(t) = -a\tau_1(t) - \alpha\tau_2(t) + \varphi(t) \\ \dot{e}(t) = -\alpha\tau_2(t) + \varphi(t) \\ \tau_1(t) \in \mathrm{sgn}(x(t)) \\ \tau_2(t) \in \mathrm{sgn}(e(t)), \end{cases} \tag{2.43}$$

where $\varphi(\cdot)$ is subject to the same magnitude upperbound (2.34) as before. Similar to Example 1.1, the system attains the sliding surface $e = 0$ in finite time where by applying the equivalent control method, it is governed by the SM equation $\dot{x}(t) = -M\tau_1(t) + \varphi(t)$. The condition $M < M_e$ ensures that the origin is not attained directly, but first the system slides on the surface $e = 0$. On this surface it is apparent from (2.43) that the dynamics in x evolves as a disturbance-free system. The discrete sliding mode system is implemented as follows:

$$\begin{cases} \tilde{x}_{k+1} = x_k - Mh\tau_{1,k+1} - M_e h\tau_{2,k+1} \\ \tilde{e}_{k+1} = e_k - M_e h\tau_{2,k+1} \\ \tau_{1,k+1} \in \mathrm{sgn}(\tilde{x}_{k+1}) \\ \tau_{2,k+1} \in \mathrm{sgn}(\tilde{e}_{k+1}), \end{cases} \tag{2.44}$$

and the update procedure representing the plant dynamics is given by:

$$\begin{cases} x_{k+1} = x_k - Mh\tau_{1,k+1} - M_e h\tau_{2,k+1} + h\varphi_{k+1} \\ e_{k+1} = e_k - M_e h\tau_{2,k+1} + h\varphi_{k+1}. \end{cases} \tag{2.45}$$

Proposition 2.2 *Consider the discrete system (2.44) with the assumptions above and with the initial conditions x_0, e_0. Then the nominal error variable $\tilde{e}_k = 0$, not straightforwardly affected by the disturbance variable φ, escapes to zero after a finite number of steps k_0 whereas $e_k = h\varphi_{k+1}$ for all $k > k_0$. Moreover, there exists $k_1 < +\infty$ such that $\tilde{x}_k = 0$ for all $k > k_0 + k_1$ and $x_k = h\varphi_k$ for all $k \geq k_0 + k_1$.*

Proof Is available in Acary et al. (2012). Since the proof is similar to that of Proposition 2.1, it is omitted for brevity.

By the above proposition, the discrete-time compensator guarantees the convergence of the nominal error variable to the origin in finite time, whereas its disturbed counterpart is equal to the disturbance, attenuated by a factor h. Summarizing (2.44) and (2.45), the discrete-time closed-loop system is as follows:

$$\begin{cases} x_{k+1} = x_k - Mh\tau_{1,k+1} - M_e h\tau_{2,k+1} + h\varphi_{k+1} \\ e_{k+1} = e_k - M_e h\tau_{2,k+1} + h\varphi_{k+1} \\ \tau_{1,k+1} = \text{proj}\left([-1, 1]; \frac{x_k - M_e h\tau_{2,k+1}}{Mh}\right) \\ \tau_{2,k+1} = \text{proj}\left([-1, 1]; \frac{e_k}{M_e h}\right). \end{cases} \tag{2.46}$$

Apparently, the above discretization is readily implementable with nested projections.

Attractive features of the proposed Euler integration method are illustrated for system (2.43), affected by $\varphi(t) = 0.1\sin(100\,t)$. In order to reproduce the continuous-time nature of the plant, its dynamics are integrated with the computer precision, whereas the input sampling time $h = 10^{-1}$s is much larger. The simulation results, reproduced from Acary et al. (2012), are presented on Fig. 2.1 with $M = 1$, $M_e = 2$ where the disturbance attenuation is clearly shown.

2.4 Hopf Bifurcation of Discontinuous Limit Cycles: Case Study

The nonlinear Van der Pol oscillator is long recognized for the generation of a stable limit cycle, independent of initial conditions. This feature explains its popular use in the model reference control. Section 1.2.3 motivates the specific modification

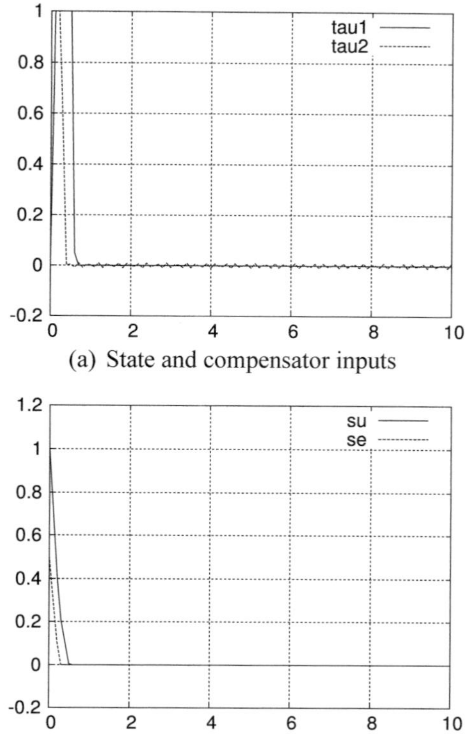

(a) State and compensator inputs

(b) Sliding variables

Fig. 2.1 Simulation of system (2.43), specified with $\varphi(t) = 0.1 \sin 100t$, $M = 1$, and $M_e = 2$. © 2012 $IEEE$. Reprinted, with permission, from Acary et al. (2012)

$$
\begin{aligned}
\dot{x}_1 &= x_2, \\
\dot{x}_2 &= -\varepsilon \left[\left(x_1^2 + \frac{x_2^2}{\mu^2} \right) - \rho^2 \right] x_2 - \mu^2 x_1
\end{aligned}
\tag{2.47}
$$

of the Van der Pol oscillator for reshaping its limit cycle to the one with harmonic behavior whose amplitude and frequency would straightforwardly rely on the oscillator parameters. Such a modification has recently been analyzed in the hybrid setting (Herrera et al. 2017) to reveal the bifurcation of the constrained Van der Pol modification to a stable limit cycle under certain parameter variations. Following Herrera et al. (2017), this phenomenon, otherwise known as a *Poincaré–Andronov–Hopf* (or simply Hopf) *bifurcation*, is further investigated via the corresponding Poincaré map to numerically carry out the set of the bifurcation parameter values and to verify the asymptotic stability of the generated hybrid limit cycle. The knowledge of the detected bifurcation parameter values is then utilized to illustrate the capability of the constrained oscillator to degenerate its limit cycle into an asymptotically stable equilibrium once the oscillator parameters are properly modified online.

2.4.1 Constrained Van der Pol Oscillator

In the sequel, the dynamics of the modified Van der Pol oscillator (2.47) are initialized and studied under the unilateral constraint

$$x_1 \geq 0. \tag{2.48}$$

The present development aims to carry out if such a constrained Van der Pol model is still capable of generating a limit cycle. Along with the numerical analysis of the limit cycle to exist, the Hopf bifurcation is revealed for the constrained Van der Pol modification under a nontrivial value of the parameter ε, yet corresponding to nonlinear dynamics. This is in contrast to the unconstrained nonlinear Van der Pol oscillator which exhibits the Hopf bifurcation just for the trivial parameter value $\varepsilon = 0$ when it degenerates to the linear oscillator.

As pointed out, starting from the very initial time instant, the modified Van der Pol model of interest operates under the unilateral position constraint $x_1 \geq 0$. Once a model trajectory hits the constraint surface

$$S = \{x \in \mathbb{R}^2 : x_1 = 0 \cup x_2 \leq 0\} \tag{2.49}$$

at a collision time instant t, the underlying model (2.47) instantaneously resets its velocity according to the elastic restitution law

$$\begin{aligned} x_1(t^+) &= x_1(t^-), \\ x_2(t^+) &= -e x_2(t^-) \text{ iff } x(t) \in S \end{aligned} \tag{2.50}$$

with the restitution parameter $e \in (0, 1)$. The above restitution rule is similar to that of impact mechanical systems (Brogliato 2016). Beyond the constraint, the continuous dynamics are governed by (2.47) whenever $x \notin S$.

2.4.2 Existence of a Constrained Limit Cycle

Similar to the constraint-free Van der Pol oscillator, the above model (2.47)–(2.50) possesses an equilibrium in the origin what is straightforwardly verified by inspection. The question then arises whether the constrained oscillator (2.47)–(2.50) generates a stable limit cycle as well. For addressing this question, the present section employs the Poincaré map to derive sufficient conditions of the constrained stable limit cycle to exist. Throughout, the standard stability concepts deal with the domain of unilaterally constrained solutions. Potential Zeno dynamics with finite impact accumulation points (similar to those of bouncing ball of Sect. 1.2.2) are actually admitted, but their finite time stability is not specifically treated as the present investigation focuses on the asymptotic stability only.

Without loss of generality, the subsequent stability analysis of the constrained Van der Pol model (2.47), (2.49), (2.50) is confined to the initial condition $x^0 = (0, v_0)^T$ with $v_0 > 0$; otherwise, one could readily re-initialize the model with such initial conditions by reconstructing the continuous dynamics in the backward time. With this in mind, let

$$x(v_0; t) = (x_1(v_0; t), x_2(v_0; t))^T \tag{2.51}$$

denote a trajectory of (2.47)–(2.50), initialized at $x_1(v_0; 0) = 0, x_2(v_0; 0) = v_0 > 0$, and let $t_k, k = 1, 2, \dots$ stand for the collision time instants when this trajectory resets its velocity.

Dependent on the oscillator parameter values, the following alternative scenarios are heuristically in order.

(S1) Inequalities

$$x_2(v_0; t_1^+) < v_0, \; x_2(v_0; t_{k+1}^+) < x_2(v_0; t_k^+) \tag{2.52}$$

hold for all $k = 1, 2, \dots$ and all $v_0 > 0$.

(S2) There exists a scalar $x^* > 0$ such that inequalities (2.52) hold for all $k = 1, 2, \dots$ and all $v_0 > x^*$ whereas the inverse inequalities

$$x_2(v_0; t_1^+) > v_0, \; x_2(v_0; t_{k+1}^+) > x_2(v_0; t_k^+) \tag{2.53}$$

hold for all $k = 1, 2, \dots$ and all $v_0 \in (0, x^*)$.

It should be pointed out that Scenario (S2) is the only possible option for the constraint-free Van der Pol oscillator (2.47) provided that $t_k, \; k = 1, 2, \dots$ stand for the time instants when the state trajectory meets the vertical semiaxis $x_1 = 0$, $x_2 > 0$. Indeed, to reproduce this conclusion it suffices to differentiate the candidate Lyapunov function

$$V(x) = \frac{1}{2}x_1^2 + \frac{1}{2\mu^2}x_2^2 \tag{2.54}$$

along the trajectories of (2.47) to obtain that

$$\dot{V}(x) = \frac{\varepsilon}{\mu^2}\left[\rho^2 - \left(x_1^2 + \frac{x_2^2}{\mu^2}\right)\right]x_2^2 \tag{2.55}$$

thereby yielding

$$\dot{V}(x)\begin{cases} > 0, \; if \; \left(x_1^2 + \frac{x_2^2}{\mu^2} < \rho^2\right), \; x_2 \neq 0 \\ < 0, \; if \; \left(x_1^2 + \frac{x_2^2}{\mu^2} > \rho^2\right), \; x_2 \neq 0 \\ = 0, \; if \; \left(x_1^2 + \frac{x_2^2}{\mu^2} - \rho^2\right)x_2^2 = 0. \end{cases} \tag{2.56}$$

Then, Scenario (S2), properly modified for the constraint-free oscillator (2.47), inevitably follows from (2.56). Remarkably, the same function (2.54) is no longer useful to detect a limit cycle of the constrained oscillator (2.47), (2.49), (2.50) because in contrast to the sign variation (2.56) of the time derivative (2.55), computed between the successive collision time instants t_k, $k = 1, 2, \ldots$, its instantaneous change $\Delta V(x(v_0; t_k)) = V(x(v_0; t_k^+)) - V(x(v_0; t_k^-))$ at the collision time instants t_k is governed by the elastic restitution law (2.50) with $e \in (0, 1)$ and therefore it remains negative definite $\Delta V(x(v_0; t_k)) = \frac{1}{2\mu^2}[x_2^2(v_0; t_k^+) - x_2^2(v_0; t_k^-)] = \frac{e^2-1}{2\mu^2} x_2^2(v_0; t_k^-) < 0$ regardless of whether the initial condition x^0 is inside or outside the ellipse, governed by $x_1^2 + \frac{x_2^2}{\mu^2} = \rho^2$.

Other possible scenarios where, for example, (2.53) holds true for all $k = 1, 2, \ldots$ and all v_0 (that would lead to instability), or a non-monotonic convergence of the sequence $x_2(v_0; t_k^+)$, $k = 1, 2, \ldots$, towards a fixed value x^* (possibly, $x^* = 0$) is in force, were ruled out due to the performed numerical tests, presented in the next sections. These tests carried out that only Scenarios (S1) and (S2) were relevant for various initial conditions and admissible combinations of the oscillator parameters.

The following result might be viewed as a counterpart of the *Poincaré–Bendixon criterion* for the constrained oscillator (2.47)–(2.50).

Theorem 2.3 *Consider the constrained Van der Pol oscillator* (2.47)–(2.50) *with a priori fixed parameters* $\rho, \mu, \varepsilon > 0$ *and* $e \in (0, 1)$. *Let Scenario (S1) be in force. Then* (2.47)–(2.50) *is globally asymptotically stable in the origin. To the contrary, let Scenario (S2) be in force. Then* (2.47)–(2.50) *possesses a stable limit cycle* γ^*, *generated by the trajectory* $x(x^*; t)$ *with* $x^0 = (0, x^*)^T$ *such that any trajectory of* (2.47)–(2.50), *initialized beyond the origin, converges to* γ^*.

Proof To begin with, let us assume that Scenario (S1) holds. Then given an initial condition $x^0 = (0, v_0)^T$ at $t = 0$ with $v_0 > 0$, the postimpact velocity values $x_2(v_0; t_k^+)$ monotonically converge to some $\tilde{x}^* \geq 0$, i.e.,

$$\lim_{k\to\infty} x_2(v_0; t_k^+) = \tilde{x}^*. \tag{2.57}$$

Due to the continuous dependence of the trajectory plots $\gamma_k(v_0) = \{x(v_0; t),\ t \in (t_k, t_{k+1})\}$, $k = 1, 2, \ldots$ on its initial values $x(v_0; t_k^+)$ at $t = t_k$, it follows that $x(v_0; t_{k+1}^-) \to x(\tilde{x}^*; t_1^-)$ as $k \to \infty$. Then by employing (2.50), one concludes that

$$\lim_{k\to\infty} x_2(v_0; t_{k+1}^+) = x_2(\tilde{x}^*; t_1^+). \tag{2.58}$$

Since (S1) ensures that $x_2(\tilde{x}^*; t_1^+) < \tilde{x}^*$ whenever $\tilde{x}^* > 0$ the latter convergence (2.58) would not contradict to the former convergence (2.57) just in the case that $\tilde{x}^* = 0$. Thus, according to (2.57), the postimpact velocity values $x_2(v_0; t_k^+)$ escape to zero as $k \to \infty$, and relying again on the continuous dependence of the trajectory plots $\gamma_k(v_0) = \{x(v_0; t),\ t \in (t_k, t_{k+1})\}$, $k = 1, 2, \ldots$ on the values $x(v_0; t_k^+)$ at $t = t_k$, the global asymptotic stability of (2.47)–(2.50) is established in the origin under Scenario (S1), which is illustrated in Fig. 2.2a.

The proof of the asymptotic stability of the limit cycle γ^* under Scenario (S2) follows the same line of reasoning. In this case, convergence (2.57) holds for some

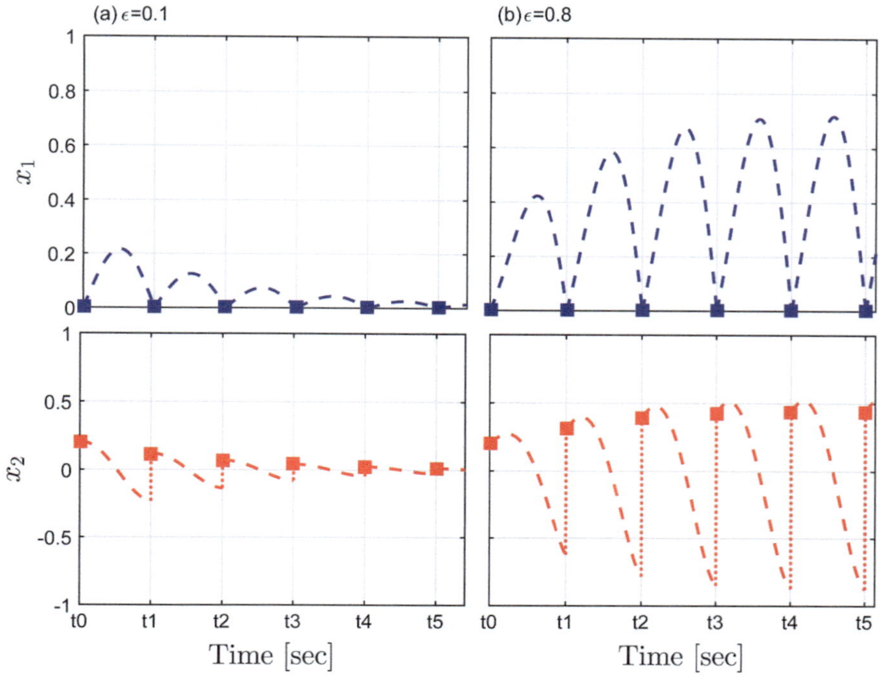

Fig. 2.2 **a** Scenario (S1) and **b** Scenario (S2) of the state dynamics of the constrained Van der Pol oscillator. The squares denote the postimpact positions and velocities at the corresponding time instants (Reprinted from Herrera et al. 2017, Copyright 2017, with permission from Elsevier)

$\tilde{x}^* \geq x^*$ if $v_0 > x^*$ and for some $\tilde{x}^* \leq x^*$ if $v_0 < x^*$. In turn, convergence (2.58) does not contradict to convergence (2.57) iff $x_2(\tilde{x}^*; t_1^+) = \tilde{x}^* = x^*$. Thus, Scenario (S2) results in a limit cycle γ^*, generated by the initial condition $x_1(0) = 0$, $x_2(0) = x^*$, and this cycle is asymptotically stable due to the continuous dependence of the trajectory plots $\gamma_k(v_0) = \{x(v_0; t),\ t \in (t_k, t_{k+1})\},\ k = 1, 2, \ldots$ on the initial values $x(v_0; t_k^+)$ at $t = t_k$. This scenario is illustrated in Fig. 2.2b. The proof is thus completed.

Theorem 2.3 admits a useful interpretation in terms of the *Poincaré map* of the trajectories of (2.47), (2.49), (2.50) where the *Poincaré section* is set at each postimpact instant. With this in mind, the velocity value of $x(t_k^+)$ after each reset can be viewed as a discrete system of the form

$$x_2(t_{k+1}^+) = F(x_2(t_k^+)),\ k = 1, 2, \ldots \tag{2.59}$$

where F represents the Poincaré map from the previous post-reset velocity to the next one. By Theorem 2.3, inequalities (2.52) and (2.53) ensure the existence of a fixed point x^* of the map F, determined by (2.59). In Sect. 2.4.4, this observation will numerically be applied to the stability analysis of a limit cycle generated by the constrained Van der Pol oscillator (2.47)–(2.50).

2.4.3 Numerical Analysis of Phenomenological Behaviors

In this section, different behaviors of the constrained Van der Pol oscillator (2.47)–(2.50) are numerically presented. Clearly, such different behaviors occur as well for the rescaled oscillator version

$$\begin{aligned} \frac{dy_1}{d\tau} &= y_2, \\ \frac{dy_2}{d\tau} &= -\alpha[(y_1^2 + y_2^2) - 1]y_2 - y_1, \quad \alpha = \frac{\varepsilon\rho^2}{\mu} > 0 \end{aligned} \tag{2.60}$$

under the velocity reset

$$\begin{aligned} y_1(\tau^+) &= y_1(\tau^-), \\ y_2(\tau^+) &= -ey_2(\tau^-) \text{ iff } y(\tau) \in S_y \end{aligned} \tag{2.61}$$

on the corresponding constraint surface

$$S_y = \{y_1 = 0 \cup y_2 \leq 0\}. \tag{2.62}$$

As opposed to the unconstrained case, where the limit cycle reshapes its amplitude, frequency and transient speed under admissible parameter variations, but it is not degenerated for any positive parameter values ρ, μ, and ε, the constrained case is shown to exhibit the Hopf bifurcation under nonzero parameter values.

It is numerically justified that given the parameter values $\rho = 1$, $\mu = 1$ and $e = 0.5$, the Hopf bifurcation of the constrained oscillator occurs for the transient speed parameter ε at some value $\varepsilon_b \approx 0.429$ that corresponds to $\alpha = \varepsilon\rho^2\mu^{-1}$ at $\alpha_b \approx 0.429$. These values are further referred to as bifurcation values. An asymptotically stable behavior of such an oscillator is shown in Fig. 2.3a for a transient speed value $\varepsilon < \varepsilon_b$ ($\alpha < \alpha_b$), whereas the capability of the same oscillator to generate an asymptotically stable limit cycle is illustrated in Fig. 2.3b for $\varepsilon > \varepsilon_b$ ($\alpha > \alpha_b$).

Furthermore, it is observed from Fig. 2.4 that when the parameter $e = 0.5$ remains fixed whereas $\alpha = \varepsilon\rho^2\mu^{-1}$ is reduced to $\alpha = 0.2$ (that corresponds, e.g., to the parameter values $\varepsilon = 0.8$, $\rho = 0.5$, $\mu = 1$), the limit cycle degenerates to the origin. To the contrary, Fig. 2.5 demonstrates that by increasing the value α to $\alpha = 1$, e.g., by decreasing $\mu = 0.2$ without modifying the parameters $\rho = 0.5$, $\varepsilon = 0.8$, $e = 0.5$, another limit cycle is generated again.

Finally, Fig. 2.6 shows that by reducing the restitution parameter to $e = 0.1$, the limit cycle degenerates to the origin once again. It is thus concluded that the Hopf bifurcation of the constrained oscillator is caused by a certain variation of the

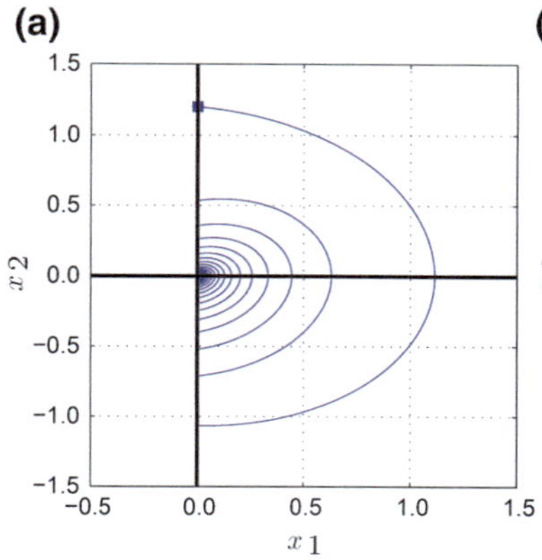

Fig. 2.3 Phase trajectories of the constrained oscillator (2.47)–(2.50), corresponding to the parameter values $\rho = 1, \mu = 1, e = 0.5$, and to the transient speed parameters $\varepsilon = 0.3$ (the left figure a) and $\varepsilon = 0.8$ (the right figure b). Squares denote the initial conditions of the corresponding trajectories (Reprinted from Herrera et al. 2017, Copyright 2017, with permission from Elsevier)

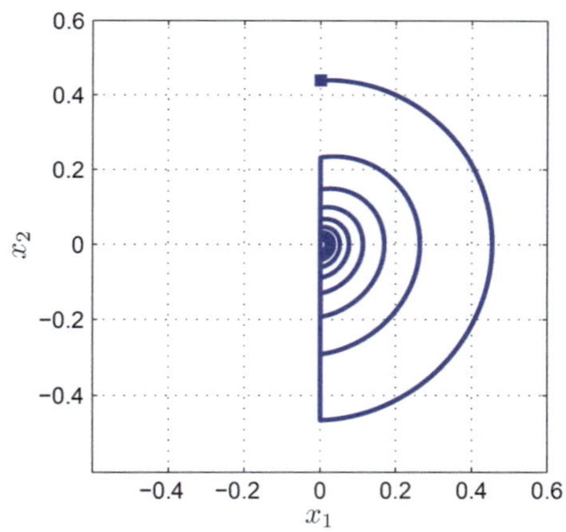

Fig. 2.4 Phase trajectory of the constrained oscillator (2.47)–(2.50), corresponding to $\varepsilon = 0.8$, $\rho = 0.5$, $\mu = 1$, $e = 0.5$ and initialized from a square on the vertical axis (Reprinted from Herrera et al. 2017, Copyright 2017, with permission from Elsevier)

oscillator parameters α and e. Relations among the bifurcation parameters α_b and e_b, resulting in the Hopf bifurcation, are numerically established next.

2.4.4 Hopf Bifurcation Analysis via Poincaré Method

In order to carry out potential relations among the bifurcation parameter values, under which a unique equilibrium of the constrained oscillator (2.60)–(2.62) bifurcates to a stable limit cycle, the method of Poincaré sections is invoked. This method allows

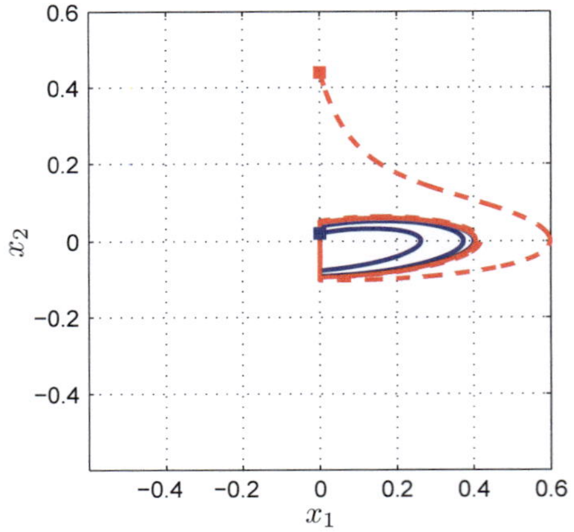

Fig. 2.5 Phase trajectories of the constrained oscillator (2.47)–(2.50) for $\varepsilon = 0.8$, $\rho = 0.5$, $\mu = 0.2$, and $e = 0.5$. Squares denote the initial conditions of two trajectories: one with small initial velocity (solid line), and one with a larger initial velocity (dashed line) (Reprinted from Herrera et al. 2017, Copyright 2017, with permission from Elsevier)

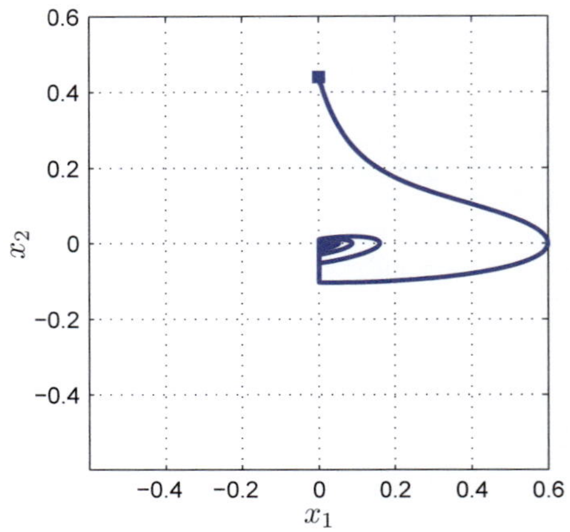

Fig. 2.6 The phase trajectory of the constrained oscillator (2.47)–(2.50), corresponding to $\varepsilon = 0.8$, $\rho = 0.5$, $\mu = 0.2$, $e = 0.1$ and initialized from a square on the vertical axis (Reprinted from Herrera et al. 2017, Copyright 2017, with permission from Elsevier)

one to get novel qualitative results on the bifurcation of nonlinear oscillators under unilateral constraints while avoiding analytical difficulties in the theoretical study.

The Poincaré section (2.59), specified immediately after each time instant when system (2.60)–(2.62) resets its velocity, involves both the continuous and discrete dynamics and it is further computed using the step-by-step numerical integration of the continuous dynamics (2.60).

The stability of the limit cycle of the constrained oscillator (2.60)–(2.62) is first analyzed for the transient speed $\alpha = 1$ and the restitution parameter $e = 0.5$. The Cobweb plot, which is well-recognized (Waugh 1964) to be a powerful tool of the qualitative study of one-dimensional maps, is depicted in Fig. 2.7 for the corresponding parameter values.

The fixed points of the Poincaré map $y_2(\tau_{k+1}^+) = F(y_2(\tau_k^+)),\ k = 1, 2, \dots$ (cf. (2.59)) are then typically obtained in the intersections of the Cobweb map with the identity map. Apart from the unstable equilibrium $y_2 = 0$, there is another

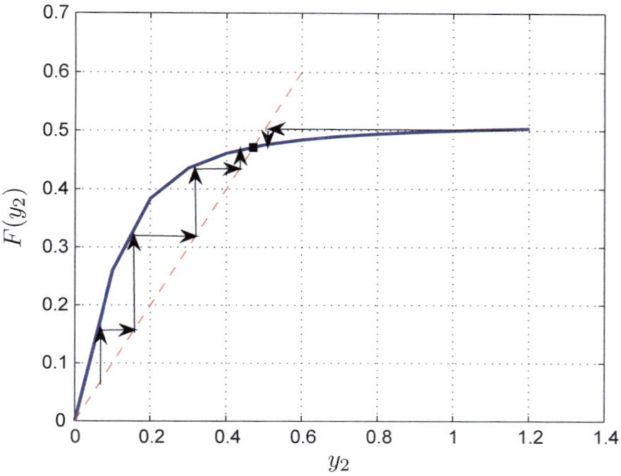

Fig. 2.7 The Cobweb plot of the Poincaré map with $e = 0.5$, and $\alpha = 1$. Solid line is for the Poincaré map $F(y_2)$, dashed line is for the identity map. The arrows illustrate the attractivity of the nonzero fixed point (black square) (Reprinted from Herrera et al. 2017, Copyright 2017, with permission from Elsevier)

intersection, that corresponds to an asymptotically stable fixed point y^*, denoted in Fig. 2.7 by the black square. Indeed, the arrows at the Cobweb plot indicate the evolution of the Poincaré map from an arbitrary initial condition of $y_2(\tau_0) = v_0 > 0$ so that all the trajectories evolve away from the origin while approaching this fixed point y^*. Thus, Scenario (S2) is in order and Theorem 2.3, applied to the constrained oscillator (2.60)–(2.62), ensures the existence and asymptotic stability of the limit cycle, which is generated by the trajectory, initialized at $y_1 = 0$, $y_2 = y^*$.

Next, the asymptotic stability of the constrained oscillator (2.60)–(2.62) is analyzed in the case where the transient speed parameter is set to $\alpha = 0.3$. The corresponding phase trajectory and Cobweb plot of the Poincaré map (2.59) are depicted in Fig. 2.3a and Fig. 2.8, respectively. In this case, the only fixed point of the Poincaré map is the origin. Since the map evolves monotonically towards the origin, Scenario (S2) is in order, and the asymptotic stability of the constrained oscillator (2.60)–(2.62), specified with $\alpha = 0.3$ and $e = 0.5$, is then verified by applying Theorem 2.3.

To reveal the bifurcation value of the parameter α under the fixed parameter $e = 0.5$ the Poincaré map (2.59) is numerically integrated for several values of α. Figure 2.9 depicts the intersections of the resulting Poincaré maps with the identity map. It is observed from the figure that while α escapes to zero, the identity map is no longer intersected with the corresponding Poincaré map that has no other fixed points than the origin. Using the same reasoning as before, the Cobweb plot straightforwardly demonstrates that the origin is attractive for low values of α. This conclusion can also be reproduced by verifying that the magnitude of ∇F is inside the unit interval. Thus, Scenario (S1) is in order, and the asymptotic stability of the origin follows from Theorem 2.3.

By plotting the fixed point y^*, computed for different values of α and fixed $e = 0.5$, the bifurcation transient speed value at $\alpha_b \approx 0.429$ was numerically shown up in Herrera et al. (2017). Moreover, by iteratively varying the restitution parameter e in the interval $(0, 1)$, the functional dependence $\alpha_b = \kappa(e)$ of the bifurcation value α_b, complying with the physically motivated conditions $\kappa(e) \to \infty$ as $e \to 0$ (no limit

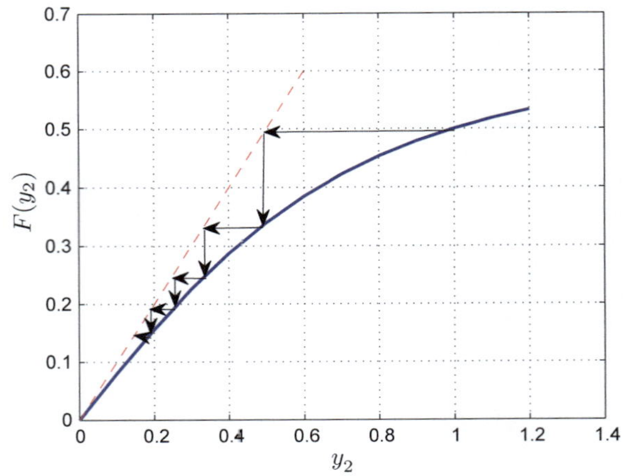

Fig. 2.8 The Cobweb plot of the Poincaré map with $e = 0.5$, and $\alpha = 0.3$. Solid line is for the Poincaré map $F(y_2)$, dashed line is for the identity map. The arrows illustrate the attractivity of the origin (Reprinted from Herrera et al. 2017, Copyright 2017, with permission from Elsevier)

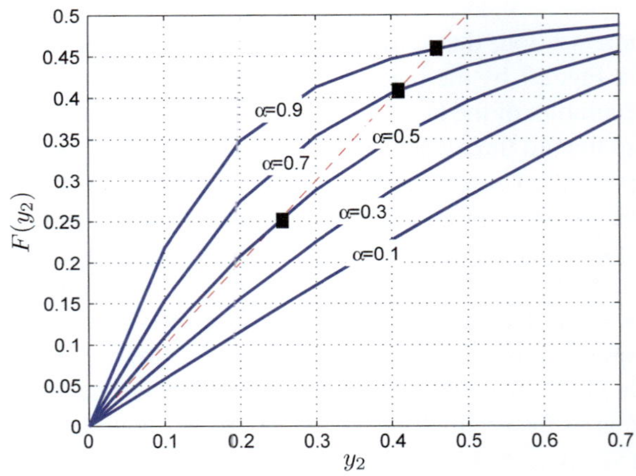

Fig. 2.9 Intersections of the Poincaré map with the identity map for different values of α and for constant $e = 0.5$. Solid lines are for the different Poincaré maps $F(y_2)$ with the corresponding values of α, dashed line is for the identity map, black square is for the fixed points of the maps (Reprinted from Herrera et al. 2017, Copyright 2017, with permission from Elsevier)

cycle for any $\alpha > 0$ and hence, no Hopf bifurcation) and $\kappa(e) \to 0$ as $e \to 1$ (a limit cycle is generated for any $\alpha > 0$), was approximately revealed in the form of the truncated Laurent series

$$\kappa(e) = \frac{k_1}{e^3} + \frac{k_2}{e^2} + \frac{k_3}{e} + k_4 + k_5 e + k_6 e^2 + k_7 e^3,$$

where the coefficients $k_1 = 0.0001$, $k_2 = -0.0053$, $k_3 = 0.0848$, $k_4 = 0.8808$, $k_5 = -1.4953$, $k_6 = 0.64$, and $k_7 = -0.1035$ were set, using the mean square fitting method (Herrera et al. 2017).

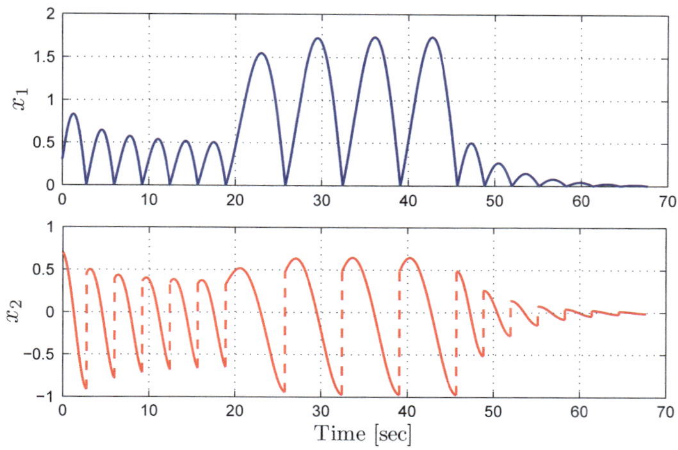

Fig. 2.10 Plot of the resulting oscillations of the constrained Van der Pol oscillator under online parameter modifications, causing three distinct behaviors (Reprinted from Herrera et al. 2017, Copyright 2017, with permission from Elsevier)

2.4.5 *Constrained Van der Pol Oscillator with Manipulated Parameters*

Motivated by the cyclic motion design in mechanical applications with impacts, the constrained Van der Pol oscillator is subsequently illustrated to be capable of generating limit cycles of manipulatable magnitude and frequency, and even of degenerating its cycle to an equilibrium point provided that the oscillator parameters are properly modified online.

For this purpose, the parameters of the constrained Van der Pol model (2.47)–(2.50) are subsequently modified in simulation runs made to illustrate phenomenological behaviors, predicted by the developed theory.

- Initially, the oscillator parameters are set to $\rho = 1$, $\mu = 1$, $e = 0.5$, $\varepsilon = 0.55$. While being specified with these parameters, the constrained oscillator (2.47)–(2.50), as it is shown in Sect. 2.4.4, generates an asymptotically stable limit cycle.
- In order to increase the amplitude of the limit cycle and to decrease its frequency the oscillator parameters are instantaneously modified to $\rho = 2$, $\mu = 0.5$, $e = 0.5$, $\varepsilon = 0.15$ right after the sixth impact.
- Finally, immediately after the tenth impact, the parameters are modified to $\rho = 1$, $\mu = 1$, $e = 0.5$, $\varepsilon = 0.05$ to cause the degeneration of the constrained oscillations to the equilibrium $x_1 = 0$, located on the constraint.

The resulting oscillator trajectory is depicted in Fig. 2.10 where the three distinct behaviors, deliberately imposed on the oscillator via appropriate parameter variations, are clearly observed. The magnitude and frequency of the resulting oscillations, produced by the constrained Van der Pol oscillator, are thus readily modified online. This feature makes the constrained oscillator extremely attractive for using it as a reference model in mechanical applications with unilateral constraints. Such a reference model is tested in Sect. 8.1.3 to generate a periodic closed-loop motion of a mass–spring system, impacting a vertical constraint.

2.5 Concluding Remarks

Basic solution concepts have been revised for VSS with resets whose ambiguous behavior has additionally been explained through the regularization technique, involving system modeling at a microscopic level. The well-posedness of the systems in question and their capability of producing Hopf bifurcation phenomena have particularly been established. Advanced mathematical tools are further presented in the infinite-dimensional setting.

References

Acary V, Brogliato B (2008) Numerical methods for nonsmooth dynamical systems. Springer Lecture notes in applied and computational mechanics, vol 35. Springer. Berlin

Acary V, Brogliato B, Orlov Y (2012) Chattering-free digital sliding-mode control with state observer and disturbance rejection. IEEE Trans. Auto Contr. 57:1087–1101

Aizerman MA, Pyatnitskii ES (1974) Foundations of the theory of discontinuous systems II. Autom Remote Control 8:39–61

Bartolini G, Zolezzi T (1985) Variable structure systems nonlinear in the control law. IEEE Trans Auto Control 30:681–684

Brogliato B (2016) Nonsmooth mechanics, 3rd edn. Springer, London

Curtain R, Zwart H (1995) An introduction to infinite-dimensional linear systems. Springer, New York

Filippov AF (1962) On certain questions in the theory of optimal control SIAM. J Control 1:76–84

Filippov AF (1988) Differential equations with discontinuous right-hand sides. Kluwer Academic Publisher, Dordrecht

Herrera L, Orlov Y, Montano O (2017) Hopf bifurcation of hybrid Van der Pol oscillators. Nonlinear Analysis: Hybrid Systems 26:225–238

Huber O, Acary V, Brogliato B (2016) Lyapunov stability and performance analysis of the implicit discrete sliding mode control. IEEE Trans Auto Control 61:3016–3030

Khalil H (2002) Nonlinear systems, 3rd edn. Prentice Hall, Upper Saddle River

Levaggi L (2002a) Infinite dimensional systems sliding motions. Eur J Control 8:508–518

Levaggi L (2002b) Sliding modes in banach spaces. Differ Integr Equ 15:167–189

Orlov Y (2009) Discontinuous systems - Lyapunov analysis and robust synthesis under uncertainty conditions. Springer, London

Utkin VI (1992) Sliding modes in control and optimization. Springer, Berlin

Waugh FV (1964) Cobweb models. J Farm Econ 46:732–750

Chapter 3
Mathematical Tools of Dynamic Systems in Hilbert Spaces

Later on, dynamic systems and solution concepts are viewed in the infinite-dimensional setting, and the modern stability paradigms (such as ISS, $\mathscr{L}_2$-gain, and FTS among others) are discussed.

3.1 Sobolev Spaces and Instrumental Inequalities

Infinite-dimensional dynamic systems, with distributed parameters in particular, are normally embedded into an appropriate Sobolev space which serves as an operational domain for such a system.

In the one-dimensional case, the *Sobolev space* $W^{l,p}(a, b)$ with an integer $l \geq 0$ is defined for $1 \leq p \leq \infty$ and $[a, b] \subseteq \mathbb{R}$ as the subset of p-integrable scalar functions $f(\cdot) \in L^p(a, b)$ such that f and its weak (also referred to as Sobolev) derivatives up to order l have finite L^p-norm on $[a, b]$. In turn, a function $g(\cdot) \in L^p(a, b)$ is said to be a *weak (Sobolev) derivative* of such a function $f(\cdot) \in L^p(a, b)$ iff

$$\int_a^b f(x)\frac{d\varphi}{dx}(x)dx = -\int_a^b g(x)\varphi(x)dx, \tag{3.1}$$

for all infinitely differentiable functions $\varphi(\cdot)$ such that $\varphi(a) = \varphi(b) = 0$.

The weak derivative concept is thus based on the well-known formula of integration by parts and it is motivated by the need of the differentiability property to be generalized for a broader class of Lebesgue integrable functions. The Sobolev space $W^{l,p}(a, b)$ becomes a Banach space if equipped with the norm

$$\|f\|_{W^{l,p}(a,b)} = \left(\sum_{i=0}^{l} \|f^{(i)}\|^p_{L^p(a,b)}\right)^{\frac{1}{p}}, \tag{3.2}$$

Y. Orlov, *Nonsmooth Lyapunov Analysis in Finite and Infinite Dimensions*, Communications and Control Engineering,
https://doi.org/10.1007/978-3-030-37625-3_3

where $f^{(i)}$ stands for the ith-order Sobolev derivative of f and

$$\|f^{(i)}\|_{L^p(a,b)} = \left(\int_a^b |f^{(i)}(x)|^p dx\right)^{\frac{1}{p}}, \quad i = 0, 1, \ldots, l \tag{3.3}$$

is the standard L_p-norm of $f^{(i)}$. The truncated summation

$$\left(\|f\|^p_{L^p(a,b)} + \|f^{(l)}\|^p_{L^p(a,b)}\right)^{\frac{1}{p}} \tag{3.4}$$

defines an alternative norm in $W^{l,p}(a,b)$ with the same induced topology as that of (3.2). For $p = 2$, the Sobolev space $W^{l,p}(a,b) = H^l(a,b)$ forms a Hilbert space with the inner product

$$\langle f, g\rangle_{H^l(a,b)} = \sum_{i=0}^{l} \int_a^b f^{(i)}(x)g^{(i)}(x)dx, \tag{3.5}$$

specified for arbitrary $f, g \in H^l(a,b)$. Below useful inequalities are listed for the PDE analysis in Sobolev spaces.

Lemma 3.1 (Cauchy–Schwartz inequality) *Given $\gamma > 0$ and $f, g \in L^2(a,b)$, the following holds*

$$\int_a^b f(x)g(x)dx \le \|f\|_{L^2(a,b)}\|g\|_{L^2(a,b)} \le \frac{\gamma}{2}\|f\|^2_{L^2(a,b)} + \frac{1}{2\gamma}\|g\|^2_{L^2(a,b)}. \tag{3.6}$$

Lemma 3.2 (Poincaré's inequality) *Let $u(x) \in H^1(a,b)$ and let $x_1, x_2 \in \mathbb{R}$ be such that $a \le x_1 \le x_2 = x_1 + h \le b$ for some $h \ge 0$. Then the following inequality*

$$\|u(\cdot)\|^2_{L^2(x_1,x_2)} \le 2h\left[u^2(x_i) + h\|u_x(\cdot)\|^2_{L^2(x_1,x_2)}\right] \tag{3.7}$$

holds for $i = 1, 2$ and $u_x(\cdot) = du/dx$.

Proof Given $u(\cdot) \in H^1(a,b)$, it is absolutely continuous and therefore,

$$u(\eta) = u(x_1) + \int_{x_1}^{\eta} u_x(x)dx \quad \forall\, \eta \in [x_1, x_2]. \tag{3.8}$$

Now squaring both sides of (3.8), exploiting the well-known inequality $2ab \le a^2 + b^2$, successively applying the Cauchy–Schwartz inequality, and taking into account that $\eta \in [x_1, x_1 + h]$ by construction, one arrives at the next chain of inequalities

$$\begin{aligned} u^2(\eta) &\le 2\left[u^2(x_1) + \left(\int_{x_1}^{\eta} u_x(x)dx\right)^2\right] \\ &\le 2\left[u^2(x_1) + (\eta - x_1)\int_{x_1}^{\eta} u_x^2(x)dx\right] \\ &\le 2\left[u^2(x_1) + h\|u_x(\cdot)\|^2_{L^2(x_1,x_2)}\right]. \end{aligned} \tag{3.9}$$

Integrating both sides of (3.9) with respect to η from x_1 to x_2 yields (3.7) with $i = 1$. The proof of (3.7) with $i = 2$ becomes identical under the change of coordinate $\zeta = x_2 - x$. ∎

In a particular case of functions $u(\cdot) \in H^1(a, b)$, vanishing at a boundary, the Poincaré inequality (3.7) is simplified to

$$\int_a^b u^2(x)dx \le C \int_a^b u_x^2(x)dx, \tag{3.10}$$

where $C = 2(b - a)^2$. Generally speaking, the Poincaré inequality (3.10) comes with a generic constant C which is chosen in the present case as in Pisano and Orlov (2017). A less conservative estimate $C = \frac{(b-a)^2}{\pi^2} < 2(b - a)^2$ is established for functions $u(x) \in H^1(a, b)$, possessing a bounded domain and vanishing at both ends.

Lemma 3.3 (Wirtinger's inequality and its generalization) *Let $u(x) \in H^1(a, b)$ be such that $u(a) = u(b) = 0$. Then*

$$\int_a^b u^2(x)dx \le \frac{(b-a)^2}{\pi^2} \int_a^b u_x^2(x)dx. \tag{3.11}$$

If additionally $u(\cdot) \in H^2(a, b)$, then the following generalization

$$\int_a^b u_x^2(x)dx \le \frac{(b-a)^2}{\pi^2} \int_a^b u_{xx}^2(x)dx \tag{3.12}$$

is in force.

Proof of Lemma 3.4 is based on the Fourier expansion of a periodic odd extension of u over $\mathbb{R}$. The interested reader can refer to Wang (1994) for the detailed proof of the lemma.

Lemma 3.4 (Agmon's inequality) *Let $u(x) \in H^1(a, b)$. Then the following holds*

$$\begin{aligned} \|u(\cdot)\|^2_{L^\infty(a,b)} &\le u^2(a) + 2\|u(\cdot)\|_{L^2(a,b)}\|u_x(\cdot)\|_{L^2(a,b)} \le u^2(a) + \|u(\cdot)\|^2_{H^1(a,b)}, \\ \|u(\cdot)\|^2_{L^\infty(a,b)} &\le u^2(b) + 2\|u(\cdot)\|_{L^2(a,b)}\|u_x(\cdot)\|_{L^2(a,b)} \le u^2(b) + \|u(\cdot)\|^2_{H^1(a,b)}. \end{aligned} \tag{3.13}$$

Proof Since

$$2\int_a^\xi u(x)u_x(x)dx = u^2(\xi) - u^2(a) \quad \forall\, \xi \in [a,b], \tag{3.14}$$

it follows that

$$u^2(\xi) \le u^2(a) + 2\int_a^b |u(x)||u_x(x)|dx. \tag{3.15}$$

By applying the Cauchy–Schwartz inequality to (3.15), one concludes that

$$\begin{aligned}\max_{\xi\in[a,b]} u^2(\xi) &\le u^2(a) + 2\|u(\cdot)\|_{L^2(a,b)}\|u_x(\cdot)\|_{L^2(a,b)} \\ &\le u^2(a) + \|u(\cdot)\|^2_{L^2(a,b)} + \|u_x(\cdot)\|^2_{L^2(a,b)},\end{aligned} \tag{3.16}$$

thereby verifying the first chain of inequalities in (3.13). The proof of the second chain of inequalities in (3.13) becomes identical to that of the first one when substituted b into (3.14) for a. ■

3.2 Linear Partial Differential Equations

3.2.1 Linear Differential Operators

A linear differential operator of interest is a multivariable polynomial

$$P\left(x, \frac{\partial}{\partial x}\right) = \sum_{|\alpha|\le m} a_\alpha(x)\left(\frac{\partial}{\partial x}\right)^\alpha \tag{3.17}$$

in the second argument $z = \frac{\partial}{\partial x}$, whose components $z_i = \frac{\partial}{\partial x_i}$, $i = 1, \dots, n$ represent a partial differentiation in the scalar variable x_i, whereas the first argument stands for the domain variable $x = (x_1, \dots, x_n)^T$. Hereinafter, a multi-index $\alpha = (\alpha_1, \dots, \alpha_n)$ consists of nonnegative integers, $|\alpha| = \sum_{i=1}^n \alpha_i$, the polynomial coefficients $a_\alpha(x)$ are smooth enough (scalar, vector, or matrix) functions on some open domain $\mathscr{D} \subseteq \mathbb{R}^n$, and $\left(\frac{\partial}{\partial x}\right)^\alpha = \frac{\partial^{|\alpha|}}{\partial x^{\alpha_1}\dots\partial x^{\alpha_n}}$, particularly, the identity operator $I = \left(\frac{\partial}{\partial x}\right)^\alpha$ comes with the trivial multi-index $\alpha = (0, \dots, 0)$. Thus, a differential operator, being applied to a sufficiently smooth function $f(x) \in C^n(\mathscr{D})$ with the domain $\mathscr{D}$, is represented as a linear combination of f and its partial derivatives, involving higher degree m. Provided that $a_\alpha(\cdot) \ne 0$ for some α such that $|\alpha| = m$, the higher degree m is typically referred to as the *order* of the differential operator.

The *theta operator*

$$\Theta = \sum_{i=1}^{n} x_i \frac{\partial}{\partial x_i} \tag{3.18}$$

is an example of a scalar differential operator of the first order. The eigenspaces of Θ are spanned by homogenous polynomials in x. For instance, in the scalar case $n = 1$, the eigenfunctions are the monomials x^k of degrees $k = 0, 1, \ldots,$ which determine the corresponding eigenvalues (indeed, $\Theta(x^k) = kx^k$). The *nabla operator*

$$\nabla = \left(\frac{\partial}{\partial x_1}, \ldots, \frac{\partial}{\partial x_n}\right)^T \tag{3.19}$$

exemplifies a vector differential (gradient) operator, also of the first order, with the gradient notation $grad = \nabla$ that has become standard in the literature.

The *Laplace operator*

$$\Delta = (\nabla)^2 = \sum_{i=1}^{n} \frac{\partial^2}{\partial x_i^2} \tag{3.20}$$

represents a popular differential operator of the second order. Differential operators of parabolic, elliptic, and hyperbolic types are discussed next.

3.2.2 Parabolic, Elliptic, and Hyperbolic Operators

A second-order differential operator

$$a\frac{\partial^2}{\partial x_1^2} + 2b\frac{\partial^2}{\partial x_1 \partial x_2} + c\frac{\partial^2}{\partial x_2} + g\frac{\partial}{\partial x_1} + f\frac{\partial}{\partial x_2} + h \tag{3.21}$$

of two independent variables $x_1, x_2 \in \mathbb{R}^1$ with constant parameters a, b, c, g, f, h is said to be of *parabolic type* iff

$$b = ac. \tag{3.22}$$

Just in this case, the differential operator (3.21) matches to a geometry equation

$$ax_1^2 + 2bx_1x_2 + cx_x^2 + gx_1 + fx_2 + h = 0 \tag{3.23}$$

of a parabola.

In turn, a differential operator (3.21) is said to be of *elliptic type* iff

$$b < ac, \tag{3.24}$$

and it is said to be of *hyperbolic type* iff

$$b > ac. \tag{3.25}$$

Similar to the parabolic case (3.22), elliptic and hyperbolic operators match to the geometry equation (3.23) of an ellipse and hyperbola, respectively.

Being specified with $a = c = 1$ and $b = g = f = h = 0$, operator (3.21) takes the form of the Laplace operator (3.20) with two independent variables and it is clearly of the elliptic type. An example of a hyperbolic operator (3.21) comes with $a = 1, c = -1$ and $b = g = f = h = 0$.

The concepts of parabolic, elliptic, and hyperbolic operators admit straightforward generalizations of a generic differential operator (3.18) for higher domain dimensions and higher orders as well as for nonconstant operator parameters. For example, a parabolic operator is hidden in

$$\nabla\big(\alpha(x)\nabla\big) + \beta(x)\nabla + \gamma(x) \tag{3.26}$$

provided that the matrix-valued function $\alpha(x)$ has a one-dimensional kernel.

3.2.3 Green Function and Mild Solutions

A *Green function* $G(x, \xi)$ of a linear differential operator $P\big(x, \frac{\partial}{\partial x}\big)$, governed by (3.18) on the domain $\mathscr{D}$ and acting on distributions over a subset of a Euclidean space, is any solution of

$$P\left(x, \frac{\partial}{\partial x}\right) G(x, \xi) = \delta(\xi - x), \tag{3.27}$$

where δ is the *Dirac distribution*, which is typically referred to as a *delta function*. Recall that the Dirac distribution is defined indirectly by specifying its effect on a continuous test function $\varphi(\xi)$ as

$$< \delta(\xi - x), \varphi(\xi) >= \varphi(x).$$

In general, Green functions are distributions. If the kernel of P is nontrivial, then the Green function of P is not unique.

If the operator $P\Big(x, \frac{\partial}{\partial x}\Big) = P\Big(\frac{\partial}{\partial x}\Big)$ is translation invariant, i.e., it has constant coefficients with respect to x, then the Green function can be taken to be a convolution operator $G(x, \xi) = G(x - \xi)$. Just in this case, a Green function is the impulse response of the inhomogeneous linear differential equation (3.27), defined on a domain with specified initial or boundary conditions. This property of Green functions is exploited to solve differential equations of the form

$$P\left(x, \frac{\partial}{\partial x}\right) u(x) = f(x). \tag{3.28}$$

Indeed, for getting a solution $u(x)$ of (3.28) in terms of the Green function it suffices to note that

$$Pu(x) = \int_{\mathscr{D}} \delta(\xi - x) f(\xi) d\xi = \int_{\mathscr{D}} PG(x, \xi) f(\xi) d\xi = P \int_{\mathscr{D}} G(x, \xi) f(\xi) d\xi, \tag{3.29}$$

which suggests the solution of (3.28) in the form

$$u(x) = \int_{\mathscr{D}} G(x, \xi) f(\xi) d\xi \tag{3.30}$$

provided that the integral in the right-hand side exists.

With the same reasoning, any solution of the nonlinear PDE

$$P\left(x, \frac{\partial}{\partial x}\right) u(x) = f(x, u) \tag{3.31}$$

satisfies its integral counterpart

$$u(x) = \int_{\mathscr{D}} G(x, \xi) f(\xi, u(\xi)) d\xi. \tag{3.32}$$

However, not any solution of (3.32) satisfies its original version (3.31). The following terminology is therefore due.

Definition 3.1 A solution of the integral equation (3.32) is said to be a *mild solution* of (3.31).

To avoid a confusion, the term of a *strong solution* is reserved for a standard solution of the original PDE (3.31).

3.2.4 Weak Solutions

Motivated by the need for considering nonsmooth solutions of (3.28) with Dirac distributions in the right-hand side, weak solutions are invoked to capture boundary and singular (point-wise) in-domain effects. The weak solution concept is in play when the meaning of the PDE (3.28) over the domain $\mathscr{D}$ is viewed in the sense of Sobolev derivatives of Sect. 3.1.

Definition 3.2 A function $u(\cdot)$ is said to be a *weak solution* of (3.28) iff

$$\int_{\mathscr{D}} u(x)Q\left(x, \frac{\partial}{\partial x}\right)\varphi(x)dx = \int_{\mathscr{D}} f(x)\varphi(x)dx \tag{3.33}$$

for all test functions $\varphi(\cdot) \in C^\infty$ with compact support

$$supp\ \varphi = \overline{\{x \in \mathbb{R}^n : \varphi(x) \neq 0\}} \subseteq \mathscr{D},$$

where the adjoint operator Q is given by

$$Q\left(x, \frac{\partial}{\partial x}\right)\varphi(x) = \sum_{\alpha \le |m|} (-1)^{|\alpha|} \left(\frac{\partial[a_\alpha(x)\varphi(x)]}{\partial x}\right)^\alpha. \tag{3.34}$$

The underlying equality (3.33) for the weak solution of (3.28) to be satisfied is argued by multiplying both sides of the PDE (3.28) by a test function φ. While integrating the resulting relation by parts to transfer all the partial derivatives from a potential solution u to a test function φ, the multiplier $(-1)^{|\alpha|}$ shows up in (3.34). It should be noted that the integration by parts yields no boundary terms because $supp\ \varphi \subseteq \mathscr{D}$.

3.3 Sturm–Liouville Operators and Their Properties

A *Sturm–Liouville operator* is a second-order differential operator $A : D \to L_r^2(0, 1)$, governed by

$$(Af)(z) = -\frac{1}{r(z)}\frac{d}{dz}\left(p(z)\frac{df}{dz}(z)\right) + \frac{q(z)}{r(z)}f(z)$$
$$\text{for all } f \in D \text{ and } z \in (0, 1). \tag{3.35}$$

It is normally accompanied with

$$\begin{aligned} q &\in C^0([0, 1]), \\ p \in C^1([0, 1]) &: p(z) > 0 \text{ for all } z \in [0, 1], \\ r \in C^0([0, 1]) &: r(z) > 0 \ \text{ for all } z \in [0, 1], \end{aligned} \tag{3.36}$$

and with the set D of all functions $f \in H^2(0, 1)$ for which

$$b_1 f(0) + b_2\frac{df}{dz}(0) = a_1 f(1) + a_2\frac{df}{dz}(1) = 0, \tag{3.37}$$

where a_1, a_2, b_1, b_2 are real constants such that

$$|a_1| + |a_2| > 0, \ |b_1| + |b_2| > 0. \tag{3.38}$$

Hereinafter, the symbol $L_r^2(0, l)$ stands for the Hilbert space of square integrable functions with the weighted inner product

$$(f, g)_r = \sqrt{\int_0^l r(z) f(z) g(z) dz}, \quad f, g, \in L_r^2(0, l)$$

whereas $L^2(0, l)$ is typically for $L_r^2(0, l)$ with the unit function $r \equiv 1$. As usual, the symbols $\phi'(\cdot)$ and $\phi''(\cdot)$ are reserved for the first- and second-order derivatives in the scalar argument.

The Sturm–Liouville operator $A : D \to L_r^2(0, 1)$, defined by (3.35) and (3.37), is well-known (Gockenbach 2011; Petrovskii 1961) to generate an orthonormal basis in $L_r^2(0, 1)$, composed of the $L_r^2(0, 1)$-orthonormalized eigenfunctions

$$\phi_n \in D \cap C^2([0, 1]) : \|\phi_n\|_{L_r^2(0,1)} = \sqrt{\int_0^1 r(z) \phi_n^2(z) dz} = 1, \; n = 1, 2, \ldots \quad (3.39)$$

of the Sturm–Liouville boundary-value problem

$$A\phi = \lambda\phi, \quad b_1\phi(0) + b_2 \frac{d\phi}{dz}(0) = a_1\phi(1) + a_2 \frac{d}{dz}(1) = 0. \quad (3.40)$$

It is worth noticing that under condition

$$\lambda_1 > 0 \quad (3.41)$$

on the smallest eigenvalue λ_1, the Sturm–Liouville operator (3.35), (3.37) is positive definite in $L_r^2(0, 1)$, i.e., $(Af, f)_r > 0$ for any nonzero $f \in L_r^2(0, 1)$.

To ensure the validity of condition (3.41) one should additionally impose a restriction on the functional coefficient $q(z)$, e.g., to be positive definite $q(z) > 0$ for all $z \in [0, 1]$. Here, such a restriction is however not assumed explicitly not to miss a valid case of sign-indefinite (particularly, semi-definite) $q(z)$. An example of such a sign semi-definite case is the Sturm–Liouville operator (3.35), (3.37) with $q(z) = 0$ under *Dirichlet boundary conditions* where $a_2 = b_2 = 0$. Just in this case, only positive eigenvalues are in play as opposed to that under *Neumann boundary conditions* with $a_1 = b_1 = 0$, whose eigenvalue $\lambda_1 = 0$ violates condition (3.41).

For sign-definite coefficients

$$a_1, a_2, b_2 \geq 0, \; b_1 \leq 0, \quad (3.42)$$

it is subsequently demonstrated that

$$\sum_{n=1}^{\infty} \lambda_n^{-1} \max_{0 \leq z \leq 1} (|\phi_n(z)|) < +\infty. \quad (3.43)$$

The above convergence is well-recognized to justify the Fourier expansion of PDE solutions, admitting separation of independent variables. It is clear that convergence (3.43) is ensured by the quadratic eigenvalue growth, admitting the evaluation of the form $\lambda_n^{-1} = O(\frac{1}{n^2})$, $n = 1, 2, \ldots$, and by the uniform boundedness of the eigenfunctions $\phi_n(z)$. These Sturm–Liouville operator properties are revealed, following the line of reasoning used in Orlov (2017).

3.3.1 Eigenvalue Estimates

For comparison, let us consider another Sturm–Liouville operator

$$(A^* f)(z) = -\frac{1}{r^*(z)}\frac{d}{dz}\left(p_*(z)\frac{df}{dz}(z)\right) + \frac{q_*(z)}{r^*(z)} f(z)$$
$$\text{for all } f \in D \text{ and } z \in (0, 1), \tag{3.44}$$

which is subject to the same boundary conditions (3.37) and which possesses coefficients $p_* \in C^1([0, 1])$, $r^*, q_* \in C^0([0, 1])$ such that

$$\begin{aligned} 0 < p_*(z) \le p(z),\ q_*(z) \le q(z), \\ r^*(z) \ge r(z) \quad \text{for all } z \in [0, 1]. \end{aligned} \tag{3.45}$$

As shown in Petrovskii (1961, Sect. 22.5), the corresponding eigenvalues $\lambda_1^* < \lambda_2^* \cdots < \lambda_n^* < \cdots$ of the Sturm–Liouville operator (3.44), (3.37) prove to be upper estimated

$$\lambda_n^* \le \lambda_n,\ n = 1, 2, \ldots \tag{3.46}$$

by those of (3.35), (3.37). The above estimates allow one to derive explicit upper/lower eigenvalues bounds as functions of n by specifying either the Sturm–Liouville operator (3.35), (3.37) or (3.44), (3.37) with spatially invariant coefficients. Indeed, let $p_*(z) = p_*, q_*(z) = q_*, r^*(z) = r^*$ for certainty, and

$$p_* = \min_{z\in[0,1]} p(z),\ q_* = \min_{z\in[0,1]} q(z),\ r^* = \max_{z\in[0,1]} r(z). \tag{3.47}$$

Then the corresponding Sturm–Liouville problem of finding eigenvalues and eigenfunctions is written in the form

$$p_*\phi''(z) - q_*\phi(z) + \lambda r^*\phi(z) = 0, \tag{3.48}$$
$$b_1\phi(0) + b_2\phi'(0) = a_1\phi(1) + a_2\phi'(1) = 0. \tag{3.49}$$

In the particular case of Neumann boundary conditions where $a_1 = b_1 = 0$, it is straightforward to compute that the eigenfunctions are multiples of $\phi_n(z) =$

$\cos \pi(n-1)z$ and the eigenvalues are explicitly given by

$$\lambda_n^* = \frac{\pi^2(n-1)^2 p_* + q_*}{r^*}, \quad n = 1, 2, \dots . \tag{3.50}$$

Respectively, in the case of Dirichlet boundary conditions with $a_2 = b_2 = 0$, the eigenfunctions are multiples of $\phi_n(z) = \sin \pi n z$ and the eigenvalues are given by

$$\lambda_n^* = \frac{(\pi n)^2 p_* + q_*}{r^*}, \quad n = 1, 2, \dots . \tag{3.51}$$

Otherwise, the explicit integration of (3.48) under boundary conditions (3.49) yields the eigenfunctions, which are multiples of

$$\phi_n(z) = b_2 \cos \mu_n z - b_1 \frac{\sin \mu_n z}{\mu_n}, \quad n = 1, 2, \dots , \tag{3.52}$$

with the corresponding eigenvalues

$$\lambda_n^* = \frac{\mu_n^2 p_* + q_*}{r^*} \tag{3.53}$$

where $\mu_n,\ n = 1, 2, \dots$ are nonnegative solutions of the algebraic equation

$$\tan \mu = \frac{(a_1 b_2 - a_2 b_1)\mu}{a_1 b_1 + a_2 b_2 \mu^2} . \tag{3.54}$$

Since the algebraic equation (3.54) cannot be solved analytically, its solutions are found numerically as intersection points of the graphs of the left-hand and right-hand sides of (3.54), which are plotted in Fig. 3.1.

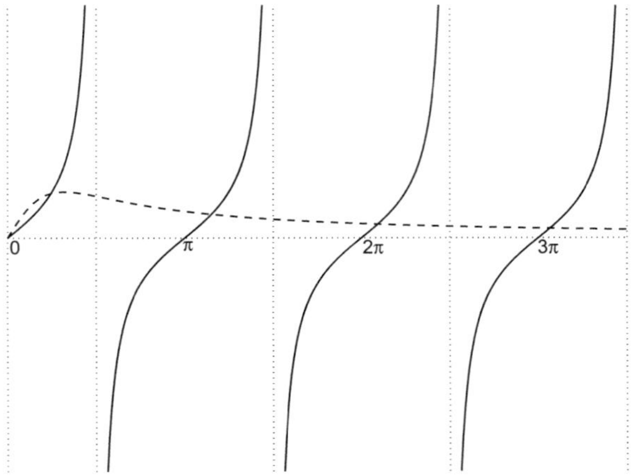

Fig. 3.1 The functions $\tan \mu$ (solid curve) and $\frac{2\mu}{1+\mu^2}$ (dashed curve). The intersection points are related to the eigenvalues of the Sturm–Liouville problem (3.48), (3.49) under Robin boundary conditions with $a_1, a_2, b_2 = 1$ and $b_1 = -1$. © 2017 *IEEE*. Reprinted, with permission, from Orlov (2017)

Due to the periodicity of the tangent function, the following conclusions are drawn from Fig. 3.1.

1. There are infinitely many nonnegative solutions $\mu_1 < \mu_2 < \cdots$ of the algebraic equation (3.54).
2. The smallest solution $\mu_1 = 0$ whereas the others are estimated as follows

$$\pi n < \mu_{n+1} < \pi(n+1), \ n = 1, 2, \ldots . \tag{3.55}$$

Taking into account (3.53), these conclusions allow one to determine the first eigenvalue $\lambda_1^* = \frac{q_*}{r^*}$ as well as to derive the subsequent eigenvalue estimates

$$\frac{\pi^2 n^2 p_* + q_*}{r^*} \leq \lambda_{n+1}^* < \frac{\pi^2 (n+1)^2 p_* + q_*}{r^*}, \ n = 1, 2, \ldots \tag{3.56}$$

where the non-strict inequality symbol has been involved to formally capture the case of Neumann boundary conditions (cf. (3.50)). By virtue of (3.46), the corresponding lower estimates

$$\lambda_n \geq \frac{\pi^2 (n-1)^2 p_* + q_*}{r^*}, \ n = 1, 2, \ldots , \tag{3.57}$$

are then obtained for the eigenvalues λ_n, $n = 1, 2, \ldots$ of the Sturm–Liouville operator (3.35) under either *Robin* or Neumann *boundary conditions* (3.37) with spatially varying coefficients subject to (3.47).

Relying on the Sturm–Liouville problem

$$p^* \phi''(z) - q^* \phi(z) + \lambda r_* \phi(z) = 0 \tag{3.58}$$

under the same (Robin or Neumann) boundary conditions (3.49), but with the coefficients

$$p^* = \max_{z \in [0,1]} p(z), \ q^* = \max_{z \in [0,1]} q(z), \ r_* = \min_{z \in [0,1]} r(z), \tag{3.59}$$

the upper eigenvalue estimates

$$\lambda_n \leq \frac{\pi^2 n^2 p^* + q^*}{r_*}, \ n = 1, 2, \ldots , \tag{3.60}$$

of the Sturm–Liouville operator (3.35), (3.37) with spatially varying coefficients p, r, q are derived in a similar way to that of deriving the lower estimates (3.57). Finally, coupling the lower and upper estimates (3.57) and (3.60) together yields

$$\frac{\pi^2 (n-1)^2 p_* + q_*}{r^*} \leq \lambda_n \leq \frac{\pi^2 n^2 p^* + q^*}{r_*}, \ n = 1, 2, \ldots . \tag{3.61}$$

In order to obtain similar eigenvalue estimates

$$\frac{\pi^2 n^2 p_* + q_*}{r^*} \le \lambda_n \le \frac{\pi^2 n^2 p^* + q^*}{r_*}, \quad n = 1, 2, \dots . \tag{3.62}$$

under the Dirichlet boundary conditions, it suffices to repeat the above derivation of (3.61), now employing both the explicit eigenvalue representation (3.51), found for the constant coefficients (3.47), and its corresponding counterpart, found for the constant coefficients (3.59).

Summarizing, the following result is obtained for a Sturm–Liouville operator with spatially varying coefficients.

Theorem 3.1 *Consider the Sturm–Liouville operator* (3.35), (3.37) *with positive functions* $p(z), r(z)$ *of class* $C^1([0, 1])$ *and* $C^0([0, 1])$, *respectively, with a continuous function* $q(z)$, *and with constant parameters* a_1, a_2, b_1, b_2 *such that* $|a_1| + |a_2| > 0$, $|b_1| + |b_2| > 0$. *Let the minimal and maximal values of these functions be determined by* (3.47) *and* (3.59) *and let* $\lambda_1 < \lambda_2 < \cdots$ *be the eigenvalues of the Sturm–Liouville operator. Then the eigenvalue estimate* (3.61) *is in force under both Robin and Neumann boundary conditions* (3.37) *with* $a_2 \neq 0$ *and/or* $b_2 \neq 0$ *whereas* (3.62) *is in force under Dirichlet boundary conditions* (3.37) *with* $a_2 = b_2 = 0$. *Moreover, the series, composed of the inverse eigenvalues of the Sturm–Liouville operator, is convergent*

$$\sum_{n=1}^{\infty} \lambda_n^{-1} < +\infty \tag{3.63}$$

for any (Robin, Neumann, Dirichlet) kind of boundary conditions, provided that the minimal eigenvalue λ_1 *is positive.*

Proof The lower and upper estimates (3.61), (3.62) have been established before. The inverse eigenvalue estimates

$$\frac{1}{\lambda_n} \le \frac{r^*}{\pi^2 (n-1)^2 p_* + q_*}, \quad n = 1, 2, \dots , \tag{3.64}$$

can then be derived from both (3.61) and (3.62) to justify the convergence

$$\sum_{n=N}^{\infty} \frac{1}{\lambda_n} \le \sum_{n=N}^{\infty} \frac{r^*}{\pi^2 (n-1)^2 p_* + q_*} < +\infty \tag{3.65}$$

where the summation starts with a sufficiently large $N = \min\{n : \pi^2 (n-1)^2 p_* + q_* > 0\}$ to avoid the appearance of an indefinite term in (3.65) that would correspond to a zero eigenvalue. Once λ_1 is positive its inverse value λ_1^{-1} is well-posed (so, all λ_n^{-1} $n = 1, 2, \dots$ are), and the series convergence (7.127) is then guaranteed by (3.65). Theorem 3.1 is thus proved. ■

3.3.2 Uniform Boundedness of the Eigenfunctions

In the case of spatially varying coefficients, the Sturm–Liouville problem in question relies on the second-order differential equation

$$\frac{d}{dz}\left(p(z)\frac{d\phi}{dz}(z)\right) - q(z)\phi(z) + \lambda r(z)\phi(z) = 0 \tag{3.66}$$

subject to the boundary conditions (3.49). In what follows, the coefficients $p(z)$ and $r(z)$ are additionally assumed to be of class $C^2([0, 1])$. Then, as shown in Petrovskii (1961, Sect. 23.2), the substitution

$$y = \int_0^z \sqrt{\frac{r(\zeta)}{p(\zeta)}}d\zeta, \; \psi = \sqrt[4]{r(y)p(y)}\phi \tag{3.67}$$

is applicable to Eq. (3.66) to represent it in terms of $\psi(y)$ in a simplified form

$$\frac{d^2\psi(y)}{dy^2} + \lambda\psi(y) = R(y)\psi(y), \; y \in (0, l) \tag{3.68}$$

where $l = \int_0^1 \sqrt{\frac{r(\zeta)}{p(\zeta)}}d\zeta$ and

$$R = \frac{\eta q - (\eta' p)'}{r\eta}, \; \eta = \frac{1}{\sqrt[4]{rp}}. \tag{3.69}$$

The boundary conditions (3.49) are respectively represented in the form

$$\beta_1\psi(0) + \beta_2\psi'(0) = \alpha_1\psi(l) + \alpha_2\psi'(l) = 0 \tag{3.70}$$

where

$$\beta_1 = \frac{1}{\sqrt[4]{r(0)p(0)}}\left[b_1 - b_2\frac{r'(0)p(0) + r(0)p'(0)}{4\sqrt{r(0)p^3(0)}}\right],$$
$$\alpha_1 = \frac{1}{\sqrt[4]{r(1)p(1)}}\left[a_1 - a_2\frac{r'(1)p(1) + r(1)p'(1)}{4\sqrt{r(1)p^3(1)}}\right],$$
$$\beta_2 = b_2\sqrt[4]{\frac{r(0)}{p^3(0)}}, \quad \alpha_2 = a_2\sqrt[4]{\frac{r(1)}{p^3(1)}}. \tag{3.71}$$

If $\phi_n(z) \in L_r^2(0, 1)$ is a normalized eigenfunction of the Sturm–Liouville problem (3.39), (3.49), (3.66) with an eigenvalue λ_n then by inspection, relation

$$\psi_n = \sqrt{r(y)p(y)}\phi_n \in L^2(0,l) \tag{3.72}$$

determines an eigenfunction of the Sturm–Liouville problem (3.68), (3.70), corresponding to the same eigenvalue λ_n and normalized in $L^2(0,l)$ according to

$$\|\psi_n\|_{L^2(0,l)} = \sqrt{\int_0^l \psi_n^2(y)dy} = 1. \tag{3.73}$$

The general solution of the linear second-order differential equation (3.68) is straightforwardly verified (Petrovskii 1961, Sect. 23.2) to satisfy the integral equation

$$\begin{aligned}\psi(y) &= \psi(0)\cos\sqrt{\lambda}y + \tfrac{\psi'(0)}{\sqrt{\lambda}}\sin\sqrt{\lambda}y\\ &+\tfrac{1}{\sqrt{\lambda}}\textstyle\int_0^y R(s)\psi(s)\sin\sqrt{\lambda}(y-s)ds.\end{aligned} \tag{3.74}$$

By letting $\lambda = \lambda_n,\ n = 1, 2, \ldots$ in (3.74), one arrives at the integral equation on an arbitrary eigenfunction $\psi_n(y)$ of the Sturm–Liouville problem (3.68), (3.70), corresponding to the eigenvalue λ_n.

The subsequent development focuses on the Robin/Neumann boundary conditions with $\beta_2 \neq 0$. The verification of the uniform boundedness of the normalized eigenfunctions for the Dirichlet boundary conditions with $b_2 = \beta_2 = 0$ may be found in Petrovskii (1961, Sect. 23.2).

Provided that $\beta_2 \neq 0$, the integral equation (3.74) is straightforwardly specified to

$$\begin{aligned}\psi_n(y) &= \psi_n(0)\cos\sqrt{\lambda_n}y - \tfrac{\beta_1\psi_n(0)}{\beta_2\sqrt{\lambda_n}}\sin\sqrt{\lambda_n}y\\ &+\tfrac{1}{\sqrt{\lambda_n}}\textstyle\int_0^y R(s)\psi_n(s)\sin\sqrt{\lambda_n}(y-s)ds\end{aligned} \tag{3.75}$$

by applying the boundary condition (3.70) at $y = 0$. The particular eigenfunction $\Psi_n(y)$, initialized with $\Psi_n(0) = \beta_2$, is then governed by the integral equation

$$\begin{aligned}\Psi_n(y) &= \beta_2\cos\sqrt{\lambda_n}y - \tfrac{\beta_1}{\sqrt{\lambda_n}}\sin\sqrt{\lambda_n}y\\ &+\tfrac{1}{\sqrt{\lambda_n}}\textstyle\int_0^y R(s)\Psi_n(s)\sin\sqrt{\lambda_n}(y-s)ds.\end{aligned} \tag{3.76}$$

Since an eigenfunction of the Sturm–Liouville problem (3.68), (3.70), corresponding to the eigenvalue λ_n, is a multiple of (3.76), its normalized representative with the unit norm (3.73) is given by

$$\psi_n(y) = \frac{\Psi_n(y)}{\|\Psi_n\|_{L^2(0,l)}},\quad n = 1, 2, \ldots. \tag{3.77}$$

Let us show that the eigenfunctions $\psi_n(y),\ n = 1, 2, \ldots$ are uniformly bounded by a constant M, independent of n. Setting

$$M_n = \max_{y \in [0,l]} \Psi_n(y), \tag{3.78}$$

one concludes from (3.76) that

$$|\Psi_n(y)| \leq \beta_2 + \frac{|\beta_1|}{\sqrt{\lambda_n}} + \frac{1}{\sqrt{\lambda_n}} M_n \int_0^l |R(s)| ds \ \forall y \in [0, 1], \tag{3.79}$$

and hence

$$M_n \leq \beta_2 + \frac{|\beta_1|}{\sqrt{\lambda_n}} + \frac{1}{\sqrt{\lambda_n}} M_n \int_0^l |R(s)| ds. \tag{3.80}$$

It follows that

$$M_n \leq \frac{\beta_2 + \frac{|\beta_1|}{\sqrt{\lambda_n}}}{1 - \frac{1}{\sqrt{\lambda_n}} \int_0^l |R(s)| ds} \leq M \tag{3.81}$$

for some constant $M > 0$ because relation (3.69) determines a continuous function $R(\cdot)$ on $[0, l]$ and $\lambda_n \to \infty$ as $n \to \infty$ by virtue of (3.61) and (3.62).

Moreover, due to (3.81), relation (3.76) leads to

$$\Psi_n^2(y) = \beta_2^2 \cos^2 \sqrt{\lambda_n} y + O\left(\frac{1}{\sqrt{\lambda_n}}\right), \tag{3.82}$$

and the straightforward integration of (3.82) yields

$$\begin{aligned} \int_0^l \Psi_n^2(y) dy &= \beta_2^2 \frac{l}{2} + \beta_2^2 \frac{\sin 2\sqrt{\lambda_n}\, l}{4\sqrt{\lambda_n}} + O\left(\frac{1}{\sqrt{\lambda_n}}\right) \\ &= \beta_2^2 \frac{l}{2} + O\left(\frac{1}{\sqrt{\lambda_n}}\right), \end{aligned} \tag{3.83}$$

thereby ensuring that for sufficiently large n, the norms $\|\Psi_n\|_{L^2(0,l)}$ are uniformly separated from zero. That is why along with the uniform boundedness (3.81) of magnitudes (3.78) of the particular eigenfunctions (3.76), the normalized functions (3.77) are uniformly bounded as well.

By inverting relation (3.72), the uniform boundedness is finally verified for the normalized eigenfunctions $\phi_n(z),\ n = 1, 2, \ldots$ of the Sturm–Liouville problem (3.39), (3.49), (3.66). Summarizing, one arrives at the following.

Theorem 3.2 *Let conditions of Theorem 3.1 be satisfied and let the functions* $p(z), r(z)$ *be additionally of class* $C^2([0, 1])$. *Then the normalized eigenfunctions* $\phi_n(z),\ n = 1, 2, \ldots$ *of the Sturm–Liouville operator* (3.35), (3.37)) *are uniformly bounded in the independent variable* $z \in [0, 1]$ *and in the number n. Moreover, the series convergence* (3.43) *is in force provided that the minimal eigenvalue* λ_1 *of the Sturm–Liouville operator is positive.*

Proof It has been shown that substitution (3.67) transforms the original Sturm–Liouville problem (3.39), (3.49), (3.66) of finding normalized eigenfunctions $\phi_n(z)$ and eigenvalues $\lambda_n,\ n = 1, 2, \ldots$ of the Sturm–Liouville operator (3.35), (3.37), to the one, governed by (3.68), (3.70), (3.73), whose eigenfunctions $\psi_n(y)$ possess the same eigenvalues, and relation (3.72) establishes the correspondence between the original and transformed eigenfunctions. Due to the invertibility of relation (3.72), the normalized eigenfunctions $\phi_n(z),\ n = 1, 2, \ldots$ of the original Sturm–Liouville problem are uniformly bounded in $z \in [0, 1]$ and n iff their transformed counterparts (3.72) are uniformly bounded in $y \in [0, l]$ and n.

In the Dirichlet case of $b_2 = \beta_2 = 0$, the uniform boundedness of the eigenfunctions in question has been established in Petrovskii (1961, Sect. 23.2). The same property is now established under $\beta_2 \neq 0$. For this purpose, the uniform boundedness (3.81) is first verified for all magnitudes (3.78) of specific non-normalized eigenfunctions (3.76), found through the general solution representation (3.74) of the transformed Sturm–Liouville problem (3.68), (3.70). Then the L_2-norms of these eigenfunctions $\Psi_n,\ n = 1, 2, \ldots$ are evaluated in the form (3.83) to demonstrate that the norms $\|\Psi_n\|_{L^2(0,l)}$ are uniformly separated from zero for sufficiently large n. Therefore, the uniform boundedness property of the normalized eigenfunctions (3.77) is inherited from that of the specific non-normalized eigenfunctions (3.76).

To complete the proof it suffices to note that the established uniform boundedness of the normalized eigenfunctions $\phi_n(z),\ n = 1, 2, \ldots$, coupled to assertion (7.127) of Theorem 3.1, results in the series convergence (3.43). ■

3.4 Separation of Variables

Separation of variables is a powerful method of solving linear PDEs in an explicit form. If applicable, the method allows one to explicitly determine the corresponding Green function. The applicability of the method is frequently reduced to an appropriate Sturm–Liouville problem of specifying an orthonormal basis with respect to which a general solution is expanded into a Fourier series. The basic idea of the method is further illustrated side by side for parabolic and hyperbolic PDEs with a special attention to determining weak solutions, matching to a point-wise right-hand side of such a PDE.

3.4.1 Parabolic Case Study

Consider a one-dimensional parabolic PDE

$$\begin{aligned} r(z)\tfrac{\partial x}{\partial z}(t,z) &= \tfrac{\partial}{\partial z}\left(p(z)\tfrac{\partial x}{\partial z}(t,z)\right) - q(z)x(t,z),\ t > 0,\ z \in (0,1) \\ b_1 x(t,0) + b_2\tfrac{\partial x}{\partial z}(t,0) &= 0,\ a_1 x(t,1) + a_2\tfrac{\partial x}{\partial z}(t,1) = 0, \end{aligned} \tag{3.84}$$

associated with the Sturm–Liouville operator (3.35), (3.37), involving the plant parameters (3.36), (3.42). The above equation describes the heat propagation $x(t, z)$ in a one-dimensional rod at the time instant t along the spatial variable $z \in [0, 1]$ with the heat capacity $r(z) > 0$, the diffusivity $p(z) > 0$, and the heat exchange $q(z)$. The sign-definite boundary coefficients (3.42) match to the physically meaningful case of the heat dynamics (3.84) which flow out of the boundaries (see, e.g., Gockenbach 2011, Sect. 10.5 for details).

The key point of separation of variables is in seeking specific solutions of (3.84), representable as a product

$$x(t, z) = T(t)Z(z) \tag{3.85}$$

of functions $T(t)$ and $Z(z)$ of independent variables t and z, respectively. Then substituting (3.85) into (3.84) yields a relation

$$r(z)Z(z)\dot{T}(t) = T(t)[p(z)Z'(z)]' - q(z)T(t)Z(z) \tag{3.86}$$

to hold true for such a solution. By collecting the T-terms and Z-terms on the opposite sides, it follows that

$$\frac{\dot{T}(t)}{T(t)} = \frac{[p(z)Z'(z)]' - q(z)Z(z)}{r(z)Z(z)}. \tag{3.87}$$

Since the left-hand side of (3.87) depends on the time variable t whereas the right-hand side depends on the spatial variable z, the latter equality proves to be in force only if it remains constant, independent of t and z. Setting this constant $-\lambda$, the following relations

$$\frac{\dot{T}(t)}{T(t)} = -\lambda, \tag{3.88}$$

$$\frac{[p(z)Z'(z)]' - q(z)Z(z)}{r(z)Z(z)} = -\lambda \tag{3.89}$$

are deduced from (3.87).

Once the same separation of variables is applied to the boundary conditions of (3.84), a hypothetical solution (3.85) is concluded to obey similar boundary conditions

$$b_1 Z(0) + b_2 Z'(0) = 0, \quad a_1 Z(1) + a_2 Z'(1) = 0. \tag{3.90}$$

Relations (3.89), (3.90), thus obtained, represent a Sturm–Liouville problem on the specific solutions (3.85) of the underlying PDE (3.84).

As noted in Sect. 3.3, the solutions to the Sturm–Liouville problem (3.89), (3.90) generate an orthonormal basis in $L_r^2(0, 1)$, composed of the $L_r^2(0, 1)$ orthonormalized eigenfunctions $\phi_n(z)$, $n = 1, 2, \ldots$, corresponding to the eigenvalues λ_n. Thus, a solution of (3.84), admitting a separation of variables (3.85), is formed by an eigenfunction $Z(z) = \phi_n(z)$, multiplied by a solution $T(t) = T_n^0 e^{-\lambda_n t}$ of (3.98), which is initialized with an arbitrary value T_n^0 and specified with the corresponding eigenvalue

$\lambda = \lambda_n$. Moreover, an arbitrary solution of (3.84) is expanded into the Fourier series

$$x(t, z) = \sum_{n=1}^{\infty} T_n^0 e^{-\lambda_n t} \phi_n(z) \tag{3.91}$$

with respect to the $L_r^2(0, 1)$-orthonormal basis $\{\phi_n(z)\}_{n=1}^{\infty}$, formed by the eigenfunctions $\phi_n(z)$, $n = 1, 2, \dots$ of the Sturm–Liouville problem (3.89), (3.90) with the corresponding eigenvalues λ_n. If the exact response to a particular initial function $x(0, z) = x_0 \in L_r^2(0, 1)$ is of interest, e.g., for determining the Green function $G(z, \zeta)$ of the parabolic PDE (3.84), one should specify (3.91) with the constants T_n^0, $n = 1, 2, \dots$ in such a manner that the initial condition

$$x_0(z) = \sum_{n=1}^{\infty} T_n^0 \phi_n(z) \tag{3.92}$$

holds true. Multiplying (3.92) by $r(z)\phi_n(z)$ and then integrating the resulting equality along the spatial domain, and finally employing the $L_r^2(0, 1)$-orthonormality of the eigenfunctions $r(z)\phi_n(z)$, yields

$$T_n^0 = \int_0^1 r(z)x_0(z)\phi_n(z)dz, \; n = 1, 2, \dots \tag{3.93}$$

are the corresponding Fourier coefficients of the initial function $x_0 \in L_r^2(0, 1)$. Relations (3.91) and (3.93), coupled together, allow one to reproduce the Green function of the parabolic PDE (3.84) in the explicit form (cf. (3.27)):

$$G(z, \zeta) = \sum_{n=1}^{\infty} \phi_n(z)\phi_n(\zeta)e^{-\lambda_n t}. \tag{3.94}$$

Let the symbol $L_\infty(0, \tau; L_r^2(0, 1))$ stand for the set of functions $f(t, z)$ such that $f(t, \cdot) \in L_r^2(0, 1)$ for almost all $t \in (0, \tau)$, and $\int_0^1 r(z)f(t, z)\phi(z)dz$ is Lebesgue measurable in t for all $\phi(\cdot) \in L_2(0, 1)$, and $ess\ \sup_{t\in(0,\tau)} \int_0^1 r(z)f^2(t, z)dz < \infty$. Then given $f(t, \cdot) \in L_{loc}^\infty\left(L_r^2(0, 1)\right) = \bigcap_{\tau>0} L^\infty\left(0, \tau; L_r^2(0, 1)\right)$, the general solution of the inhomogeneous parabolic PDE

$$\begin{aligned} r(z)\tfrac{\partial x}{\partial z}(t, z) = \tfrac{\partial}{\partial z}\left(p(z)\tfrac{\partial x}{\partial z}(t, z)\right) - q(z)x(t, z) + f(t, z), \; t > 0, \; z \in (0, 1) \\ b_1 x(t, 0) + b_2\tfrac{\partial x}{\partial z}(t, 0) = 0, \; a_1 x(t, 1) + a_2\tfrac{\partial x}{\partial z}(t, 1) = 0 \end{aligned} \tag{3.95}$$

is represented in accordance with (3.30) as

$$x(t,z) = \sum_{n=1}^{\infty} T_n^0 e^{-\lambda_n t}\phi_n(z)$$
$$+\sum_{n=1}^{\infty}\int_0^t e^{-\lambda_n(t-\tau)}\int_0^1 r(\zeta)f(\tau,\zeta)\phi_n(\zeta)d\zeta d\tau\phi_n(z). \qquad (3.96)$$

It is clear that the solution representation (3.96) can straightforwardly be derived by substituting the Fourier series

$$f(t,z) = \sum_{n=1}^{\infty}\int_0^1 r(\zeta)f(t,\zeta)\phi_n(\zeta)d\zeta\phi_n(z) \qquad (3.97)$$

into (3.95) for $f(t,z)$ and by applying the separation-of-variables method to the resulting inhomogeneous PDE (3.95). Just in this case, (3.98) is modified to

$$\dot{T}_n(t) = -\lambda_n T_n(t) + \int_0^1 r(\zeta)f(t,\zeta)\phi_n(\zeta)d\zeta, \qquad (3.98)$$

thereby validating (3.96).

It is worth noticing that in a particular case of the spatially uniform heat capacity $r(z)=r$, diffusivity $p(z)=p$, and heat exchange $q(z)=q$, the Green function (3.94) of (3.84) is simplified to

$$G(t,z,\zeta) = 2\sum_{n=1}^{\infty}\sin\pi nz\sin\pi n\zeta\, e^{-\frac{q+p(\pi n)^2}{r}t} \qquad (3.99)$$

under the Dirichlet boundary conditions with $a_2=b_2=0$, and it is

$$G(t,z,\zeta) = 1+2\sum_{n=1}^{\infty}\cos\pi nz\cos\pi n\zeta\, e^{-\frac{q+p(\pi n)^2}{r}t} \qquad (3.100)$$

under the Neumann boundary conditions with $a_1=b_1=0$. The interested reader may refer to the handbook (Butkovskii 1982) for specific Green functions of parabolic PDEs.

Remarkably, the solution representation (3.96) remains in force for weak solutions of (3.95), matching to point-wise distributions

$$f(t,z) = u(t)\delta(z-z_0) \qquad (3.101)$$

with locally integrable $u(t)$ and $z_0 \in (0,1)$. Since being specified with (3.101), the right-hand side of (3.95) contains the Dirac distribution $\delta(z-z_0)$, the meaning of such a boundary-value problem (3.95) is defined indirectly according to the *weak solution* concept of Sect. 3.2.4.

Definition 3.3 A continuous function $x(t, \cdot) \in H^1(0, 1)$ is said to be a *weak solution* of (3.95), (3.101) on $[0, \tau)$ if it satisfies the corresponding boundary conditions and for every test function $\varphi(z) \in H^1(0, 1)$ with compact support $supp\ \varphi \subseteq [0, 1]$, the function $\int_0^1 x(t, z)\varphi(z)dz$ is absolutely continuous on $[0, \tau)$ and relation

$$\frac{d}{dt}\int_0^1 r(z)x(t, z)\varphi(z)dz =$$
$$-\int_0^1 p(z)x'(t, z)\varphi'(z)dz - \int_0^1 q(z)x(t, z)\varphi(z)dz + u(t)\varphi(z_0) \quad (3.102)$$

holds for almost all $t \in [0, \tau)$.

The weak solution concept (3.102) relies on the well-defined action $\int_0^1 \varphi(z)\delta(z - z_0)dz = \varphi(z_0)$ of the shifted Dirac distribution $\delta(z - z_0)$, $z_0 \in [0, 1]$ on an arbitrary test function $\varphi(\xi) \in H^1(0, 1)$ with compact support $supp\ \varphi \subseteq [0, 1]$ and it is based on the integration-by-parts property

$$\int_0^1 [p(z)x'(t, z)]'\varphi(z)dz = -\int_0^1 p(z)x'(t, z)\varphi'(z)dz$$

of the Sobolev derivatives of the $H^1(0, 1)$-valued functions. Since the Fourier coefficients of the point-wise distribution (3.101) with respect to the $L_r^2(0, 1)$-orthonormal eigenfunctions $\phi_n(z)$, $n = 1, 2, \ldots$ of the Sturm–Liouville problem (3.89), (3.90) is given by

$$\int_0^1 r(\zeta)f(\tau, \zeta)\phi_n(\zeta)d\zeta = u(\tau)r(z_0)\phi_n(z_0), \quad (3.103)$$

the solution representation (3.96) is simplified in the present case to

$$x(t, z) = \sum_{n=1}^{\infty} T_n^0 e^{-\lambda_n t}\phi_n(z)$$
$$+r(z_0)\sum_{n=1}^{\infty}\phi_n(z_0)\int_0^t e^{-\lambda_n(t-\tau)}u(\tau)d\tau\ \phi_n(z). \quad (3.104)$$

Due to the uniqueness of the Green function, relation (3.104) may be viewed as an *alternative definition of the weak solution* of the PDE (3.95) with the Dirac distribution (3.101).

It is worth noticing that the solution representation (3.104) remains in force for $z_0 = 0$ as well as for $z_0 = 1$ provided that $b_2 \neq 0$ and, respectively, $a_2 \neq 0$. Just in the case, the point-wise distribution (3.101) corresponds to the non-homogenous boundary condition $b_1 x(t, 0) + b_2 \frac{\partial x}{\partial z}(t, 0) = -u(t)$ for $z_0 = 0$ and to $a_1 x(t, 1) + a_2 \frac{\partial x}{\partial z}(t, 1) = u(t)$ for $z_0 = 1$.

More precisely (Butkovskii 1982), an arbitrary mild solution of such a boundary-value problem (3.84), accompanied with the above non-homogeneous boundary conditions, coincides with the weak solution of the boundary-value problem (3.95), satisfying the same initial condition and the corresponding homogeneous boundary conditions, the non-homogenous right-hand side of which is specified with (3.101) and either $z_0 = 0$ or $z_0 = 1$, respectively. To reproduce these conclusions it suffices to introduce the auxiliary function $\tilde{x}(t, z) = x(z, t) + 1 - \chi(z) + \chi(z - 1)$ where $x(z, t)$ is a mild solution of the former boundary-value problem and $\chi(z)$ is the Heaviside step function. It is then straightforward to verify that $\tilde{x}(t, z)$ coincides with the mild solution $x(t, z)$ inside the interval $(0, 1)$, whereas it proves to be a weak solution of the latter boundary-value problem by inspection.

In the case of the non-homogeneous Dirichlet boundary conditons $x(t, 0) = -u(t)$ and $x(t, 1) = u(t)$ at either end when $b_2 = 0$ or $a_2 = 0$, the corresponding point-wise distribution (3.101) should be modified to Butkovskii (1982)

$$f(t, z) = u(t)\delta'(z - i), \quad i = 0 \quad \text{or} \quad i = 1 \tag{3.105}$$

where the δ-function derivative $\delta'(z)$ is defined on compactly supported smooth test functions $\phi(z)$ by $\delta'(z - z_0)\phi(z) = -\delta\phi'(z) = -\phi(z_0)$ in the sense of distributions. With this in mind, the solution representation under the above non-homogeneous Dirichlet boundary conditions takes the form

$$\begin{aligned} x(t, z) = &\sum_{n=1}^{\infty} T_n^0 e^{-\lambda_n t} \phi_n(z) \\ &- \sum_{n=1}^{\infty} [r(z)\phi_n(z)]' \Big|_{z=i} \int_0^t e^{-\lambda_n (t-\tau)} u(\tau) d\tau \; \phi_n(z), \end{aligned} \tag{3.106}$$

where $i = 0$ stands for the non-homogeneity $-u(t)$, located at the left boundary and corresponding to $z_0 = 0$, and $i = 1$ is for the non-homogeneity $u(t)$, located at the right boundary and corresponding to $z_0 = 1$.

Clearly, relation (3.106) may be viewed as an *alternative definition of the weak solution* of the PDE (3.95) with the distributional Dirac derivative (3.105).

3.4.2 Hyperbolic Case Study

Let us now consider a one-dimensional hyperbolic PDE

$$\begin{aligned} r(z)\tfrac{\partial^2 x}{\partial z^2}(t, z) &= \tfrac{\partial}{\partial z}\left(p(z)\tfrac{\partial x}{\partial z}(t, z)\right) - q(z)x(t, z), \; t > 0, \; z \in (0, 1) \\ b_1 x(t, 0) + b_2 \tfrac{\partial x}{\partial z}(t, 0) &= 0, \; a_1 x(t, 1) + a_2 \tfrac{\partial x}{\partial z}(t, 1) = 0, \end{aligned} \tag{3.107}$$

associated with the same Sturm–Liouville operator (3.35)–(3.37), (3.42). The above equation is referred to as a wave equation because it describes the string oscillation $x(t, z)$ at the time instant t along the position location $z \in [0, 1]$ with the mass density $r(z)$, the elasticity $p(z)$, and the stiffness $q(z)$.

As in the parabolic case of Sect. 3.4.1, the separation-of-variables substitution (3.85) into (3.107) results in the Sturm–Liouville problem (3.89), (3.90) on the spatially varying solution component $Z(z)$ of the wave PDE (3.107) whereas its time-varying component $T(t)$ meets the second-order ODE

$$\frac{\ddot{T}(t)}{T(t)} = -\lambda, \tag{3.108}$$

where λ is an eigenvalue of (3.89), (3.90).

Thus, similar to the parabolic PDE (3.84), the general solution of the wave PDE (3.107) can be expanded into the Fourier series with respect to the $L_r^2(0,1)$-orthonormal basis $\{\phi_n(z)\}_{n=1}^{\infty}$, formed by the eigenfunctions $\phi_n(z)$, $n = 1, 2, \ldots$ of the Sturm–Liouville problem (3.89), (3.90) with the corresponding eigenvalues λ_n. For instance, under the physically reasonable restriction

$$q(z) \geq 0 \text{ for all } z \in [0, 1] \tag{3.109}$$

on the string stiffness, all the eigenvalues $\lambda_1, \lambda_2, \ldots$ turn out to be nonnegative, and the general solution of (3.107) is then represented by the Fourier series

$$x(t,z) = \sum_{n=1}^{\infty} \left[T_n' \sin\sqrt{\lambda_n}\, t + T_n'' \cos\sqrt{\lambda_n}\, t \right] \phi_n(z) \tag{3.110}$$

where the constants T_n', T_n'', $n = 1, 2, \ldots$ are determined by the initial functions $x(0,z)$ and $\dot{x}(0,t)$. In a particular case of the unit density $r(z) \equiv 1$ and elasticity $p(z) \equiv 1$, no-value stiffness $q(z) \equiv 0$, and Dirichlet boundary conditions, which are specified with $a_2 = b_2 = 0$, the Green function of the wave PDE (3.107) is readily deduced from (3.110) in the form

$$G(z,\zeta) = \frac{4}{\pi} \sum_{k=0}^{\infty} \frac{1}{2k+1} \sin\frac{(2k+1)\pi z}{2} \sin\frac{(2k+1)\pi\zeta}{2} \sin\frac{(2k+1)\pi t}{2}. \tag{3.111}$$

Other specific Green functions of wave PDEs may be found in Butkovskii (1982). As a matter of fact, Green functions are applicable in the hyperbolic case as well to explicitly represent solutions of the inhomogeneous wave PDE

$$\begin{aligned} r(z)\tfrac{\partial^2 x}{\partial z^2}(t,z) &= \tfrac{\partial}{\partial z}\left(p(z)\tfrac{\partial x}{\partial z}(t,z)\right) - q(z)x(t,z) + f(t,z),\ t > 0,\ z \in (0,1) \\ b_1 x(t,0) + b_2 \tfrac{\partial x}{\partial z}(t,0) &= 0,\ a_1 x(t,1) + a_2 \tfrac{\partial x}{\partial z}(t,1) = 0, \end{aligned} \tag{3.112}$$

in the Fourier series form provided that $f(t,\cdot) \in L_{loc}^{\infty}\left(L_r^2(0,1)\right)$. The same is also true for weak solutions of the inhomogeneous wave PDE (3.112), which is specified with a point-wise distribution (3.101) and the meaning of which is viewed along the weak solution concept to straightforwardly be extended from the parabolic case of Definition 3.3 to the present hyperbolic case.

3.5 Nonlinear First-Order Partial Differential Equations

A scalar nonlinear PDE of the first order is of the form

$$F(x, V(x), \nabla V(x)) = 0, \tag{3.113}$$

where F is a continuous scalar function of the arguments $x = (x_1, \ldots, x_n)^T \in \mathbb{R}^n$, $V(\cdot) \in \mathbb{R}$, $\nabla V(\cdot) \in \mathbb{R}^n$, and $\nabla = \left(\frac{\partial}{\partial x_1}, \ldots, \frac{\partial}{\partial x_n}\right)^T$ is the nabla operator.

It should be pointed out that (3.113) may be viewed in the time-varying setting as well when the argument $(x_1, \ldots, x_n, t)^T \in \mathbb{R}^{n+1}$ is formally substituted into (3.127) for x and the operator nabla reads $\nabla = \left(\frac{\partial}{\partial x_1}, \ldots, \frac{\partial}{\partial x_n}, \frac{\partial}{\partial t}\right)^T$.

In general, the above PDE is not solvable in the class of smooth functions which is why the meaning of such a PDE should be revisited to admit continuous solutions of (3.113).

3.5.1 Viscosity Solutions of First-Order PDEs

This section focuses on the viscosity solution concept of Crandall and Lions (1983) that addresses continuous solutions of PDEs.

Definition 3.4 A continuous function $V(x)$ is said to be a *viscosity solution* of (3.113) iff for all x, whenever W is a smooth function such that $V - W$ possesses a local minimum at x, one has

$$F(x, V, \nabla W) \geq 0, \tag{3.114}$$

and whenever W is a smooth function such that $V - W$ possesses a local maximum at x, one has

$$F(x, V, \nabla W) \leq 0. \tag{3.115}$$

The set $\partial^D V(x)$ of values $\nabla W(x)$, where W is a smooth function such that $V - W$ possesses a local maximum at x is referred to as the *Dini (or viscosity) superdifferential* of V at x. The corresponding concept with minimum instead of maximum defines the *Dini (or viscosity) sub-differential* $\partial_D V(x)$. This terminology is due to the fact (see Clarke 1983 for details) that in the superdifferential case, one has $\zeta \in \partial^D V(x)$ iff the *directional Dini superderivative*

$$D^+V(x; \nu) = \limsup_{\mu \to \nu;\ \tau \to 0+} \frac{1}{\tau} [V(x + \tau\mu) - V(x)] \tag{3.116}$$

satisfies

$$D^+V(x; \nu) \leq \zeta^T \nu \text{ for any } \nu \in \mathbb{R}^n. \tag{3.117}$$

Respectively, in the sub-differential case, one has $\zeta \in \partial_D V(x)$ iff the *directional Dini subderivative*

$$D_+ V\,(x;\,\nu) = \liminf_{\mu \to \nu;\ \tau \to 0+} \frac{1}{\tau}\,[V\,(x + \tau\mu) - V\,(x)] \tag{3.118}$$

satisfies

$$D_+ V(x;\,\nu) \geq \zeta^T \nu \text{ for any } \nu \in \mathbb{R}^n. \tag{3.119}$$

In the set-valued analysis (Aubin and Cellina 1984), the directional Dini superderivative (3.116) and subderivative (3.118) are called the *contingent hypoderivative* and, respectively, *contingent epiderivative*.

In terms of Dini sub- and superdifferentials, a viscosity solution of (3.113) is redefined as a continuous function $V(x)$ that satisfies the following inequalities

$$F(x, V, \partial_D V(x)) \geq 0 \tag{3.120}$$

$$F(x, V, \partial^D V(x)) \leq 0 \tag{3.121}$$

for all $x \in \mathbb{R}^n$.

It is of interest to note that specific linear and quadratic functions $W(x)$ that appear in the viscosity solution definition are perfectly addressed in terms of proximal sub- and supergradients.

Recall (Clarke 1983) that a vector $\zeta(y) \in \mathbb{R}^n$ is said to be a *proximal supergradient* of a scalar function $f(x)$ at $y \in \mathbb{R}^n$ iff there exists some $\sigma(y) \geq 0$ such that

$$f(x) \leq f(y) + \zeta^T(y)(x - y) + \sigma(y)\|x - y\|^2 \tag{3.122}$$

for all x in some neighborhood $U(y) \subset \mathbb{R}^n$ of y. Complimentary to the above is the proximal sub-gradient concept. A vector $\zeta(y) \in \mathbb{R}^n$ is said to be a *proximal sub-gradient* of $f(x)$ at $y \in \mathbb{R}^n$ iff $-\zeta(y)$ is a proximal supergradient of $-f(x)$ at y, i.e., there exists some $\sigma(y) \geq 0$ such that

$$f(x) \geq f(y) + \zeta^T(y)(x - y) - \sigma(y)\|x - y\|^2 \tag{3.123}$$

for all x in some neighborhood $U(y) \subset \mathbb{R}^n$ of y. The sets $\partial^P f(y)$ and $\partial_P f(y)$ of proximal supergradients and proximal sub-gradients of f at y are referred to as the *proximal superdifferential* and, respectively, *proximal sub-differential* of f at y. These sets are definitely convex, but they may be empty, closed or open as well as bounded or unbounded.

Thus, given $\zeta \in \partial_P V(x)$, the function $V(y) - W(y)$ with a particular $W(y) = \zeta^T y - \sigma\|y - x\|^2$ and some positive constant σ possesses a local minimum at x and ∇W at x equals ζ. Hence, the inequality $F(x, V(x), \zeta) \geq 0$ holds for any $\zeta \in \partial_P V(x)$, thereby yielding the multivalued inequality

$$F(x, V, \partial_P V(x)) \geq 0 \text{ for all } x \in \mathbb{R}^n. \tag{3.124}$$

Similar to the above, the multivalued inequality

$$F(x, V, \partial^P V(x)) \leq 0 \text{ for all } x \in \mathbb{R}^n \tag{3.125}$$

is obtained.

Summarizing, a viscosity solution of (3.113) necessarily satisfies its proximal bilateral version (3.124)–(3.125). Under appropriate assumptions, the sufficiency of (3.124)–(3.125) for V to be a viscosity solution is established (Clarke 1983) by applying Subbotin's theorem (Subbotin 1995) according to which any Dini subgradient can be, roughly speaking, approximated by a proximal sub-gradient.

Alternatively, viscosity solutions of (3.113) can be interpreted in terms of the limit of classical (smooth) solutions $V_\varepsilon(x)$ to the parameterized family of Laplacian PDEs

$$F(x, V_\varepsilon(x), \nabla V_\varepsilon(x)) = \varepsilon \Delta V_\varepsilon(x) \tag{3.126}$$

as the positive parameter ε goes to zero. Since in fluid mechanics, the PDE (3.126) describes the viscous fluid motion, it was a motivation of the viscosity solution terminology, introduced in Crandall and Lions (1983). The basic idea behind the motivating approximation of a viscosity solution $V(x)$ of (3.113) by classical solutions $V_\varepsilon(x)$ of the viscous fluid PDE (3.126) is as follows.

Let $\xi \in \partial^D V(x)$ for some x. Then it is well-known (see, e.g., Liberzon 2012, p. 173) that there exists a scalar C^1 function $\Phi(x)$ such that $\nabla\Phi(x) = \xi$, $\Phi(x) = V(x)$, and $\Phi(y) \geq V(y)$ for all y in a vicinity of x, i.e., $\Phi - V$ possesses a local minimum at x. Approximating $\Phi(x) \in C^1$ by a C^2 function (whenever necessary) one concludes that given V_ε, which is close to V for small $\varepsilon > 0$, the function $\Phi - V_\varepsilon$ possesses a local minimum at some x_ε near x, thereby ensuring that $\nabla\Phi(x_\varepsilon) = \nabla V_\varepsilon(x_\varepsilon)$ and $\Delta\Phi(x_\varepsilon) \geq \Delta V_\varepsilon(x_\varepsilon)$. Since V_ε satisfies (3.126), it follows that $F(x_\varepsilon, V_\varepsilon(x_\varepsilon), \nabla\Phi(x_\varepsilon)) \leq \varepsilon\Delta\Phi(x_\varepsilon)$. Taking the limit as $\varepsilon \to 0$, by continuity of F, one arrives at $F(x, V(x), \nabla\Phi(x)) \leq 0$ where $\nabla\Phi(x) = \xi \in \partial^D V(x)$. Thus, (3.125) is verified. Its sub-gradient counterpart (3.124) is verified in a similar manner. Relations (3.125) and (3.124), coupled together, ensure that the limit $V(x)$ of the parameterized solutions $V_\varepsilon(x)$ of (3.126) as $\varepsilon \to 0$ is a viscosity solution of (3.113).

To conclude this section, the viscous solution concept is illustrated with a simple example.

Example 3.1 Let $x \in \mathbb{R}$ and let the PDE (3.113) take the form $1 - |\nabla V| = 0$. By inspection, the functions $V(x) = x$ and $V(x) = -x$ are classical (strong) solutions of (3.113). Let us verify that $V(x) = |x|$ is a viscosity solution of the same PDE. Indeed, for all $x \neq 0$, the functions $V(x)$ is differentiable and (3.113) holds true. Moreover, $\partial^D V(0) = \emptyset$ by the superdifferential definition (3.116)–(3.117), and hence (3.124) is true whereas $\partial_D V(0) = [-1, 1]$ by the sub-differential definition (3.118)–(3.119) and the inequality $1 - |\xi| \geq 0$ holds for all $\xi \in \partial_D V(0) = [-1, 1]$, thereby ensuring (3.125) as well. It is worth noticing that the function $V(x) = |x|$ is no longer a viscosity solution of the above PDE, rewritten as $|\nabla V| - 1 = 0$. Thus, the sign convention, used in the PDE representation, becomes critical in the definition of a viscosity solution.

3.5.2 Discontinuous Strict Hamilton–Jacobi Inequality and Its Proximal Solutions

Viscosity solutions are just a way of defining nonsmooth solutions of the first-order PDE (3.113) through bilateral approximations (3.114) and (3.115). Alternative solution concepts are due to Clarke (1983), which are particularly well-suited to ϵ-Nash equilibrium strategies, proposed in Mylvaganam et al. (2014) for nonzero differential games. Apparently, while dealing with the corresponding partial differential inequality

$$F(x, V, \nabla V) \leq 0, \tag{3.127}$$

only the unilateral approximation (3.125) that uses the proximal superdifferential is relevant. With this in mind, proximal solutions are next brought into play for a particular form of (3.127), referred to as a Hamilton–Jacobi inequality and used in the $\mathscr{L}_2$-gain analysis of nonsmooth systems.

The Hamilton–Jacobi inequality of interest is associated with a variable structure system of the form

$$\dot{x} = f(x, t) + g(x, t)w(t), \tag{3.128}$$

$$z = h(x, t), \tag{3.129}$$

and in the corresponding time-varying setting, it is given by

$$\frac{\partial V}{\partial t} + \frac{\partial V}{\partial x} f(x, t) + \frac{1}{4\gamma^2} \frac{\partial V}{\partial x} g(x, t) g^T(x, t) \left(\frac{\partial V}{\partial x}\right)^T + h^T(x, t) h(x, t) < 0. \tag{3.130}$$

Hereinafter, the state vector $x(t) \in \mathbb{R}^n$, the time variable $t \in \mathbb{R}$, the unknown disturbance $w(t) \in \mathbb{R}^r$, the performance output $z(t) \in \mathbb{R}^p$, the vector function $h(x, t) : \mathbb{R}^n \mapsto \mathbb{R}^p$, the matrix function $g(x, t) : \mathbb{R}^n \mapsto \mathbb{R}^{n \times r}$, the system nonlinearity $f(x, t) : \mathbb{R}^n \mapsto \mathbb{R}^n$ undergoes discontinuities on the time-varying surface

$$S(t) = \{x \in \mathbb{R}^n : s(x, t) = 0\}, \tag{3.131}$$

determined by a smooth function $s(x, t) : \mathbb{R}^{n+1} \mapsto \mathbb{R}$, and it is switched according to

$$f(x, t) = \begin{cases} f^+(x, t) & \text{if } s(x, t) > 0 \\ f^-(x, t) & \text{if } s(x, t) < 0\,. \end{cases} \tag{3.132}$$

It is assumed that the functions $w(t)$, $f^+(x, t)$, $f^-(x, t)$, $g(x, t)$, and $h(x, t)$ are piecewise continuous in t and locally Lipschitz continuous in x on their domains.

The precise meaning of the differential equation (3.128) with a piecewise continuous right-hand side is defined in the sense of Filippov as that of the corresponding differential inclusion, which is obtained from the right-hand side of (3.128) by passing to the minimal convex hull, spanned by the vectors $f^+(x,t)$ and $f^-(x,t)$.

As established in Sect. 2.2, a sliding mode on the switching surface (3.131), if any, is governed by

$$\dot{x} = f_0(x,t) + g_0(x,t)w \tag{3.133}$$

where

$$f_0(x,t) + g_0(x,t)w = \mu(x,t)f^+(x,t) + \big[1 - \mu(x,t)\big]f^-(x,t) + g(x,t)w \tag{3.134}$$

with

$$\mu(x) = \frac{\text{grad}^T\, s(x)[\varphi^-(x) + \psi(x)\omega]}{\text{grad}^T\, s(x)[\varphi^-(x) - \varphi^+(x)]}, \tag{3.135}$$

being found from the condition

$$\text{grad}^T\, s(x,t)\cdot\Big[\mu(x,t)f^+(x,t) + \big(1 - \mu(x,t)\big)f^-(x,t) + g(x,t)w\Big] = 0 \tag{3.136}$$

that the velocity vector (3.134) is in the plane T, tangential to S. Summarizing, the following relations

$$\begin{aligned} f_0(x,t) = &\left(\frac{\text{grad}^T\, s(x,t)f^-(x,t)}{\text{grad}^T\, s(x,t)[f^-(x,t) - f^+(x,t)]}\right)f^+(x,t) \\ &- \left(\frac{\text{grad}^T\, s(x,t)f^+(x,t)}{\text{grad}^T\, s(x,t)[f^-(x,t) - f^+(x,t)]}\right)f^-(x,t), \end{aligned} \tag{3.137}$$

$$\begin{aligned} g_0(x,t)w = g(x,t)w + &\left(\frac{\text{grad}^T\, s(x,t)g(x,t)w}{\text{grad}^T\, s(x,t)[f^-(x,t) - f^+(x,t)]}\right)f^+(x,t) \\ + &\left(1 - \frac{\text{grad}^T\, s(x,t)g(x,t)w}{\text{grad}^T\, s(x,t)[f^-(x,t) - f^+(x,t)]}\right)f^-(x,t) \end{aligned} \tag{3.138}$$

are derived to determine the functions $f_0(x,t)$ and $g_0(x,t)$.

The Hamilton–Jacobi inequality (3.130) is introduced as is outside the switching surface (3.131) whereas on the switching surface (3.131), (3.130) is specified according to (3.134) with

$$f(x,t) = f_0(x,t),\; g(x,t) = g_0(x,t) \;\text{ provided that }\; s(x,t) = 0. \tag{3.139}$$

In other words, the Hamilton–Jacobi inequality (3.130), if confined to the discontinuity surface (3.131), takes the form

$$\frac{\partial V}{\partial t} + \frac{\partial V}{\partial x} f_0(x,t) + \frac{1}{4\gamma^2} \frac{\partial V}{\partial x} g_0(x,t) g_0^T(x,t) \left(\frac{\partial V}{\partial x}\right)^T + h^T(x,t)h(x,t) < 0. \tag{3.140}$$

The solution concept for such a discontinuous PDI is inherited from Osuna et al. (2018).

Definition 3.5 A locally Lipschitz continuous function $V(x,t)$ is said to be a (local) *proximal solution* of the PDI (3.130) subject to (3.137)–(3.139) iff its proximal superdifferential $\partial^P V(x,t)$ is everywhere non-empty and (3.130) holds with $V(x,t)$ beyond the discontinuity surface (3.131) (locally around the origin) for all proximal supergradients $\left(\frac{\partial V}{\partial x}, \frac{\partial V}{\partial t}\right)^T \in \partial^P V(x,t)$ whereas the sliding mode Hamilton–Jacobi inequality (3.140) is satisfied on the discontinuity surface (3.131) (locally around the origin) for all $\left(\frac{\partial V}{\partial x}, \frac{\partial V}{\partial t}\right)^T \in \partial^P V(x,t)$.

It should be pointed out that if confined to continuous vector fields, the above definition coincides with that of proximal solutions by Clarke (1983).

Given a proximal solution $V(x,t)$ of the Hamilton–Jacobi inequality (3.130), which is associated with the VSS (3.128), (3.129) and which is specified to (3.140) whenever $s(x,t) = 0$, the following result is instrumental to estimate the time derivative $\dot{V}(x(t),t)$ along Filippov solutions of (3.128).

Lemma 3.5 *Let $x(t) : \mathbb{R} \to \mathbb{R}^n$ be an absolutely continuous function of the time variable t and let $V(x,t) : \mathbb{R}^n \to \mathbb{R}$ be a locally Lipschitz continuous function. Then the composite function $V(x(t),t)$ is absolutely continuous and its time derivative, being expressed in terms of the Dini derivative $DV(x,t;v,1) = \lim_{h\to 0} \frac{V(x+hv,t+h)-V(x,t)}{h}$, is given by*

$$\frac{\mathrm{d}}{\mathrm{d}t} V(x(t),t) = DV(x(t),t;\dot{x}(t),1) \tag{3.141}$$

almost everywhere. Furthermore,

$$DV(x(t),t;\dot{x}(t),1) \le \frac{\partial V}{\partial t} + \frac{\partial V}{\partial x}\dot{x}(t) \tag{3.142}$$

for almost all t and for all supergradients $\left(\frac{\partial V}{\partial x}, \frac{\partial V}{\partial t}\right)^T \in \partial^P V(x,t)$, if any.

Proof Since the function $V(x,t)$ is Lipschitz continuous whereas $x(t)$ is absolutely continuous, the composite function $V(x(t))$ is absolutely continuous as well and (3.141) is satisfied for almost all t. Indeed,

$$\begin{aligned}\frac{d}{dt}V(x(t),t) &= \lim_{h\to 0}\frac{V(x(t+h),t+h)-V(x(t),t)}{h}\\ &= \lim_{h\to 0}\frac{V(x(t)+h\dot{x}(t),t+h)-V(x(t),t)}{h}\\ &\quad - \lim_{h\to 0}\frac{V(x(t+h),t+h)-V(x(t)+h\dot{x}(t),t+h)}{h}\\ &= \frac{d}{dh}V(x(t)+h\dot{x}(t),t+h)\Big|_{h=0} = DV(x(t),t;\dot{x}(t),1)\end{aligned} \tag{3.143}$$

provided that the derivatives $\dot{x}(t)$ and $dV(x(t),t)/dt$ exist at the time instant t. In order to reproduce (3.143), one should take into account that the last limit in (3.143) equals to zero because the function $V(x,t)$ locally satisfies the Lipschitz condition and $x(t+h) = x(t) + h\dot{x}(t) + o(h)$ where $o(h)$ is such that $\lim_{h\to 0}\frac{o(h)}{h} = 0$.

To complete the proof it remains to note that inequality (3.142) is validated by the time-varying counterpart $DV(x,t;\nu,1) \le \frac{\partial V}{\partial t} + \left(\frac{\partial V}{\partial x}\right)^T \nu$ of (3.117) that particularly holds true with $\nu = \dot{x}(t)$, whenever $\dot{x}(t)$ exists, and for all supergradients $\left(\frac{\partial V}{\partial x}, \frac{\partial V}{\partial t}\right)^T \in \partial^T V(x,t)$. ■

3.6 Stability in Euclidean and Hilbert Spaces

Stability concepts to subsequently be addressed are first introduced for abstract dynamic systems, then extended in the presence of uncertainties, and finally specified in terms of ODEs, PDEs, integrodifferential equations, and differential inclusions. Discontinuous vector fields are either implicitly or explicitly involved. Special attention is called to homogeneous differential inclusions, whose asymptotic stability is typically accompanied with the finite time stability provided their homogeneity degree is negative. Once the asymptotic stability remains uniform for certain disturbances, the finite time stability of such a perturbed system is shown to persist as well.

3.6.1 *Abstract Dynamic Systems and Relevant Stability Concepts*

A generic dynamic system, evolving in a Hilbert (particularly, Euclidean) space X, is characterized by a (possibly set-valued) shift operator (also called translation operator) $S(t,t_0,x_0)$ along the system trajectories, initialized within an operational domain $D \subseteq X$ with $x_0 \in D$ at time instant t_0 and specified with $x(t) = S(t,t_0,x_0) \subset X$ at the current time instant $t \ge t_0$. The following properties should normally be postulated for such an operator to match specific dynamic systems, governed by ODEs and differential inclusions, by PDEs and integrodifferential equations:

- $S(t, t_0, x_0)$ is continuous in t for all $t_0 \in \mathbb{R}, x_0 \in D$ (continuity of the dynamics);
- $S(t, t_0, x_0)$ is continuous in (t_0, x_0) for all $t \geq t_0$ (continuous dependence on the initial data);
- $S(t_0, t_0, x_0) = x_0$ for all initial data $t_0 \in \mathbb{R}, x_0 \in D$ (no shift at $t = t_0$);
- $S(t, t_1, S(t_1, t_0, x_0)) = S(t, t_0, x_0)$ for all $t \geq t_1 \geq t_0$ and all $x_0 \in D$ (semigroup property).

An operator $S(t, t_0, x_0)$, possessing the above properties, is *referred to as a dynamic system*. It *is said to be autonomous* iff

$$S(t, t_0, x_0) = S(t - t_0, 0, x_0) \quad \text{for all} \quad t \geq t_0, x_0 \in D,$$

i.e., given an arbitrary $x_0 \in D$, the operator S is actually a function of the one variable $t - t_0$ rather than of two variables t and t_0.

It is worth noticing that for a single-valued operator S and fixed initial data t_0, x_0, a trajectory $S(t, t_0, x_0)$ of the dynamic system, associated with S, is uniquely defined for $t > t_0$, whereas S is generally speaking non-invertible for $t < t_0$ (as typically happens for sliding mode dynamics and retarded systems). A set-valued operator S is in turn invoked to deal with dynamic systems, which generate multiple trajectories, originated from the same initial data (as is the case, e.g., of systems, governed by differential inclusions).

As noted, examples of dynamic systems are shift operators along ODE (classical, Caratheodory, or Filippov) solutions, along PDE (classical, strong, mild, or weak) solutions, along solutions of integrodifferential and differential inclusions. In what follows, well-known stability concepts are addressed within the adopted framework of abstract dynamic systems.

Suppose that $x_0 = 0$ is an equilibrium point of a dynamic system $S(t, t_0, x_0)$, i.e., $0 \in D$ and $S(t, t_0, 0) = 0$ for all $t \geq t_0$. Let $B_\delta = \{x \in D : \|x\| \leq \delta\}$ stands for the intersection of the domain D and a ball of the radios $\delta > 0$, which is centered in the origin.

Definition 3.6 The equilibrium point $x = 0$ of the dynamic system $S(t, t_0, x_0)$ is *(uniformly) stable* iff for each $t_0 \in \mathbb{R}$, $\varepsilon > 0$, there is $\delta = \delta(\varepsilon, t_0) > 0$, dependent on ε and possibly dependent on t_0 (respectively, independent of t_0), such that the inequality

$$\|S(t, t_0, x_0)\| < \varepsilon \tag{3.144}$$

holds for all $t \geq t_0$ and all $x_0 \in B_\delta$.

Definition 3.7 The equilibrium point $x = 0$ of the dynamic system $S(t, t_0, x_0)$ is *(uniformly) asymptotically stable* iff it is (uniformly) stable and the convergence

$$lim_{t\to\infty} \|S(t, t_0, x_0)\| = 0 \tag{3.145}$$

holds for all $t_0 \in \mathbb{R}$, $x_0 \in B_\delta$ with some $\delta > 0$ (uniformly in the initial data t_0 and x_0). If this convergence remains in force regardless of the choice of the initial data (respectively, the convergence is uniform in $t_0 \in \mathbb{R}$ and $x_0 \in B_\delta$ for any $\delta > 0$), the equilibrium point is said to be *globally (uniformly) asymptotically stable*.

Definition 3.8 The equilibrium point $x = 0$ of the dynamic system $S(t, t_0, x_0)$ is *(globally) exponentially stable* with a positive decay rate $\beta > 0$ iff the exponential decay

$$\|S(t, t_0, x_0)\| \leq e^{-\beta(t-t_0)}\|x_0\| \tag{3.146}$$

is in force for all $t \geq t_0$ and all $x_0 \in B_\delta$ with some $\delta > 0$ (respectively, with any $\delta > 0$).

Definition 3.9 The equilibrium point $x = 0$ of the dynamic system $S(t, t_0, x_0)$ is *(uniformly) finite time stable* iff it is (uniformly) asymptotically stable, and the limiting relation

$$S(t, t_0, x_0) = 0 \tag{3.147}$$

holds for all $t \geq t_0 + T(t_0, x_0)$ and for all $x_0 \in B_\delta$ with some $\delta > 0$ where the settling time function

$$T(t_0, x_0) = \inf\{T \geq 0 : S(t, t_0, x_0) = 0 \text{ for all } t \geq t_0 + T\} < \infty \tag{3.148}$$

is well-defined for all $t_0 \in \mathbb{R}$ and $x_0 \in B_\delta$ (and it is respectively such that the upper bound

$$\mathcal{T}(B_\delta) = sup_{t_0 \in \mathbb{R},\ x_0 \in B_\delta} T(t_0, x_0) < \infty \tag{3.149}$$

is also well-defined). If along with the settling time function $T(t_0, x_0) < \infty$, its upper bound $\mathcal{T}(B_\delta) < \infty$ remains well-defined for all $\delta > 0$ the equilibrium point is said to be *globally finite time stable*.

Definition 3.10 The equilibrium point $x = 0$ of the dynamic system $S(t, t_0, x_0)$ is *fixed time stable* iff it is globally finite time stable, and the upper bound (3.149) is such that $sup_{\delta>0}\mathcal{T}(B_\delta) < \infty$.

More stability notions are further discussed in the presence of uncertainties.

3.6.2 Robust Stability of Uncertain Dynamic Systems: Basic Definitions

An uncertain dynamic system is not characterized by an a priori known operator S, satisfying the postulates of Sect. 3.6.1. Instead, it belongs to a family $\{S_f\}_{f \in F}$ of such operators $S_f(t, t_0, x_0)$ where the index f runs through a subset $F \subset L^\infty_{loc}(\mathcal{F})$ of locally integrable functions $f(t)$ with values in a Hilbert space $\mathcal{F}$.

Recall that $L^\infty_{loc}(\mathcal{F}) = \bigcap_{T>0} L^\infty(0, T; \mathcal{F})$ where $L^\infty(0, T; \mathcal{F})$ stands for the Banach space of functions $f(t)$ such that $f(t) \in \mathcal{F}$ for almost all $t \in (0, T)$ and $ess\ \sup_{t \in (0,T)} \|f(t)\| < \infty$.

Throughout, $f \equiv 0$ is supposed to belong to F and it matches to the nominal system S_0 whereas F is interpreted as a set of admissible disturbances f, which may affect the nominal system S_0.

Consider a family of uncertain dynamic systems $S_f(t, t_0, x_0)$ with $f \in F$ and suppose that $x_0 = 0$ is an equilibrium of S_f for any $f \in F$. The following definitions are in order to address stability concepts for uncertain dynamic systems regardless of whichever admissible disturbance affects the nominal system.

Definition 3.11 The equilibrium point $x = 0$ of the nominal dynamic system S_0 is *equiuniformly stable* against admissible disturbances $f \in F$ (or simply equiuniform stable) iff for each $t_0 \in \mathbb{R},\ \varepsilon > 0$, there is $\delta = \delta(\varepsilon)$, dependent on ε and independent of t_0 and f, such that the inequality

$$\|S_f(t, t_0, x_0)\| < \varepsilon \tag{3.150}$$

holds for all $t \geq t_0$, all $x_0 \in B_\delta$, and all $f \in F$.

Definition 3.12 The equilibrium point $x = 0$ of the nominal dynamic system S_0 is *equiuniformly asymptotically stable* if it is equiuniformly stable and the convergence

$$lim_{t\to\infty}\|S_f(t, t_0, x_0)\| = 0 \tag{3.151}$$

is uniform in disturbances $f \in F$ and in the initial data $t_0 \in \mathbb{R},\ x_0 \in B_\delta$ for some $\delta > 0$. If this uniform convergence remains in force with any $\delta > 0$, the equilibrium point is said to be globally equiuniformly asymptotically stable.

Definition 3.13 The equilibrium point $x = 0$ of the nominal dynamic system S_0 is *equiuniformly finite time stable* iff it is equiuniformly asymptotically stable, and the limiting relation

$$S_f(t, t_0, x_0) = 0 \tag{3.152}$$

holds for all $f \in F$, for all $t \geq t_0 + T_f(t_0, x_0)$, and for all $x_0 \in B_\delta$ with some $\delta > 0$ where the settling time function

$$T_f(t_0, x_0) = \sup_{f\in F} \inf\{T \geq 0 : S_f(t, t_0, x_0) = 0 \text{ for all } t \geq t_0 + T\} < \infty, \tag{3.153}$$

matching to S_f is such that the upper bound

$$\mathscr{T}_F(B_\delta) = sup_{f\in F,\ t_0\in\mathbb{R},\ x_0\in B_\delta} T_f(t_0, x_0) < \infty \tag{3.154}$$

is well-defined. If along with the settling time function $T_f(t_0, x_0) < \infty$, its upper bound $\mathscr{T}_F(B_\delta) < \infty$ remains well-defined for all $\delta > 0$ the equilibrium point is said to be *globally equiuniformly finite time stable*.

Definition 3.14 The equilibrium point $x = 0$ of the nominal dynamic system S_0 is *equiuniformly fixed time stable* iff it is globally equiuniformly finite time stable, and the upper bound (3.154) is such that $sup_{\delta>0}\mathscr{T}_F(B_\delta) < \infty$.

Recently, the ISS and integral ISS (iISS) concepts were introduced for ODEs to quantify a robustness degree versus disturbances (see survey Sontag 2008 and references therein). The ISS meaning is in that the state norm is upper bounded by a continuous disturbance-dependent function, escaping to zero when the disturbance magnitude is nullified, whereas the effect of an arbitrary initial condition is captured by an additional term, which depends on time and on the norm of the initial state, and which asymptotically decays as time goes to infinity. Later on Dashkovskiy and Mironchenko (2013), Prieur and Mazenc (2012), the ISS was revisited for DPS. The iISS of DPS was tackled as well (Mironchenko and Ito 2015), and particularly, within the linear $\mathscr{H}_\infty$ framework (Morris 2001). The ISS concept is now reworked for uncertain dynamic systems S_f, $f \in F$ whereas its integral $\mathscr{H}_\infty$ counterpart is given in terms of $\mathscr{L}_2$ gain.

The next definition is inspired from Dashkovskiy and Mironchenko (2013), Pisano and Orlov (2017) and it specifies ISS uncertain dynamic systems with exponential decay rates and linear disturbance attenuation.

Definition 3.15 The equilibrium point $x = 0$ of the nominal dynamic system S_0 is *(globally) exponentially ISS* iff the exponential ISS inequality

$$\|S_f(t, t_0, x_0)\|^2 \leq \|x_0\|^2 e^{-\beta(t-t_0)} + \gamma \|f\|^2_{L_\infty(0,t;\mathscr{F})} \tag{3.155}$$

holds with some decay rate $\beta > 0$ and attenuation level $\gamma > 0$ for all $t \geq t_0$, for all $f \in F$ such that $\|f(t)\| \leq \delta$, and for all $x_0 \in B_\delta$ with some $\delta > 0$ (respectively, with any $\delta > 0$).

For uncertain dynamic systems with an output operator

$$z_f(t, t_0, x_0) = h(S_f(t, t_0, x_0)) : X \to Z \tag{3.156}$$

where Z is a Hilbert space, the dissipation-flavored $\mathscr{L}_2$ gain notion is inspired from Willems (1972), and it is as follows.

Definition 3.16 Given $\gamma > 0$, the nominal dynamic system S_0 with the output operator (3.156) has *$\mathscr{L}_2$ gain less than γ (locally around the origin)* iff for some positive definite function $\alpha(\cdot)$ of a scalar argument, the system response (3.156), resulting from $f \in F$, satisfies

$$\int_{t_0}^{t_1} \|z_f(t, t_0, x_0)\|^2 dt < \alpha(\|x_0\|) + \gamma^2 \int_{t_0}^{t_1} \|f(t)\|^2 dt \tag{3.157}$$

for all $t_1 \geq t_0$, for all $x_0 \in B_\delta$ with any $\delta > 0$ (respectively, with some $\delta > 0$), and for all $f \in F$ such that $S_f(t, t_0, x_0)$ remains in B_δ.

In the sequel, uncertain dynamic systems are exemplified with Hilbert space-valued sliding modes and retarded processes as well as with homogeneous dynamics, governed by differential inclusions.

3.6.3 Sliding Mode Dynamics in Hilbert Space

As in Sect. 2.2.5, consider a Cauchy initial-value problem

$$\dot{x} = Ax + f(x,t) + bu(x,t), \ t > t_0, \ x(t_0) = x^0 \in \mathscr{D}(A) \tag{3.158}$$

where the state variable $x(t)$ and the input signal $u(x, t)$ are abstract functions with values in Hilbert spaces H and U, respectively, the infinitesimal operator A with domain $\mathscr{D}(A)$ generates a strongly continuous semigroup $T_A(t)$ on H, the operator function $f(x, t)$ with values in H is of class C^1 in all arguments, and b is a linear bounded operator, acting from U to H.

Recall (Curtain and Zwart 1995) that a family $\{T(t)\}_{t\geq 0}$ of linear bounded operators $T(t), \ t \geq 0$ forms a strongly continuous semigroup on a Hilbert space H iff $T(0)$ is the identity operator and $T(t+\tau) = T(t)T(\tau)$ is satisfied for all $t, \tau \geq 0$, and the functions $T(t)x$ are continuous in $t \geq 0$ for all $x \in H$ and continuous in $x \in H$ for all $t \geq 0$. Thus, the strongly continuous semigroup $T(t)$ generates an autonomous dynamic system $S(t, x) = T(t)x$ in the Hilbert space H that can be viewed as a shift operator along the classical solutions of the linear homogeneous system.

The induced operator norm $\|T(t)\|$ of the semigroup is well-known (Curtain and Zwart 1995) to respect the inequality $\|T(t)\| \leq \omega e^{\beta t}, t \geq 0$ with some growth bound β and some $\omega > 0$. The domain of an operator A, generating a strongly continuous semigroup, forms the Hilbert space $\mathscr{D}(A)$ with the graph inner product $\langle \cdot, \cdot \rangle_{\mathscr{D}(A)}$ defined by means of the inner product $\langle \cdot, \cdot \rangle_H$ of the underlying Hilbert space H:

$$\langle x, y \rangle_{\mathscr{D}(A)} = \langle x, y \rangle_H + \langle Ax, Ay \rangle_H, \ x, y \in \mathscr{D}(A).$$

If β is a growth bound of the semigroup, then given $\lambda > \beta$, there holds $(A - \lambda I)^{-1}H = \mathscr{D}(A)$ where I is the identity operator, and the norm of $x \in \mathscr{D}(A)$ given by $\|(A - \lambda I)x\|_H$ is equivalent to the graph norm $\|x\|_{\mathscr{D}(A)}$ of $\mathscr{D}(A)$. In particular, $\|x\|_{\mathscr{D}(A)} = \|Ax\|_H$ if A is exponentially stable and possesses a growth bound $\beta < 0$. It should be noted that $\mathscr{D}(A) \hookrightarrow H$ is continuously embedded into H, i.e., $\mathscr{D}(A) \subset H, \ \mathscr{D}(A)$ is dense in H and the inequality $\|x\|_H \leq \omega_0 \|x\|_{\mathscr{D}(A)}$ holds for all $x \in \mathscr{D}(A)$ and some constant $\omega_0 > 0$.

If the input function u meets the same smoothness conditions as that imposed on the system nonlinearity f, Eq. (3.158) locally has a unique strong solution $x(t)$ which is defined as follows.

Definition 3.17 A continuous function $x(t)$, defined on $[0, T)$, is a *strong solution* of the initial-value problem (3.158) with a continuously differentiable input $u(x, t)$ iff $\lim_{t\downarrow 0} \|x(t) - x^0\|_H = 0$, and $x(t)$ is continuously differentiable and satisfies the equation for $t \in (0, T)$.

The precise meaning of the solutions of (3.158), for inputs which are only piecewise continuously differentiable, is defined as a limiting result obtained through the regularization procedure, similar to that proposed for finite-dimensional systems.

Let the input $u(x, t)$ be continuously differentiable beyond a linear manifold

$$cx = 0 \tag{3.159}$$

with $c \in \mathscr{L}(H, S)$ being a linear bounded operator from H to some Hilbert space S, and let $u(x, t)$ undergo discontinuities on this manifold. Then the strong solutions of (3.158) are only considered whenever they are beyond the discontinuity manifold (3.159) whereas in the vicinity of this manifold, the original system is replaced by a related system, which takes into account all possible imperfections in the new input function $u^\delta(x, t)$ (e.g., delay, hysteresis, and saturation) and for which there exists a strong solution. A generalized solution of system (3.158) is then obtained by making the characteristics of the new system approach those of the original one. As in the finite-dimensional case, a motion along the discontinuity manifold is referred to as a sliding mode.

To recall a rigorous sliding mode introduction in the infinite-dimensional setting of Sect. 2.2.5, let us complement the subspace

$$H_1 = ker\ c = \{x_1 \in H : cx_1 = 0\} \subseteq H$$

by the subspace $H_2 \subseteq H$ such that

$$H = H_1 \oplus H_2.$$

Clearly, the discontinuity manifold (3.159), written through the new coordinates $x_1(t) = P_1 x(t) \in H_1$ and $x_2(t) = P_2 x(t) \in H_2$, takes the form $x_2 = 0$. Hereinafter, P_i is the projector on the subspace H_i, $A_i = A|_{H_i}$ is the operator restriction on H_i, $i = 1, 2$.

Definition 3.18 (Orlov 2009) An absolutely continuous function $x^\delta(t)$, defined on some interval $[0, \tau)$, is said to be an *approximate δ-solution* of system (3.158) if it is a strong solution of

$$\dot{x}^\delta = Ax^\delta + f(x^\delta, t) + bu^\delta(x^\delta, t) \tag{3.160}$$

with some $u^\delta(x, t)$ such that

$$\|u^\delta(x, t) - u(x, t)\| \le \delta \tag{3.161}$$

for all $t \ge 0$ and for all $x = (x_1, x_2) \in H = H_1 \oplus H_2$ subject to $\|x_2\|_{\mathscr{D}(A_2)} \ge \delta$.

Definition 3.19 (Orlov 2009) An absolutely continuous function $x(t)$, defined on some interval $[0, \tau)$, is said to be a *generalized solution* of system (3.158) if there exists a family of approximate δ-solutions $x^\delta(t)$ of the system such that

$$\lim_{\delta \to 0} \|x^\delta(t) - x(t)\|_{\mathscr{D}(A)} = 0\ uniformly\ in\ t \in [0, \tau).$$

Although beyond the discontinuity manifold the above definitions are confined to strong solutions of the initial-value problem, an extension to the case, where such a solution is defined in a mild sense as a solution to a corresponding integral equation, is possible.

By definition, the sliding motion, which is in general nonunique, does not depend on the precise specification of the discontinuous input on the discontinuity manifold. Moreover, an equivalent control value $u^{eq}(x,t)$, maintaining the system motion on this manifold, is imposed by the original system itself.

To describe sliding modes in the infinite-dimensional system, let us rewrite Eq. (3.158) in terms of variables $x_1(t) \in H_1$ and $x_2(t) \in H_2$:

$$\dot{x_1} = A_{11}x_1 + A_{12}x_2 + f_1(x_1, x_2, t) + b_1u(x_1, x_2, t), \; x_1(0) = x_1^0, \quad (3.162)$$

$$\dot{x_2} = A_{21}x_1 + A_{22}x_2 + f_2(x_1, x_2, t) + b_2u(x_1, x_2, t), \; x_2(0) = x_2^0, \quad (3.163)$$

where $A_{ij} = P_iA_j$, $i, j = 1, 2$ are the operators from H_j to H_i, and $f_i = P_if$, $b_i = P_ib$. If the operator b_2 is non-singular and the inverse operator b_2^{-1} is bounded there exists a unique solution of the algebraic equation $A_{21}x_1 + f_2(x_1, 0, t) + b_2u(x, t) = 0$ with respect to u. This solution

$$u^{eq}(x_1) = -b_2^{-1}[A_{21}x_1 + f_2(x_1, 0, t)] \quad (3.164)$$

is accepted as the equivalent control value because (3.164) is the only input ensuring that $\dot{x}_2 = 0$ on the discontinuity manifold $x_2 = 0$, thereby maintaining system (3.158) with appropriate initial conditions on the manifold $x_2 = 0$. Setting $\tilde{A} = A_{11} - b_1b_2^{-1}A_{21}$ and $f_0(x_1, x_2, t) = f_1(x_1, x_2, t) - b_1b_2^{-1}f_2(x_1, x_2, t)$, the sliding mode equation

$$\dot{x} = \tilde{A}x_1 + f_0(x_1, 0, t), \quad (3.165)$$

governing the system motion on the discontinuity manifold $cx = 0$, is then obtained by substituting the equivalent control value (3.164) into (3.163) for u.

By Theorem 2.2, the initial value problem (3.158) possesses a unique generalized solution, and this solution is governed by the sliding mode equation (3.165).

Thus, the shift operator $S(t, t_0, x_0) = x(t)$ is well-defined along the generalized solutions $x(t)$ of the Cauchy initial-value problem (3.158), and hence $S(t, t_0, x_0)$ determines a Hilbert space-valued dynamic system with potential motions on the discontinuity manifold (3.159).

3.6.4 *Hilbert Space-Valued Dynamics with Delay*

Now consider a linear infinite-dimensional system

$$\dot{x}(t) = Ax(t) + A_1x(t - \tau(t)), \quad t \geq t_0, \quad (3.166)$$

evolving in a Hilbert space $\mathscr{H}$ where $x(t) \in \mathscr{H}$ is the instantaneous state of the system. Let the following assumptions be satisfied

A1 The operator A generates a strongly continuous semigroup $T(t)$ and the domain $\mathscr{D}(A)$ of the operator A is dense in $\mathscr{H}$;
A2 The linear operator A_1 is bounded in $\mathscr{H}$;
A3 The function $\tau(t)$ is piecewise-continuous of class C^1 on the closure of each continuity subinterval and it satisfies

$$\inf_t \tau(t) \geq 0, \ \sup_t \tau(t) \leq h \tag{3.167}$$

with some constant $h > 0$ for all $t \geq t_0$.

Let the initial conditions

$$x_{t_0} = \varphi(\theta), \ \theta \in [-h, 0], \phi \in W \tag{3.168}$$

be specified for $x_{t_0}(\theta) = x(t_0 + \theta)$ in the space

$$W = C([-h, 0], \mathscr{D}(A)) \cap C^1([-h, 0], \mathscr{H}). \tag{3.169}$$

A function $x(t) \in C([t_0 - h, t_0 + \eta], \mathscr{D}(A))$ *is said to be a solution* of the initial-value problem (3.166), (3.168) on $[t_0 - h, t_0 + \eta]$ if $x(t)$ is initialized with (3.168), it is absolutely continuous for $t \in [t_0, t_0 + \eta]$, and it satisfies (3.166) for almost all $t \in [t_0, t_0 + \eta]$.

The initial-value problem (3.166), (3.168) turns out to be well-posed on the semi-infinite time interval $[t_0, \infty)$ and due to the lemma given below, its solutions can be found as mild solutions, i.e., as those of the integral equation

$$\begin{aligned} x(t) &= T(t - t_0)x(t_0) \\ &+ \textstyle\int_{t_0}^t T(t - s)A_1 x(s - \tau(s))ds, \ t \geq t_0. \end{aligned} \tag{3.170}$$

The following result is in order.

Lemma 3.6 (Fridman and Orlov 2009) *Under assumptions A1–A3 there exists a unique solution of the initial-value problem* (3.166)*)*, (3.168) *on* $[t_0, \infty)$. *This solution is also a unique solution of the integral initial-value problem* (3.168), (3.170).

Proof To begin with, let us choose a positive η_0 small enough to ensure that $\eta_0 < \inf_t \tau(t)$ and the first discontinuity point $t_0^1 > t_0$ of $\tau(t)$ is such that the difference $t_0^1 - t_0$ is multiple to η_0, i.e., $t_0^1 = t_0 + k_0\eta_0$ for some integer $k_0 > 0$. While being viewed over the time segment $[t_0, t_0 + \eta_0]$, the initial-value problem (3.166) is equivalent to

$$\dot{x}(t) = Ax(t) + A_1\phi(t - t_0 - \tau(t)), \ x(t_0) = \phi(0) \tag{3.171}$$

where the inhomogeneous term $A_1\phi(t - t_0 - \tau(t))$ is of class C^1 on $[0, \eta_0]$. By Theorem 3.1.3 of Curtain and Zwart (1995), there exists a unique local solution of (3.171) and this solution satisfies the integral equation (3.170) on $[t_0, t_0 + \eta_0]$.

The same line of reasoning is step-by-step applied to the time segments $[t_{i-1}, t_{i-1} + \eta_0]$, $i = 1, \ldots, k_0$ with $t_i = t_{i-1} + \eta_0$ and $t_{k_0} = t_0^1$. Following this line, the initial-value problem is established to possess a unique solution $x(t, t_0, \phi)$ for $t \in [t_0, t_0^1]$, which satisfies the integral equation (3.170) on $[t_0, t_0^1]$. The assertion of Lemma 3.6 is then concluded by iteration on the time segments $[t_0^j, t_0^{j+1}]$, $j = 1, 2, \ldots$ where $t_0^1 < t_0^2 < \cdots$ are the successive discontinuity points of the function $\tau(t)$. ∎

As in the delay-free case of Sect. 3.6.3, the shift operator $S(t, t_0, \phi) = x(t, t_0, \phi)$ is well-defined along the solutions $x(t)$ of the Cauchy initial-value problem (3.166), (3.168), and hence $S(t, t_0, \phi)$ determines a Hilbert space-valued dynamic system with delay.

3.6.5 *Homogeneous Differential Inclusions and Their Finite Time Stability*

Homogeneous systems form an important class of nonlinear systems, featuring trajectory scaling that allows one to particularly simplify ISS and finite time stability analysis. The original homogeneity definition of a class of dilations where each state is dilated with a different weight is due to Zubov (1964). Asymptotic stabilization of continuous homogeneous systems, thus defined, was studied in Kawski (1989), Rousier (1992). The notion of geometric homogeneity and its application to stabilization was then developed in references (Kawski 1991, 1995). A detailed literature review on the topic of geometric homogeneity was presented in Bhat and Bernstein (2005). Later on, homogeneous approximations led to the development of tools to establish global asymptotic (and in some cases finite time) stability of nonlinear systems. This result used previous results on the so-called homogeneous domination approach (see Andrieu et al. 2008, Sect. 5; Moulay and Perruquetti 2006; Qian 2005 and references therein for a detailed literature review). Very recently (Polyakov et al. 2016, 2018; Polyakov 2018), linear dilations were generalized in Banach and Hilbert spaces to address the homogeneity in the infinite-dimensional setting.

Below, the homogeneity concepts are introduced for dynamic systems, generated by differential inclusions and discontinuous vector fields in particular. Consider a differential inclusion

$$\dot{x} \in \Phi(x, t) \tag{3.172}$$

with a multivalued right-hand side $\Phi : \mathbb{R}^{n+1} \to 2^{\mathbb{R}^n}$. In a particular case

$$\dot{x} = \varphi(x, t) \tag{3.173}$$

of discontinuous vector fields of Sect. 2.2, $\Phi(x, t)$ is constituted by the Filippov set, generated by a piecewise continuous function $\varphi(x, t)$.

Recall that an absolutely continuous function $x(t)$ *is said to be a solution* of (3.172) on an interval $\mathscr{I} \in \mathbb{R}^n$ iff it satisfies (3.172) for almost all $t \in \mathscr{I}$. Pretty general sufficient conditions of such a solution to exist (e.g., the set-valued function Φ to be upper bounded and the set $\Phi(x, t)$ to be non-empty, compact and convex for all x, t) may be found in Aubin and Cellina (1984). It is however clear, a solution $x(t, t_0, x_0)$ of (3.172), initialized at $t_0 \in \mathbb{R}$ with $x(t_0) = x_0 \in \mathbb{R}^n$, is not unique so that the shift operator $S(x(t, t_0, x_0) = x(t, t_0, x_0)$ along the solutions of (3.172) generates a multivalued dynamic system.

Definition 3.20 (*Homogeneity of differential inclusions* Orlov 2004) The differential inclusion (3.172) is called homogeneous of degree $q \in R$ with respect to dilation $(r_1, r_2, \ldots, r_n)$, where $r_i > 0, i = 1, \ldots, n$, if there exists a constant $c_0 > 0$, called a lower estimate of the homogeneity parameter such that any solution $x(\cdot)$ of (3.172) generates a parameterized set of solutions $x^c(\cdot)$ with components

$$x_i^c(t) = c^{r_i} x_i(c^q t) \tag{3.174}$$

and any parameter $c \geq c_0$.

It should be noted that the above definition relies on solutions of the differential inclusion. However, similar to the Lyapunov function method, there is no need to explicitly compute the solutions as it is only required to verify that the parameterized set $x^c(\cdot)$ with components (3.174) satisfies (3.172) provided that $x(\cdot)$ is a solution of (3.172). More explicit homogeneity definition comes for discontinuous vector fields, generated by a piecewise continuous function.

Definition 3.21 (*Homogeneity of piecewise continuous vector fields* Orlov 2004) A piecewise continuous vector field $\phi(x, t) : \mathbb{R}^{n+1} \to \mathbb{R}^n$ of (3.173) is called homogeneous of degree $q \in \mathbb{R}$ with respect to dilation $(r_1, r_2, \ldots, r_n)$, where $r_i > 0, i = 1, \ldots, n$, if there exists a constant $c_0 > 0$ such that

$$\phi_i(c^{r_1} x_1, c^{r_2} x_2, \ldots, c^{r_n} x_n, c^{-q} t) = c^{q+r_i} \phi_i(x_1, x_2, \ldots, x_n, t) \tag{3.175}$$

for all $c \geq c_0$, for all $(x_1, x_2, \ldots, x_n)^T \in \mathbb{R}^n$, and for almost all $t \in \mathbb{R}$.

It is worth noticing that the above definitions are consistent in the sense that homogeneity of a piecewise continuous vector field $\phi(x, t)$ ensures homogeneity of the corresponding differential inclusion (3.172).

Lemma 3.7 *Let a piecewise continuous vector field ϕ be homogeneous of degree $q \in \mathbb{R}$ with respect to dilation $(r_1, \ldots, r_n)$. Then the differential inclusion* (3.172) *with the right-hand side, corresponding to the Filippov set $\Phi(x, t)$ of the function $\phi(x, t)$, is homogeneous of the same degree $q \in \mathbb{R}$ with respect to the same dilation $(r_1, \ldots, r_n)$.*

Proof Let $x(\cdot)$ be a solution of the differential inclusion (3.172), specified with the Filippov set $\Phi(x, t)$ of the function $\phi(x, t)$. Then for all $c \geq c_0$ it is straightforward to verify that due to (3.175), the function $x^c(\cdot)$ with components (3.174) is also a solution of (3.172), thus specified. Hence, the differential inclusion (3.172) with the right-hand side $\Phi(x, t)$, being the Filippov set of the homogeneous function $\phi(x, t)$ in the sense of Definition 3.21, is homogeneous in the sense of Definition 3.20. ■

The above homogeneity definitions may not be suitable far away from the origin when the differential inclusion (3.172) (respectively, the function ϕ) is not homogeneous in the large. Formally, the need for the homogeneity localization can be argued as follows. Let a system be governed by a globally homogeneous differential inclusion and along with this inclusion, let an auxiliary system be considered, which coincides with the underlying homogeneous inclusion inside a ball B_δ. Beyond this homogeneity ball, however, it is appropriately extended in an arbitrary non-homogeneous manner. In this case, a localized homogeneity definition could be invoked to admit that the parameterized solutions $x^c(\cdot)$ evolve homogeneously only within the ball B_δ. Certainly, in such a case, there should be an implicit upper bound on c that limits $x^c(\cdot)$ to within B_δ whereas an arbitrarily large c can thus be chosen for globally homogeneous systems only. This motivates the local homogeneity to be defined as follows.

Definition 3.22 (*Local homogeneity* Orlov 2004) A piecewise continuous vector field $\phi(x, t) : \mathbb{R}^{n+1} \to \mathbb{R}^n$ of (3.173) is said to be *locally homogeneous* of degree $q \in \mathbb{R}$ with respect to dilation $(r_1, r_2, \ldots, r_n)$, where $r_i > 0,\ i = 1, \ldots, n$ (as well as the differential inclusion (3.172) is) if there exists a constant $c_0 > 0$ and a ball $B_\delta \in \mathbb{R}^n$ such that relation (3.175) holds true (respectively, any solution $x(\cdot)$ of (3.172) generates a parameterized set of solutions $x^c(\cdot)$ with components (3.174)) for all $(x_1, x_2, \ldots, x_n)^T \in B_\delta$, for almost all $t \in \mathbb{R}$, and for all $c \geq c_0$ such that $(c^{r_1}x_1, c^{r_2}x_2, \ldots, c^{r_n}x_n)^T \in B_\delta$.

A locally homogeneous differential inclusion (3.172) of degree $q < 0$ with respect to dilation $(r_1, \ldots, r_n)$ proves to be globally (locally) finite time stable whenever it is globally (locally) asymptotically stable. This fact is closely related to rescaling of the time and state variables. Going through this route allows one to additionally obtain an upper estimate

$$T(t_0, x^0) \leq \tau(x^0, E_R) + \frac{1}{1 - 2^q}(\delta R^{-1})^q s(\delta) \tag{3.176}$$

of the settling time function

$$T(t_0, x^0) = \sup_{x(\cdot, t_0, x^0)} \inf\{T \geq 0 : x(t, t_0, x^0) = 0 \ for\ all\ t \geq t_0 + T\} \tag{3.177}$$

via the reaching time functions

$$\tau(x^0, E_R) = \sup_{x(\cdot, t_0, x^0)} \inf\{T \geq 0 : x(t, t_0, x^0) \in E_R \ for\ all\ t_0 \in \mathbf{R},\ t \geq t_0 + T\} \tag{3.178}$$

and

$$s(\delta) = \sup_{x^0 \in E_\delta} \tau(x^0, E_{\frac{1}{2}\delta}) \tag{3.179}$$

where E_R denotes an ellipsoid of the form

$$E_R = \left\{ x \in \mathbf{R}^n : \sqrt{\Sigma_{i=1}^n \left(\frac{x_i}{R^{r_i}}\right)^2} \leq 1 \right\}, \tag{3.180}$$

to be located within a homogeneity ball, $\delta \geq c_0 R$, and $c_0 > 0$ is a lower estimate of the homogeneity parameter.

Theorem 3.3 (Homogeneity Principle Orlov 2009) *Let the differential inclusion* (3.172) *be locally homogeneous of degree $q < 0$ with respect to dilation $(r_1, \ldots, r_n)$ where $r_i > 0$, $i = 1, \ldots, n$ and let the equilibrium point $x = 0$ of* (3.172) *be globally (locally) uniformly asymptotically stable. Then the differential inclusion* (3.172) *is globally (locally) uniformly finite time stable and an upper estimate of the settling time function* (3.177) *is given by* (3.176).

Proof Due to the global (local) uniform asymptotic stability of (3.172), all the trajectories of the differential inclusion, initialized within a compact set, are uniformly steered toward an arbitrarily small ellipsoid (3.180). Then the condition

$$x(t) \in E_R \ for\ t \geq t_0 + \tau(x^0, E_R) \tag{3.181}$$

holds for an arbitrary solution $x(t)$ of (3.172), properly initialized with $x(t_0) = x^0$, where the ellipsoid E_R has been assumed to be small enough to be located within a homogeneity ball.

Apart from this, given an a priori fixed $\delta \geq c_0 R$ where $c_0 > 0$ is a lower estimate of the homogeneity parameter, there exists $s(\delta) > 0$ such that for each initial time moment $\tilde{t}_0$ and all the solutions $\tilde{x}(t)$ with $\tilde{x}(\tilde{t}_0) \in E_\delta$ one has $\tilde{x}(t) \in E_{\frac{1}{2}\delta}$ for $t \geq \tilde{t}_0 + s(\delta)$. Since the function $x^c(t)$, whose components (3.174) are specified with $c = \delta R^{-1} \geq c_0$, is a solution of (3.172) by homogeneity, and in addition, $x^c(\tilde{t}_0) \in E_\delta$ at $\tilde{t}_0 = c^{-q}(t_0 + \tau(x^0, E_R))$, it follows that

$$x^c(t) \in E_{\frac{1}{2}\delta} \ for\ t \geq c^{-q}(t_0 + \tau(x^0, E_R)) + s(\delta). \tag{3.182}$$

The latter relation, rewritten in terms of $x(t)$ by means of (3.174) subject to $c = \delta R^{-1}$, is represented as follows:

$$x(t) \in E_{\frac{1}{2}R} \ for\ t \geq t_1 = t_0 + \tau(x^0, E_R) + (\delta R^{-1})^q s(\delta). \tag{3.183}$$

Now, by applying the same derivation to a solution of (3.172) with $x(t_1) \in E_{\frac{1}{2}R}$, one obtains

$$x(t) \in E_{\frac{1}{4}R} \ for \ t \geq t_2 = t_1 + 2^q(\delta R^{-1})^q s(\delta). \tag{3.184}$$

In general, the following relations are derived

$$x(t) \in E_{2^{-(i+1)}R} \ for \ t \geq t_{i+1} = t_i + 2^{qi}(\delta R^{-1})^q s(\delta), \ i = 1, 2, \dots \tag{3.185}$$

by iterating on i. Since $\lambda = 2^q < 1$ by virtue of $q < 0$, the convergence of the time instants $t_k,\ k = 1, 2, \dots$ to a finite limit takes place:

$$\lim_{k\to\infty} t_k = t_0 + \tau(x^0, E_R) + \lim_{k\to\infty} \Sigma_{i=0}^{k-1}\lambda^i(\delta R^{-1})^q s(\delta) =$$
$$t_0 + \tau(x^0, E_R) + \lim_{k\to\infty} \frac{1-\lambda^k}{1-\lambda}(\delta R^{-1})^q s(\delta) =$$
$$t_0 + \tau(x^0, E_R) + \frac{1}{1-\lambda}(\delta R^{-1})^q s(\delta) < \infty. \tag{3.186}$$

Hence, relations (3.185) result in

$$x(t) \in \cap_{i=1}^{\infty} E_{2^{-i}R} = \{0\} \ for \ t \geq t_0 + \tau(x^0, E_R) + \frac{1}{1-2^q}(\delta R^{-1})^q s(\delta), \tag{3.187}$$

thereby establishing both the required finite time convergence property for the locally homogeneous differential inclusion (3.172) and the upper estimate (3.176) of the settling time function (3.177). Theorem 3.3 is thus proved. ■

Remark 3.1 For a globally homogeneous differential inclusion (3.172) one can choose $\delta_0 = c_0 R_0$ and R_0 sufficiently large to guarantee that $x(t, t_0, x^0) \in E_{R_0}$ for all $t \geq t_0$. Then $\tau(x^0, E_{R_0}) = 0$ and the upper estimate (3.176) of the settling time function (3.177) is simplified to

$$T(t_0, x^0) \leq \frac{c_0^q}{1-2^q} s(\delta_0) \tag{3.188}$$

with $\delta_0 = \delta_0(x^0)$ dependent on x_0.

An important corollary of the homogeneity principle is obtained if the differential inclusion (3.172) is generated by an uncertain differential equation

$$\dot{x} = \varphi(x, t) + \psi(x, t) \tag{3.189}$$

with a locally homogeneous piecewise continuous functions $\varphi(x, t)$ and locally uniformly bounded $\psi(x, t)$ within a ball B_δ such that

$$|\psi_i(x, t)| \leq M_i, \ i = 1, \dots, n \tag{3.190}$$

for almost all $(x, t) \in B_\delta \times \mathbb{R}$ and some constants $M_i \geq 0$, fixed a priori.

An absolutely continuous function $x(t)$, defined on an interval I, *is said to be a solution* of the uncertain differential equation (3.189) with the rectangular uncertainty constraint (3.190) iff it is a Filippov solution of (3.189) on the interval I for some piecewise continuous function $\psi(x,t)$, satisfying (3.190). As a matter of fact, the differential equation (3.173) with no uncertain term ψ is representable in the form of (3.189), (3.190) with $M_i = 0$ for all $i = 1, \ldots, n$.

Although such an uncertain discontinuous system (3.189) with a rectangular constraint (3.190) is inhomogeneous for a particular $\psi(x,t)$, it might be interpreted (Orlov 2009) as a homogeneous differential inclusion of the form

$$\dot{x} \in \Phi(x,t) + \Psi \tag{3.191}$$

where $\Phi(x,t)$ is the Filippov set of the function $\varphi(x,t)$ and Ψ is the Cartesian product of the intervals $\Psi_i = [-M_i, M_i]$, $i = 1, \ldots, n$, whereas

$$\Phi(x) + \Psi = \{\phi + \psi : \phi \in \Phi(x),\ \psi \in \Psi\} \tag{3.192}$$

is the union of these sets.

It is worth noticing that if $\varphi(x,t) = \varphi(x)$ is time-independent then the corresponding Filippov set $\Phi(x,t) = \Phi(x)$ in (3.191) is time-independent, too, and the uncertain system (3.189), (3.190) in this case would thus be governed by the autonomous differential inclusion, in spite of the presence of the uncertain time-varying term $\psi(x,t)$ in (3.189).

Playing with the above interpretation (3.191) of the uncertain system (3.189), (3.190), the following specification of the homogeneity principle is obtained.

Theorem 3.4 (Quasi-homogeneity Principle Orlov 2009) *Let the following conditions be satisfied*

1. *The right-hand side of an uncertain differential equation* (3.189) *consists of a locally homogeneous piecewise continuous function φ of degree $q < 0$ with respect to dilation $(r_1, \ldots, r_n)$ and a piecewise continuous function ψ whose components ψ_i, $i = 1, \ldots, n$ are locally uniformly bounded by constants $M_i \geq 0$ within a homogeneity ball;*
2. *$M_i = 0$ whenever $q + r_i > 0$;*
3. *The uncertain system* (3.189), (3.190) *is globally (locally) equiuniformly asymptotically stable around the origin.*

Then the differential inclusion (3.191), (3.192) *is globally (locally) equiuniformly finite time stable and its settling time function* (3.148) *is estimated as* (3.176).

Proof Conditions 1 and 2 of the theorem guarantee that the differential inclusion (3.191), (3.192) is locally homogeneous of degree $q < 0$ with respect to dilation $(r_1, \ldots, r_n)$.

Indeed, let $x(t) = (x_1(t), \ldots, x_n(t))^T$ be a solution of (3.189) under some piecewise continuous function $\psi(x,t)$, satisfying (3.190), and let $x(t)$ evolve within a ball B_δ where the homogeneity condition (3.175) holds almost everywhere for all $c \geq c_0$. Then for arbitrary $c \geq \max(1, c_0)$ such that the function $x^c(t)$ with components $x_i^c(t) = c^{r_i} x_i(c^q t),\ i = 1, \ldots, n$ evolves within B_δ, it is straightforward to verify the following. The function $x^c(t)$, thus constructed, is a solution of (3.189) with the piecewise continuous function $\psi(x, t) = \psi^c(x, t)$ whose components are

$$\psi_i^c(x, t) = c^{q+r_i} \psi_i(c^{-r_1} x_1, \ldots, c^{-r_n} x_n, c^q t).$$

Since by Condition 2 of the theorem one has $c^{q+r_i} \leq 1$ whenever $M_i > 0$, the function $\psi^c(x, t)$ is also admissible in the sense of (3.190). Hence, any solution of the uncertain differential equation (3.189), evolving within a homogeneity ball B_δ, generates a parameterized set of solutions $x^c(t)$ with the parameter c reasonably large. Taking into account Lemma 3.7, the differential inclusion (3.191), (3.192) is therefore locally homogeneous of degree $q \in \mathbf{R}$ with respect to dilation $(r_1, \ldots, r_n)$.

Thus, Conditions 1–3, coupled together, ensure that Theorem 3.3 is applicable to the differential inclusion (3.191), (3.192). By applying Theorem 3.3, the validity of Theorem 3.4 is established. ■

Apparently, Remark 3.1 remains valid for an uncertain system (3.189), (3.190), the right-hand side of which consists of a globally homogeneous piecewise continuous function φ of degree $q < 0$ and a globally uniformly bounded piecewise continuous function ψ. Recently, an interesting specification of the quasi-homogeneity principle for sectorial disturbances, vanishing in the origin, was proposed in Oza et al. (2015).

3.7 Concluding Remarks

Mathematical tools of dynamic systems are revised in the infinite-dimensional setting. Important instrumental inequalities (such as Cauchy–Schwartz, Agmon, Wirtinger, Poincaré inequalities) are recalled in Sobolev spaces, and useful eigenvalue/eigenfunction estimates are derived for Sturm–Liouville operators. Various (strong, mild, weak, viscosity, and proximal) solution concepts are addressed for parabolic and hyperbolic PDEs. Abstract dynamic systems, evolving in a Hilbert space, particularly, infinite-dimensional sliding mode dynamics and Hilbert space-valued dynamics with delays are introduced to specify stability concepts of interest. Special attention is given to dynamic systems, governed by differential inclusions, where stability equiuniformity with respect to disturbances is emphasized. It is shown that if along with global (local) equiuniform asymptotic stability, such a dynamic system possesses a weighted homogeneity of some negative degree it exhibits a stronger feature of (local) finite time stability, which is also equiuniform with respect to disturbances. The proposed arsenal of dynamic systems aims to develop universal stability tools in finite and infinite dimensions.

References

Andrieu V, Praly L, Astolfi A (2008) Homogenous approximation, recursive oserver design, and output feedback. SIAM J Control Optim 47:1814–1850

Aubin JP, Cellina A (1984) Differential inclusions - set valued maps and viability theory. Springer, Berlin

Bhat SP, Bernstein DS (2005) Geometric homogeneity with applications to finite-time stability. Math Control Signals Syst 17:101–127

Butkovskii AG (1982) Greens functions and transfer functions handbook. Horwood Ltd., Chichester

Clarke FH (1983) Optimization and nonsmooth analysis. Wiley Interscience, New York

Crandall MG, Lions PL (1983) Viscosity solutions of Hamilton–Jacobi equations. Trans Am Math Soc 277:1–42

Curtain R, Zwart H (1995) An introduction to infinite-dimensional linear systems. Springer, New York

Dashkovskiy S, Mironchenko A (2013) Input-to-state stability of infinite-dimensional control systems. Math Control Signals Syst 25:1–35

Fridman E, Orlov Y (2009) Exponential stability of linear distributed parameter systems with time-varying delays. Automatica 45:194–201

Gockenbach MS (2011) Partial differential equations-analytical and numerical methods, 2nd edn. SIAM, Philadeplhia

Kawski M (1989) Stabilization of nonlinear systems in the plane. Syst Control Lett 12:169–175

Kawski M (1991) Families of dilations and asymptotic stability. In: Analysis of controlled dynamical systems (Lyon, 1990). Progress in systems and control theory, vol 8, pp 285–294. Birkhäuser, Boston

Kawski M (1995) Geometric homogeneity and stabilization. In: Preprints of IFAC nonlinear control systems design symposium (NOLCOS'95), pp 164–169

Liberzon D (2012) Calculus of variations and optimal control theorey - a coincise introduction. Princeton University Press, Princeton

Mironchenko A, Ito H (2015) Construction of Lyapunov functions for interconnected parabolic systems: an iISS approach. SIAM J Control Optim 53:3364–3382

Morris K (2001) H_∞ output feedback of infinite-dimensional systems via approximation. Syst Control Lett 44:211–217

Moulay E, Perruquetti W (2006) Finite time stability and stabilization of a class of continuous systems. J Math Anal Appl 323:1430–1443

Mylvaganam T, Sassano M, Astolfi A (2014) Constructive ε-Nash equilibria for nonzero-sum differential games. IEEE Trans Auto Control 60(4):950–965

Orlov Y (2000) Discontinuous unit feedback control of uncertain infinite-dimensional systems. IEEE Trans Auto Control 45:834–843

Orlov Y (2004) Finite time stability and robust control synthesis of uncertain switched systems. SIAM J Control Optim 43:1253–1271

Orlov Y (2009) Discontinuous systems - Lyapunov analysis and robust synthesis under uncertainty conditions. Springer, London

Orlov Y (2017) On General properties of eigenvalues and eigenfunctions of a Sturm–Liouville operator: Comments on "ISS with respect to boundary disturbances for 1-D parabolic PDEs". IEEE Trans. Auto. Ctrl. 62(11): 5970–5973

Osuna T, Orlov Y, Aguilar L (2018) $\mathscr{L}_2$-gain tuning of variable structure SISO systems of relative degree n. Int J Control 81:2422–2444

Oza HB, Orlov Y, Spurgeon SK (2015) Continuous uniform finite time stabilization of planar controllable systems SIAM. J Control Optim 53:1154–1181

Petrovskii IG (1961) Lectures on partial differential equations, 3rd edn. Fizmatgiz, Moscow

Pisano A, Orlov Y (2017) On the ISS properties of a class of parabolic DPS with discontinuous control using sampled-in-space sensing and actuation. Automatica 81:447–454

Polyakov A, Efimov D, Fridman E, Perruquetti W (2016) On homogeneous distributed parameters equations. IEEE Trans Auto Control 61:3657–3662
Polyakov A, Coron JM, Rosier L (2018) On homogeneous finite-time control for linear evolution equation in hilbert space. IEEE Trans Auto Control 63:3143–3150
Polyakov A (2018) Sliding mode control design using canonical homogeneous norm. Int J Robust Nonlinear Control 29:682–701
Prieur C, Mazenc F (2012) ISS-Lyapunov functions for timevarying hyperbolic systems of balance laws. Math Control Signals Syst 24:111–134
Qian C (2005) A homogeneous domination approach for global output feedback stabilization of a class of nonlinear systems. In: Proceedings of the American control conference, pp 4708–4715
Rousier L (1992) Homogeneous Lyapunov function for homogeneous continuous vector field. Syst Control Lett 19:467–473
Sontag E (2008) Input to state stability - basic concepts and results. In: Nistri P, Stefani G (eds) Nonlinear and optimal control theory. Springer, Berlin, pp 163–220
Subbotin AI (1995) Generalized solutions of first-order PDE's. Birkhäuser, Boston
Wang T (1994) Stability in abstract functional-differential equations. J Math Anal Appl 186:835–861
Willems J (1972) Dissipative dynamical systems. Part I: general theory. Arch Rat Mech Anal 45:321–351
Zubov VI (1964) Methods of A.M. Lyapunov and their applications. Noordhoff, Leiden

Part II
Construction of Nonsmooth Lyapunov Functions

The Lyapunov function method, being initially proposed for smooth dynamic systems in $\mathbb{R}^n$, remains a classical analysis tool for differential inclusions and particularly for Filippov's dynamics in discontinuous systems (Utkin 1992; Yakubovich et al. 2004; Orlov 2009). It is additionally well-recognized to be useful for robust synthesis in the presence of disturbances. The relevant robust synthesis relies on having a continuous positive-definite control Lyapunov function (CLF) $V(x,t) : \mathbb{R}^n \times \mathbb{R} \to \mathbb{R}_{\geq 0}$ such that there is some control input that ensures the time derivative $\dot{V}(x(t),t)$, computed along the closed-loop system trajectories x(t), to be negative definite regardless of whichever admissible disturbance affects the system. Since such a CLF $\dot{V}(x,t)$ is admitted to be continuous only its time derivative $\dot{V}(x(t),t)$ is viewed in the sense of directional Dini superderivatives (contingent hypoderivatives), earlier addressed in Sect. 3.5. Involving nonsmooth CLFs allows one to arrive at a stabilizing discontinuous feedback law, which is actually motivated by a well-known fact that a stabilizing continuous feedback may fail to exist even for simple controllable systems.

To begin with, various kinds of strict and nonstrict Lyapunov functions are introduced for addressing different stability concepts such as ISS and finite time stability among others. To add universality, the Lyapunov approach is revisited in the infinite-dimensional setting. Constructive methods are then proposed to design nonsmooth Lyapunov functions. First, generalized Lyapunov forms are introduced to specify FTS Lyapunov functions for SM algorithms such as twisting and supertwisting ones. Next, Lyapunov minmax approach and speed gradient method are presented to illustrate the resulting discontinuous feedback synthesis in the presence of matched disturbances. After that, the construction methodology, relying on appropriate solutions of the so-called Lyapunov gradient equation, is proposed for multiple Lyapunov functions of VSS. Finally, positive definite proximal solutions of Hamilton–Jacobi PDIs are invoked to construct robust Lyapunov functions of uncertain VSS with resets.

References

Orlov Y (2009) Discontinuous systems - Lyapunov analysis and robust synthesis under uncertainty conditions. Springer, London

Utkin VI (1992) Sliding modes in control and optimization. Springer, Berlin
Yakubovich VA, Leonov GA, Gelig AK (2004) Stability of stationary sets in control systems with discontinuous nonlinearities. World Scientfic, Singapore

Chapter 4
Modern Lyapunov Tools

This chapter deals with a dynamic system $S(t, t_0, x_0)$ to analyze its stability based on the Lyapunov approach. Throughout, such a system is assumed to evolve in a Euclidean/Hilbert space X, whichever specified, with initial conditions $t_0 \in \mathbb{R}$ $x_0 \in D$, located in an operational domain $D \subseteq X$. The state space X is equipped with a standard norm $\|\cdot\|$, generated by the inner product in X.

While being well-suited for the standard stability analysis, the existing arsenal of Lyapunov functions presumes a more delicate application where along with the asymptotic stability, complement features (e.g., ISS and/or finite time stability) are of interest. Strict Lyapunov functions, satisfying a certain differential inequality, and particularly homogeneous Lyapunov functions, which are always available for homogeneous dynamics, suffice in many cases. Alternatives in a situation, where a strict Lyapunov function cannot be identified, are non-strict and multiple Lyapunov functions. Non-strict Lyapunov functions possess only negative semi-definite time derivatives along the system trajectories, which is why invoking advanced methods such as an appropriate extension of the invariance principle and semi-global Lyapunov function strictification becomes critical to finally establish the asymptotic stability. In turn, multiple Lyapunov functions, independently constructed in the operational sub-domains, additionally call for verifying certain consistency conditions along the sub-domain boundaries to ensure the desired asymptotic stability of the system in question. Once the asymptotic stability is established, the finite time convergence of trajectories might be verified by using semi-global and homogeneous Lyapunov functions. Later on, various kinds of strict and non-strict Lyapunov functions are addressed side by side.

4.1 Strict Lyapunov Functionals

The Lyapunov approach to establishing asymptotic stability of a Hilbert space-valued dynamic system $S(t, t_0, x_0)$ consists in finding a positive definite functional of the state and possibly of time variable such that its time derivative, computed along the

Y. Orlov, *Nonsmooth Lyapunov Analysis in Finite and Infinite Dimensions*,
Communications and Control Engineering,
https://doi.org/10.1007/978-3-030-37625-3_4

system trajectories, is negative definite. The sign-definiteness is further adopted in such a manner to conceptually suit it to an appropriate introduction of strict and non-strict Lyapunov functionals.

As usual, a function $v(\cdot) : \mathbb{R}_{\geq 0} \to \mathbb{R}_{\geq 0}$ is said to be positive definite iff it is continuous, $v(0) = 0$, and $v(s) > 0$ for all $s > 0$; a $v(\cdot)$ is said to be of class $\mathcal{K}$ iff it is continuous, strictly increasing, and $v(0) = 0$; it is moreover said to be of class $\mathcal{K}_\infty$ iff in addition, it escapes to infinity as its argument goes to infinity (i.e., $v(s) \to \infty$ as $s \to \infty$).

A continuous functional $V(x, t) : D \times \mathbb{R} \to \mathbb{R}_{\geq 0}$ is said to be *positive semi-definite* iff $V(x, t) \geq v_0(\|x\|) \geq 0$ for some continuous comparison function $v_0(\cdot)$ and for all $(x, t) \in D \times \mathbb{R}$, and *positive definite* iff $v_0(\cdot)$ is so. Moreover, it is said to be *strictly positive definite* iff $v_0(\cdot) \in \mathcal{K}$, *radially unbounded* iff $v_0(\cdot) \in \mathcal{K}_\infty$, and *decrescent* iff $V(x, t) \leq v_1(\|x\|)$ for another comparison function $v_1(\cdot) \in \mathcal{K}$ and for all $(x, t) \in D \times \mathbb{R}$. It is said to be *negative definite (semi-definite)* iff $-V(x, t)$ is positive definite (semi-definite).

Definition 4.1 (*Lyapunov functional*) A *strict Lyapunov functional* of the dynamic system $S(t, t_0, x_0)$ is a continuous, strictly positive definite, decrescent functional $V(x, t) : X \times \mathbb{R} \to \mathbb{R}_{\geq 0}$ such that

$$\dot{V}(x, t) = \limsup_{\tau \to 0+} \frac{V(S(t + \tau, t, x), t + \tau) - V(x, t)}{\tau} \tag{4.1}$$

is negative definite. If (4.1) is negative semi-definite rather than negative definite, the Lyapunov functional $V(x, t)$ is said to be *non-strict*.

Under certain conditions, a dynamic system proves to be asymptotically stable provided it possesses a strict Lyapunov functional.

Theorem 4.1 *Suppose that*

1. *A dynamic system $S(t, t_0, x_0)$ is uniformly continuous in t, i.e., it generates uniformly continuous trajectories $x(t) = S(t, t_0, x_0)$, $t \geq t_0$ for all $t_0 \in \mathbb{R}$ and for all $x_0 \in D$;*
2. *The system possesses a strict Lyapunov functional $V(x, t)$, locally defined on a ball $B_\delta = \{x \in D : \|x\| \leq \delta,\ \delta > 0\}$ for all t.*

Then the underlying dynamic system is asymptotically stable.

Proof Since $V(x, t)$ is a strict Lyapunov functional of the system in question, there are comparison functions $v_0(\cdot), v_1(\cdot), v(\cdot) : \mathbb{R}_{\geq 0} \to \mathbb{R}_{\geq 0}$ such that $v_0, v_1 \in \mathcal{K}$, $v(\cdot)$ is positive definite,

$$v_0(\|x\|) \leq V(x, t) \leq v_1(\|x\|) \text{ for all } (x, t) \in B_\delta \times \mathbb{R}, \tag{4.2}$$

and the inequality

$$\dot{V}(x(t), t) \leq -v(\|x(t)\|) \tag{4.3}$$

holds true for almost all $t \in \mathbb{R}$ with the time derivative of the composite functional $V(x(t), t)$, computed along the system trajectories $x(t)$, which are initialized within B_δ.

Subject to $r \in (0, \delta)$ and $c \in (0, \min_{\|x\|=r} v_0(\|x\|))$, the set $\{x \in B_\delta : v_0(\|x\|) \leq c\}$ is within the interior of the ball B_r. Let a time-dependent set be given by $\Omega_{c,t} = \{x \in B_r : V(x, t) \leq c\}$. Then by virtue of (4.2), one concludes that

$$\{x \in B_r : v_1(\|x\|) \leq c\} \subset \Omega_{c,t} \subset \{x \in B_r : v_0(\|x\|) \leq c\} \subset B_r \subset B_\delta \qquad (4.4)$$

for all $t \geq t_0$.

Due to (4.3), each solution $x(t) = S(t, t_0, x_0)$ is a priori estimated via

$$\sup_{t\in[t_0,\infty)} V(x(t), t) \leq c \qquad (4.5)$$

if it is initialized at $t_0 \in \mathbb{R}$ with $x(t_0) = x_0 \in \Omega_{c,t_0}$. It follows that $x(t) \in \Omega_{c,t}$ for all $t \geq t_0$. Coupled to (4.4), this ensures that any solution $x(t)$ starting in $\{x \in B_r : v_1(\|x\|) \leq c\}$ stays in $\{x \in B_r : v_0(\|x\|) \leq c\}$ for all $t \geq t_0$. Hence, $x(t)$ is uniformly bounded in t. Moreover, by choosing r arbitrarily small, $x = 0$ is shown to be a stable equilibrium point of the system.

It remains to demonstrate that

$$\lim_{t\to\infty} \|x(t)\| = 0. \qquad (4.6)$$

For this purpose, let us integrate (4.3) on a solution $x(t) = S(t, t_0, x_0)$, initialized at $t_0 \in \mathbb{R}$ with $x(t_0) = x_0$ within a compact set $\{x \in B_r : v_1(\|x\|) \leq c\}$. The resulting inequality

$$\int_{t_0}^{\infty} v(\|x(t)\|)dt \leq \lim_{t\to\infty} [V(x(t_0), t_0) - V(x(t), t)] \leq V(x(t_0), t_0) \leq c \qquad (4.7)$$

is thus obtained where the integrand $v(\|x(t)\|)$ is uniformly continuous in t. Indeed, $v(\|x\|)$ is uniformly continuous on the compact set $\{x \in B_r : v_0(\|x\|) \leq c\}$ whereas $x(t)$ is uniformly continuous in t by the first condition of the theorem.

Thus, the following convergence

$$\lim_{t\to\infty} v(\|x(t)\|) = 0 \qquad (4.8)$$

is straightforwardly established by applying Barbalat's Lemma 2.2 to the integral inequality (4.7). Taking into account that $v(\cdot)$ is positive definite, convergence (4.6) is then guaranteed by (4.8).

Indeed, the final conclusion can be shown by contradiction, for if it were not true, there would exist a solution $x(t) = S(t, t_0, x_0)$, time instants t_k, $k = 1, 2, \ldots,$ and $\varepsilon > 0$ such that $t_k \to \infty$ as $k \to \infty$ and $\|x(t_k)\| > \varepsilon$ for all $k = 1, 2, \ldots$. In accordance with (4.4), it follows that the solution magnitudes $\|x(t_k)\|,\ k = 1, 2, \ldots$ belong to the compact set

$$K = \{s \in \mathbb{R} : s \geq \varepsilon \ and \ v_0(s) \leq c\}.$$

Since the function v is continuous and such that $v(\cdot) : \mathbb{R}_{>0} \to \mathbb{R}_{>0}$, it is verified that

$$\inf_k v(\|x(t_k)\|) \geq \min_{s \in K} v(s) > 0. \tag{4.9}$$

The proof of Theorem 4.1 is then completed by noting that (4.9) contradicts the limiting relation (4.8). ■

It is worth noticing that in a particular case of a strict Lyapunov functional $V(x, t)$, globally defined on the entire operational domain $D = X$, Theorem 4.1 ensures the global asymptotic stability of the system in question provided that the functional $V(x, t)$ is radially unbounded.

4.1.1 Multiple Lyapunov Functionals

While analyzing switched, hybrid, and variable structure dynamic systems $S(t, t_0, x_0)$, multiple strict Lyapunov functionals $V_i(x, t),\ i = 1, 2, \ldots N < \infty$ (e.g., a family of strict Lyapunov functionals of each individual subsystem, forming the overall system) may be involved (Branicky 1989). If it so happens that the values of the functionals $V_i(x, t),\ i = 1, 2, \ldots N$ coincide on their mutual boundaries of the corresponding sub-domains $D_i(x, t) = D(V_i(x, t)),\ i = 1, 2, \ldots N$, which possess disjoint interiors, then there exists a continuous common Lyapunov functional $V(x, t)$ of the underlying VSS such that $V(x, t) = V_i(x, t)$ for all $(x, t) \in D_i(x, t),\ i = 1, 2, \ldots N$. Just in the case, the stability analysis may be reduced to the standard one, based on the common Lyapunov functional $V(x, t)$.

In general, the resulting functional $V(x, t)$ is however discontinuous. Thus, one should presume that while a specific functional $V_i(x, t)$ is decreasing when the corresponding system structure is active, it may increase when the structure becomes inactive. To prevent such a situation, an extra condition should be imposed on the functionals $V_i(x, t),\ i = 1, 2, \ldots N$ to form a decreasing sequence of their values at the beginning of each interval when the individual structure becomes active. The following result (Liberzon 2003, Theorem 3.1) is in force for finite-dimensional switched systems composed of the individual subsystems

$$\dot{x} = f_i(x),\ i = 1, 2, \ldots, N \tag{4.10}$$

with locally Lipschitz functions $f_i(x) : \mathbb{R}^n \to \mathbb{R}^n$.

Theorem 4.2 *Let* (4.10) *be a finite family of globally asymptotically stable systems and let $V_i(x),\ i = 1, 2, \ldots, N$ be a family of corresponding radially unbounded* *strict Lyapunov functions. Suppose that there exists a family of positive definite continuous functionals $W_i(x)$, $i = 1, 2, \ldots, N$ with the property that for every pair of switching times t_k, t_j, $k < j$ such that the ith subsystem is active at these time instants and it is inactive between them, one has*

$$V_i(x(t_j)) - V_i(x(t_k)) \leq -W_i(x(t_k)). \tag{4.11}$$

Then the switched system (4.10) *is globally asymptotically stable.*

Proof The validation of Theorem 4.2 is rather technical and can be found in Liberzon (2003, pp. 33–34). ■

It is worth noticing that the above result is actually pretty universal and admits a particular interpretation in terms of VSS with impulse effects incorporated. Section 4.3 presents a similar result with an alternative reasoning for Hilbert space-valued dynamic systems subject to state restitutions under unilateral constraints.

4.1.2 Semi-global Lyapunov Functionals

Let us now consider a parameterized family of *strict semi-global Lyapunov functionals* $V^\delta(x, t)$, $\delta > 0$ such that each $V^\delta(x, t)$ is well-posed on the corresponding domain $x \in B_\delta \subset X$ for all $t \in \mathbb{R}$. It is clear that the domains of the semi-global Lyapunov functionals are nested and $\bigcap_{\delta>0} B_\delta = \{0\}$ whereas their union $\bigcup_{\delta>0} B_\delta = X$ covers the entire state space as $\delta \to \infty$.

By definition, the strict semi-global Lyapunov functionals $V^\delta(x, y)$, $\delta > 0$ respect the relations

$$v_0^\delta(\|x\|) \leq V^\delta(x, t) \leq v_1^\delta(\|x\|) \tag{4.12}$$

for all $(x, T) \in B_\delta \times \mathbb{R}$ and for some comparison functions $v_0^\delta(\cdot), v_1^\delta(\cdot) \in \mathscr{K}$. Moreover, their time derivatives $\dot{V}^\delta(x(t), t)$, computed along the system trajectories $x(t) = S(t, t_0, x_0)$, which are initialized within B_δ, are negative definite, i.e., the inequality

$$\dot{V}^\delta(x(t), t) \leq -v^\delta(\|x(t)\|) \tag{4.13}$$

holds almost everywhere for all $x(t) = S(t, t_0, x_0)$, $x_0 \in B_\delta$, and for some positive definite $v^\delta(\cdot)$. Apart from this, the family $\{V^\delta(x, t)\}_{\delta>0}$ is assumed to be radially unbounded in the sense that

$$\lim_{\delta\to\infty} \min_{\|x\|=\delta} v_0^\delta(x) = \infty. \tag{4.14}$$

The following result is then in order.

Theorem 4.3 *Suppose that*

1. *A dynamic system $S(t, t_0, x_0)$ is uniformly continuous in t, i.e., it generates uniformly continuous trajectories $x(t) = S(t, t_0, x_0)$, $t \geq t_0$ for all $t_0 \in \mathbb{R}$ and for all $x_0 \in D$;*
2. *There exist strict Lyapunov functionals $V^\delta(x, t)$, parameterized by $\delta > 0$, such that* (4.12), (4.13) *hold for all $(x, t) \in B_\delta \times \mathbb{R}$, and for some comparison functions $v_0^\delta(\cdot), v_1^\delta(\cdot) \in \mathscr{K}$ and positive definite $v^\delta(\cdot)$;*
3. *the family $\{V^\delta(x, t)\}_{\delta>0}$ is radially unbounded in the sense of* (4.14).

Then the underlying dynamic system is globally asymptotically stable.

Proof The asymptotic stability of the dynamic system in question is straightforwardly obtained by applying Theorem 4.1. As the family $\{V^\delta(x, t)\}_{\delta>0}$ is radially unbounded, the constant c, used in the proof of Theorem 4.1, can be chosen to be arbitrarily large to include any initial state in the set $\{x \in \mathbb{R} : v_0^\delta(\|x\|) \leq c\}$ for some $\delta > 0$. Hence, convergence (4.6) holds for any initial conditions. The validity of Theorem 4.3 is thus established. ■

In the sequel, stronger conditions are imposed on the time derivative of a strict Lyapunov functional to ensure its finiteness and ISS features.

4.1.3 Finite Time Stable Lyapunov Functionals

Suppose that a dynamic system $S(t, t_0, x_0)$ possesses a strict Lyapunov functional $V(x, t) : X \times \mathbb{R} \to \mathbb{R}_{\geq 0}$ with the time derivative $\dot{V}(x(t, t))$ along the system trajectories $x(t) = S(t, t_0, x_0)$ such that the inequality

$$\frac{dV(x(t), t)}{dt} \leq -2\gamma V^\beta(x(t), t) \tag{4.15}$$

holds for almost all $t \geq t_0$ and for some constants $\gamma > 0$, $\beta \in [0, 1)$. Then the equality $V(x(t), t) = 0$ holds true for all $t \geq t_0 + T_{ste}$ where

$$T_{ste} = [2\gamma(1 - \beta)]^{-1} V^{1-\beta}(x_0) \tag{4.16}$$

is a settling time estimate. The finiteness of the composite functional $V(x(t), t)$ is straightforwardly concluded from Orlov (2009, Lemma 4.3) which is reproduced herein as follows.

Lemma 4.1 *Let an everywhere nonnegative functional $\mathscr{V}(t)$ meet the differential inequality*

$$\dot{\mathscr{V}}(t) \leq -2\gamma \mathscr{V}^\beta(t) \tag{4.17}$$

for all $t \geq t_0$ *and for some constants* $\gamma > 0$ *and* $\beta \in (0, 1)$. *Then* $\mathscr{V}(t) = 0$ *for all*

$$t \geq t_0 + [2\gamma(1-\beta)]^{-1}\mathscr{V}^{1-\beta}(t_0). \tag{4.18}$$

Proof By Comparison Lemma 2.1, an arbitrary nonnegative solution $\mathscr{V}(t)$ of inequality (4.17) is dominated $\mathscr{V}(t) \leq \mathscr{W}(t)$ by the solution

$$\mathscr{W}(t) = \begin{cases} [\mathscr{V}^{(1-\beta)}(t_0) - 2\gamma(1-\beta)(t-t_0)]^{\frac{1}{1-\beta}} & \text{if } t \in [t_0, t_0 + \frac{\mathscr{V}^{(1-\beta)}(t_0)}{2\gamma(1-\beta)}] \\ 0 & \text{if } t \geq t_0 + \frac{\mathscr{V}^{(1-\beta)}(t_0)}{2\gamma(1-\beta)} \end{cases} \tag{4.19}$$

of the differential equation

$$\dot{\mathscr{W}} = -2\gamma\mathscr{W}^{\beta},$$

specified with the same initial condition $\mathscr{W}(t_0) = \mathscr{V}(t_0)$. Since $\mathscr{W}(t) = 0$ for all $t \geq t_0 + [2\gamma(1-\beta)]^{-1}\mathscr{V}^{1-\beta}(t_0)$, the functional $\mathscr{V}(t)$ vanishes after a finite time moment $T \leq t_0 + [2\gamma(1-\beta)]^{-1}\mathscr{V}^{1-\beta}(t_0)$, i.e., $\mathscr{V}(t) = 0$ for all $t \geq t_0 + [2\gamma(1-\beta)]^{-1}\mathscr{V}^{1-\beta}(t_0)$. Lemma 4.1 is thus proved. ∎

The strict Lyapunov functional $V(x, t)$, satisfying (4.16), is therefore referred to as a *finite time stable Lyapunov functional*, or simply FTS Lyapunov functional.

Another differential inequality

$$\frac{dV(x(t), t)}{dt} \leq -2k\gamma(t)V(x(t), t), \ t \in [t_0, t_0 + T), \tag{4.20}$$

constituting a nonautonomous FTS Lyapunov functional $V(x, t)$ is specified with a time-varying function $\gamma(t)$ of class $\Gamma(t_0, t_0 + T)$ and some finite $T > 0$, and it is confined to the semi-open interval $[t_0, t_0 + T)$.

A function $\gamma(t)$ *is said to be of class* $\Gamma(t_0, t_0 + T)$ iff it possesses the properties, listed below:

1. $\gamma(t)$ is a monotonically increasing function with finite values on $[t_0, t_0 + T)$;
2. $\gamma(t_0) = 1$;
3. $\lim_{t \to t_0 + T} \int_{t_0}^{t} \gamma(\tau)d\tau = \infty$.

Examples

$$\gamma(t) = \frac{T^{m+n}}{(T + t_0 - t)^{m+n}} \tag{4.21}$$

of such a function $\gamma(t) \in \Gamma(t_0, t_0 + T)$ are inherited from Song et al. (2017a, b) with positive integers m, n.

The finiteness of a nonnegative solution of (4.20) is guaranteed by the following result, inspired from Song et al. (2017a, Lemma 1).

Lemma 4.2 *Let an everywhere nonnegative function* $\mathscr{V}(t)$ *meet the differential inequality*

$$\dot{\mathscr{V}}(t) \le -2k\gamma(t)\mathscr{V}(t) \tag{4.22}$$

for some $\gamma(t) \in \Gamma(t_0, t_0 + T)$ *and* $k > 0$*, and for all* $t \in [t_0, t_0 + T)$*. Then*

$$\mathscr{V}(t) \le e^{-2k\int_{t_0}^{t}\gamma(\tau)d\tau}\mathscr{V}(t_0), \quad t \in [t_0, t_0 + T), \tag{4.23}$$

and $\mathscr{V}(t)$ *escapes to zero as* $t \to t_0 + T$*.*

Proof As in the proof of Lemma 4.1, solving (4.22) yields (4.23). Moreover, since $\gamma(t) \in \Gamma(t_0, t_0 + T)$ the right-hand side of (4.23) escapes to zero as $t \to t_0 + T$, and hence $\mathscr{V}(t_0 + T) = 0$. Lemma 4.2 is thus proved. ■

A potential generalization of the differential inequality (4.20) could deal with a directional Dini superderivative (contingent hypoderivative) in its left-hand side. This would allow one to involve nonsmooth FTS Lyapunov functionals $V(x, t)$. A very recent generalization in this regard may be found in Matusik et al. (2018).

4.1.4 Homogeneous Lyapunov Functions

As established in the finite-dimensional setting of Sect. 3.6.5, a locally homogeneous system of a negative homogeneity degree is necessarily (globally) finite time stable provided it is (globally) asymptotically stable. An important result, carried out in Rosier (1992), is that a finite-dimensional autonomous asymptotically stable homogeneous system possesses an autonomous homogeneous strict Lyapunov function. It is worth noticing that the homogeneity definition for scalar functions $V(x)$ of a vector argument $x \in \mathbb{R}^n$ is different from Definition 3.21 of vector fields, and it is as follows.

Definition 4.2 (*Homogeneity of scalar functions* Kawski 1995) A continuous function $V(x) : \mathbb{R}^n \to \mathbb{R}$ is called homogeneous of degree $p \in \mathbb{R}$ with respect to dilation $(r_1, r_2, \ldots, r_n)$, where $r_i > 0,\ i = 1, \ldots, n$, if

$$V(c^{r_1}x_1, c^{r_2}x_2, \ldots, c^{r_n}x_n) = c^p V(x_1, x_2, \ldots, x_n) \tag{4.24}$$

for all $(x_1, x_2, \ldots, x_n)^T \in \mathbb{R}^n$ and for all $c > 0$.

Thus, a homogeneous function is characterized by the property that any scaled level set of the function is also a level set whereas a homogeneous vector field is featured by the property that any scaled orbit of the vector field is also an orbit. An interested reader may refer to Kawski (1995) for more distinct details on the homogeneity concepts for functions and vector fields.

The aforementioned existence of a strict homogeneous Lyapunov function was first established for homogeneous vector fields of class C^1 (Hahn 1963; Zubov 1964) and then extended to continuous homogeneous vector fields (Rosier 1992, Theorem 2). The latter result was established based on a construction of a strict homogeneous Lyapunov function, using a strict (not necessarily homogeneous) Lyapunov function, the existence of which is guaranteed by the converse of Lyapunov's second theorem, proven in Kurzweil (1956) for continuous vector fields. Since the Lyapunov's converse has become available for discontinuous systems (Bacciotti and Rosier 1998) a strict homogeneous Lyapunov function appears to be constructible for discontinuous vector fields as well.

A stronger version of Rosier (1992, Theorem 2), given below, is extracted from Bhat and Bernstein (2005, Theorems 7.1 and 7.2) to bring into play homogeneous FTS Lyapunov functions.

Theorem 4.4 *Let continuous $f(x) : \mathbb{R}^n \to \mathbb{R}^n$ be homogeneous of degree q with respect to dilation $(r_1, r_2, \ldots, r_n)$ and let $\beta \in (0, 1)$. If the origin is an FTS equilibrium of the vector field f, then $q < 0$, and there exist $\gamma > 0$ and continuous positive definite homogeneous function $V(x) : \mathbb{R}^n \to \mathbb{R}_{\geq 0}$ of degree $p = -\frac{q}{1-\beta}$ which is of class C^1 on $\mathbb{R}^n \setminus \{0\}$ and it is such that the Lie derivative $L_f V(x) := \nabla^T V(x) f(x)$ is continuous on $\mathbb{R}^n$ and*

$$L_f V(x) \leq -2\gamma V^{\beta}(x). \tag{4.25}$$

Coupling Theorem 4.4 to Lemma 4.1 results in a settling time estimate (4.16), which remains in force for the underlying vector field $f(x)$.

4.1.5 Input-to-State Stable Lyapunov Functions

In the finite-dimensional setting, ISS Lyapunov functions were invoked in Sontag and Wang (1995) to characterize the ISS property for a general nonlinear system

$$\dot{x} = f(x, w) \tag{4.26}$$

with a continuously differentiable function $f(x, w) : \mathbb{R}^{n+m} \to \mathbb{R}^n$ such that $f(0, 0) = 0$. It was established that such a system is ISS iff it admits an ISS Lyapunov function which is defined as follows.

Definition 4.3 (*ISS Lyapunov functions* Sontag and Wang 1995) A smooth function $V(x) : \mathbb{R}^n \to \mathbb{R}_{\geq 0}$ is called an ISS Lyapunov function for system (4.26) if inequalities

$$\alpha_1(\|x\|) \leq V(x) \leq \alpha_2(\|x\|) \tag{4.27}$$

hold for all $x \in \mathbb{R}^n$ and for some functions α_1, α_2 of class $\mathscr{K}_\infty$, and given any input $w \in \mathbb{R}^n$, another inequality

$$\nabla^T V(x) f(x, w) \leq -\alpha_3(\|x\|) \tag{4.28}$$

holds for all sufficiently large $x \in \mathbb{R}^n$ so that $\|x\| \geq \alpha_4(\|w\|)$ and for some functions α_3, α_4 of class $\mathscr{K}$.

In the infinite-dimensional setting, a pretty straightforward extension of the ISS Lyapunov function concept was developed in Dashkovskiy and Mironchenko (2013) to demonstrate that for a wide class of dynamic systems, evolving in a Banach space, the existence of an ISS Lyapunov functional ensures the system to be ISS. However in contrast to the finite-dimensional setting, a converse result that an ISS infinite-dimensional system possesses an ISS Lyapunov functional has not been established yet.

4.2 Non-strict Lyapunov Functionals

An alternative to establishing asymptotic stability through strict Lyapunov functionals is via using non-strict Lyapunov functionals, which are normally easier to guess. However, the knowledge of a non-strict Lyapunov functional is not so efficient tool because it is in general capable of analyzing a partial asymptotic stability in a reduced state space only. The following result, representing an infinite-dimensional extension of LaSalle–Yoshizawa Theorem (Krstic et al. 1995, Theorem A.8) that has been inherited from the finite-dimensional treatment (LaSalle 1960; Yoshizawa 1963), is due.

Theorem 4.5 *Suppose that*

(C1) *a dynamic system $S(t, t_0, x_0)$ is uniformly continuous in t;*

(C2) *$S(t, t_0, x_0)$ possesses a non-strict Lyapunov functional $V(x, t)$, locally defined on a ball $B_\delta = \{x \in D : \|x\| \leq \delta, \ \delta > 0\}$ for all t so that inequalities* (4.2) *and* (4.3) *hold for some comparison functions $v_0(\cdot), v_1(\cdot)$ of class $\mathscr{K}$ and for positive semi-definite $v(\cdot)$;*

(C3) *$S(t, t_0, x_0)$ is initialized within $\{x \in B_\delta : v_1(\|x\|) \leq c\}$ where*

$$c \in (0, \min_{\|x\|=r} v_0(\|x\|)) \textit{ for some } r \in (0, \delta).$$

Then the corresponding trajectories $x(t) = S(t, t_0, x_0) \in \{x \in B_\delta : v_0(\|x\|) \leq c\}$ for all $t \geq t_0$ and $\lim_{t\to\infty} v(\|x(t)\|) = 0$.

Proof Analyzing the proof of Theorem 4.1, it is straightforward to verify that the positive semi-definiteness of the function $v(\cdot)$ would suffice to establish both boundedness (4.5) of properly initialized system trajectories and the limiting relation (4.8) along them. Thus, the line of reasoning, used in the proof of Theorem 4.1, applies here to validate Theorem 4.5. ■

It should be pointed out that under the conditions of Theorem 4.5, the dynamic system proves to be partially asymptotically stable in a subspace $X_1 \subseteq X$ provided that inequality (4.3) holds with a positive definite function $v(\|x_1\|)$, which depends on the state component $x_1 \in X_1$ only. In other words, the convergence $\lim_{t\to\infty} \|x_1(t)\| = 0$ is additionally guaranteed in this case. To reproduce this conclusion, it suffices to note that as in the proof of Theorem 4.1, the limiting relation $\lim_{t\to\infty} \|x_1(t)\| = 0$ follows from the convergence $\lim_{t\to\infty} v(\|x_1(t)\|) = 0$ with a positive definite function $v(\cdot)$.

4.2.1 Invariance Principle

For autonomous dynamic systems, once a non-strict Lyapunov functional is identified, the asymptotic stability can be carried out by applying the well-known Krasovskii–LaSalle invariance principle, initially proven for Euclidian continuous vector fields (Krasovskii 1959; LaSalle 1960) and later on extended to a broader class of dynamic systems such as discontinuous, unilaterally constrained, and infinite-dimensional systems (Alvarez et al. 2000; Brogliato and Goeleven 2005; Henry 1981) to name a few.

Consider an autonomous dynamic system $S(t, x_0)$, evolving in a Hilbert space X and initialized in an operational domain $D \subseteq X$ at $t = 0$ with $x_0 \in D$.

A set $M \in D$ is said to be *an invariant set* of the dynamic system $S(t, x_0)$ iff

$$x_0 \in M \ \Rightarrow S(t, x_0) \in M \ \forall\, t \geq 0. \tag{4.29}$$

For an arbitrary $x_0 \in D$, let

$$\gamma(x_0) = \{S(t, x_0),\ t \geq 0\} \tag{4.30}$$

denote the orbit through x_0 and let

$$\omega(x_0) = \{x \in D : x = \lim_{n\to\infty} S(t_n, x_0) \text{ for some } t_n \to \infty \text{ as } n \to \infty\} \tag{4.31}$$

denote the ω-limit set of x_0. An important property of ω-limit sets, extracted from Hale (1969, Lemma 3) and given below, constitutes a basis of the invariance principle in the infinite-dimensional setting.

Theorem 4.6 *Suppose $x_0 \in D$ and the orbit $\gamma(x_0)$ of the dynamic system $S(t, x_0)$ belongs to a compact set $\Omega \in X$. Then the ω-limit set $\omega(x_0)$ is non-empty, compact, and invariant. Moreover, the corresponding trajectory $S(t, x_0)$ approaches $\omega(x_0)$ as $t \to \infty$ in the sense that*

$$\lim_{t\to\infty} \inf_{x\in\omega(x_0)} \|S(t, x_0) - x\| = 0. \tag{4.32}$$

Proof Since under the conditions of the theorem, the closures $\overline{\gamma(S(\tau, x_0)}$ of the ω-limit sets $\gamma(S(\tau, x_0))$ are compact for all $\tau \geq 0$, the ω-limit set

$$\omega(x_0) = \cap_{\tau \geq 0} \overline{\gamma(S(\tau, x_0))}$$

is thus representable in the form of the intersection of nested non-empty compact sets. Hence, the set $\omega(x_0)$ is non-empty and compact. To show that it is invariant, it suffices to verify that $S(t, x) \in \omega(x_0)$ for any $x \in \omega(x_0)$ and $t \geq 0$. Then given $x \in \omega(x_0)$ and $t \geq 0$, there exists $t_n \to \infty$ as $n \to \infty$ such that $S(t_n, x_0) \to x$. By the semigroup property, one has $S(t + t_n, x_0) = S(t, S(t_n, x_0))$, and by continuity, it follows that

$$\lim_{n\to\infty} S(t + t_n, x_0) = \lim_{n\to\infty} S(t, S(t_n, x_0)) = S(t, x) \in \omega(x_0). \tag{4.33}$$

Finally, convergence (4.32) is established by contradiction, for if it were not true, there would exist an $\epsilon > 0$ and a sequence $\{t_n\}_{n=1}^{\infty}$, escaping to infinity as $n \to \infty$ such that

$$\|S(t_n, x_0) - x\| > \epsilon \text{ for all } n = 1, 2 \ldots \text{ and for all } x \in \omega(x_0). \tag{4.34}$$

Since $S(t_n, x_0)$ is pre-compact, it contains a convergent subsequence $S(t_{n_k}, x_0) \to y$ as $k \to \infty$. By definition of the ω-limit set, $y \in \omega(x_0)$ that contradicts to (4.34), resulting in $\|x - y\| \geq \epsilon$ for all $x \in \omega(x_0)$. The proof of Theorem 4.6 is thus completed. ■

An infinite-dimensional generalization of the Krasovskii–LaSalle invariance principle, summarizing that of Henry (1981), Luo et al. (1999), Walker (1980), is now readily deduced from Theorem 4.6.

Theorem 4.7 (Invariance Principle) *Suppose an autonomous dynamic system $S(t, x_0)$ possesses a non-strict Lyapunov functional $V(x) : X \to \mathbb{R}_{\geq 0}$ with a nonpositive definite time derivative $\dot{V}(x(t))$ along the system trajectories $x(t) = S(t, x_0)$. Let M be the maximal invariant subset of $E = \{y \in X : \dot{V}(y) = 0\}$. Then $S(t, x_0)$ approaches M as $t \to \infty$ in the sense that*

$$\lim_{t\to\infty} \inf_{y \in M} \|S(t, x_0) - y\| = 0. \tag{4.35}$$

provided that the corresponding orbit $\gamma(x_0)$ lies in a compact set of X.

Proof Since under the conditions of the theorem, $V(S(t, x_0))$ is bounded from below by zero and nonincreasing for all $t \geq 0$, there exists a limit $v = \lim_{t\to\infty} V(S(t, x_0))$. If $x \in \omega(x_0)$ then $V(x) = v$ and by virtue of (4.33), it is established that $V(S(t, x)) = v$ for all $t \geq 0$. Thus, $\dot{V}(x) = 0$ for any $x \in \omega(x_0)$, and hence $\omega(x_0) \subseteq M$. By applying Theorem 4.6, the validity of Theorem 4.7 is then verified. ■

Although in its present form, the invariance principle is confined to autonomous dynamic systems, extensions are possible, e.g., to asymptotically autonomous and periodic systems (Khalil 2002; Sastry 1999) as well as to discontinuous systems with a nonambiguous behavior to the right (Alvarez et al. 2000). However in general, it remains applicable neither to nonautonomous systems nor to differential inclusions (Michel and Wang 1995).

4.2.2 *Invariance Principle Extension*

In order to analyze the asymptotic stability of nonautonomous dynamic systems based on a non-strict Lyapunov functional, a complementary technique was developed in Orlov (2003). Following this technique, a nonautonomous dynamic systems $S(t, t_0, x_0)$ is presently analyzed in a Hilbert space X, using a non-strict Lyapunov functional and an auxiliary sign-indefinite functional. While being coupled together, the functionals lead to a sign-definite functional that converges to zero along the system trajectories. The system is thus concluded to be asymptotically stable in the entire state space. Such a strictification procedure of coupling a non-strict Lyapunov functional and an indefinite functional is formalized as follows.

Theorem 4.8 (Extended Invariance Principle Orlov 2003) *Consider a continuous (possibly, sign-indefinite) functional $W(x, t) : X \times \mathbb{R} \to \mathbb{R}$. Let along with Conditions C1–C3 of Theorem 4.5, the following hold true for the system trajectories $x(t) = S(t, t_0, x_0)$, initialized in $\{x \in B_\delta : v_1(\|x\|) \le c\}$:*

(C4)

$$\dot{W}(x(t), t) \le -w(\|x(t)\|) + \kappa v_1(\|x(t)\|) \tag{4.36}$$

for a positive constant κ and for some continuous, positive semi-definite function $w(\cdot)$, whose combination $v + w$ with $v(\cdot)$ is positive definite.

Then the dynamic system is asymptotically stable.

Proof By Theorem 4.5, an arbitrary system trajectory $S(t, t_0, x_0)$, initialized within $\{x \in B_\delta : v_1(\|x\|) \le c\}$, remains uniformly bounded

$$x(t) = S(t, t_0, x_0) \in \{x \in B_\delta : v_0(\|x\|) \le c\} \tag{4.37}$$

for all $t \ge t_0$, and

$$\lim_{t\to\infty} v(\|x(t)\|) = 0. \tag{4.38}$$

Thus, it remains to prove that any motion of the system within (4.37) approaches the origin as $t \to \infty$. For this purpose, let us preliminarily demonstrate that

$$\lim_{t\to\infty} w(\|x(t)\|) = 0 \tag{4.39}$$

for each trajectory within (4.37).

To begin with, let us note that relation (4.5), verified in the proof of Theorem 4.5, is certainly in force. Furthermore, integrating inequality (4.36) yields

$$\int_{t_0}^{t} w(\|x(t)\|)dt \leq [W(x(t_0), t_0) - W(x(t), t)] + c_1 \int_{t_0}^{\infty} v(\|x(t)\|)dt \leq C, \tag{4.40}$$

where the latter inequality with some positive constant C is guaranteed by (4.5) and by the boundedness of the continuous function $W(x)$ on the compact domain (4.37). By the same reasoning and by Condition C1 of the theorem, which ensures the uniform continuity of any solution $x(t) = S(t, t_0, x)$, the integrand $w(\|x(t)\|)$ is uniformly continuous in t.

Hence, Barbalat's Lemma 2.2 is applicable to (4.40), and by applying this lemma, the limiting relation (4.39) is obtained.

Moreover, since the linear combination $u(\|x\|) = v(\|x\|) + w(\|x\|)$ is positive definite by Condition C4, then relations (4.38), (4.39), coupled together, result in

$$\lim_{t\to\infty} u(\|x(t)\|) = 0, \tag{4.41}$$

thereby concluding the desired state convergence $\lim_{t\to\infty} x(t) = 0$. The latter conclusion can be shown by contradiction similar to that of the proof of Theorem 4.1. The dynamic system is therefore asymptotically stable, and the proof of Theorem 4.8 is completed. ■

4.3 Lyapunov Functionals Under Unilateral Constraints

The Lyapunov approach remains an effective analysis tool for dynamic systems $S(t, t_0, x_0)$, operating in a Hilbert space X. Let such a constraint

$$F(x, t) \geq 0, \tag{4.42}$$

be specified with a continuous functional $F(x, t) : X \times \mathbb{R}^1 \to \mathbb{R}^1$.

Suppose that at impact-time instants of hitting the constraint, the system is enforced to instantaneous state resets

$$x(t+) = \sigma(x(t-), t), \tag{4.43}$$

determined by a continuous operator $\sigma(x, t) : X \times \mathbb{R}^1 \to X$ where $x(t-) = \lim_{s\to t-} x(s), x(t+) = \lim_{s\to t+} x(s)$. Then, having a strict Lyapunov functional for a constraint-free dynamic system $S(t, t_0, x_0)$ is still capable of ensuring the asymptotic stability under the unilateral constraint (4.42) if in addition, the functional proves to be nonincreasing at the resetting time instants.

The following result is inspired by its finite-dimensional counterpart (Haddad et al. 2006, Theorem 2.4).

Theorem 4.9 *Suppose an unconstrained dynamic system $S(t, t_0, x_0)$, evolving in a Hilbert space X, is uniformly continuous and possesses a strict Lyapunov functional $V(x, t) : X \times \mathbb{R} \to \mathbb{R}_{\geq 0}$ such that its time derivative $\dot{V}(\tilde{x}(t), t)$ along the system trajectories $\tilde{x}(t)$ is locally Lipschitz continuous. Consider the composite functional $V(x(t), t)$, computed along the system trajectories $x(t)$ which are subject to the state reset* (4.43) *at the unilateral constraint* (4.42). *Let $V(x(t), t)$ be nonincreasing, i.e.,*

$$V(x(t+), t) - V(x(t-), t) \leq 0 \tag{4.44}$$

for all resetting time instants $t : F(x(t), t) = 0$ when the trajectory $x(t)$ hits the constraint. Then the constrained dynamic system is (globally) asymptotically stable (provided that the functional V is globally defined on the entire state space).

Proof By Comparison Lemma 2.1, the composite functional $V(x(t), t)$, computed along the trajectories $x(t)$ of the constrained system is upper estimated

$$V(x(t), t) \leq V(\tilde{x}(t), t) \quad \forall t \geq t_0 \tag{4.45}$$

by the composite functional $V(\tilde{x}(t), t)$, computed along the trajectories of the unconstrained system provided that the trajectories $x(t)$ and $\tilde{x}(t)$ are initialized with the same initial condition $x(t_0) = \tilde{x}_0(t_0)$. Indeed, by virtue of (4.44), this is true not only before the first resetting time instant but also between the first two consecutive resetting time instants. By iteration, this conclusion is then straightforwardly extended until the first accumulation point (if any) of resettling time instants, and so on for all $t \geq t_0$. Since by Theorem 4.1, the unconstrained dynamic system $S(t, t_0, x)$ is asymptotically stable, then $\lim_{t\to\infty} V(\tilde{x}(t), t) = 0$. Due to (4.45), it follows that $\lim_{t\to\infty} V(x(t), t) = 0$. As in the proof of Theorem 4.1, the latter convergence with the positive definite functional V ensures the asymptotic stability of the constrained dynamic system. The global asymptotic stability is respectively ensured if V is globally defined on the entire state space. Theorem 4.9 is thus proved. ∎

4.4 Concluding Remarks

The modern arsenal of strict and non-strict Lyapunov functionals presented is shown to be capable of analyzing not only specific stability features of dynamic systems (such as FTS properties) but also their robustness (e.g., ISS) in the presence of external disturbances. Although the majority of the proposed Lyapunov-based stability tools are well-established to be applicable to a wide class of dynamic systems, some of these tools call for further development to make them applicable in the infinite-dimensional setting. For instance, the homogeneity concept, just recently introduced for the PDE setting (Polyakov 2016), is expected to burst research interest on its applications to distributed parameter and time delay systems. A potential extension of the invariance principle to periodic infinite-dimensional systems is also among open problems of interest. Theoretical aspects of the constructive design of various kinds of Lyapunov functionals are addressed in the next chapter.

References

Alvarez J, Orlov Y, Acho L (2000) An invariance principle for discontinuous dynamic systems with applications to a Coulomb friction oscillator. J Dyn Syst Meas Control 74:190–198

Bacciotti A, Rosier L (1998) Liapunov and Lagrange stability: inverse theorems for discon- tinuous systems. Math Control Signals Syst 11:101–128

Bhat SP, Bernstein DS (2005) Geometric homogeneity with applications to finite-time stability. Math Control Signals Syst 17:101–127

Branicky M (1989) Multiple Lyapunov functions and other analysis tools for switched and hybrid systems. IEEE Trans Auto Control 43:475–482

Brogliato B, Goeleven D (2005) The Krasovskii–LaSalle invariance principle for a class of unilateral dynamical systems. Math Control Signals Syst 17:57–76

Dashkovskiy S, Mironchenko A (2013) Input-to-state stability of infinite-dimensional control systems. Math Control Signals Syst 25:1–35

Haddad WM, Chellaboina V, Nersesov S (2006) Impulsive and hybrid dynamical systems - stability, dissipativity, and control. Princeton University Press, Princeton

Hahn W (1963) Theory and applications of Liapunov's direct method. Prentice-Hall, Englewood Cliffs

Hale JK (1969) Dynamical systems and stability. J Math Anal Appl 26:39–59

Henry D (1981) Geometric theory of semilinear parabolic equations. Lecture notes in mathematics. Springer, Berlin

Kawski M (1995) Geometric homogeneity and stabilization. In: Proceedings of the 3rd IFAC Nonlinear Control Systems Design Symposium, pp 2799–2805

Khalil H (2002) Nonlinear systems, 3rd edn. Prentice Hall, Upper Saddle River

Krasovskii NN (1959) Problems of the theory of stability of motion. Fizmatgiz, Moscow (in Russian); English translation (1963) Stanford University Press, Stanford

Krstic H, Kokotovic PV, Kanellakopoulos I (1995) Nonlinear and adaptive control design. Wiley, New York

Kurzweil J (1956) On the inversion of Lyapunov's second theorem on stability of motion. Transl Am Math Soc 24:19–77

LaSalle JP (1960) Some extensions of Lyapunov's second method. IRE Trans Circuit Theory CT-7:520–527

Liberzon D (2003) Switching in systems and control. Birkhäuser, Boston

Luo ZH, Guo BZ, Morgul M (1999) Stability and stabilization of infinite dimensional systems with applications. Springer, London

Matusik R, Nowakowaski A, Plaskazc S, Rogowski A (2018) Finite-time stability for differential inclusions with applications to neural networks. SIAM J Control Optim (to be published)

Michel A, Wang K (1995) Qualitative theory of dynamical systems: the role of stability preserving mappings. Marcel Dekker Inc, New York

Orlov Y (2003) Extended invariance principle for nonautonomous switched systems. IEEE Trans Auto Control 48:1448–1452

Orlov Y (2009) Discontinuous systems - Lyapunov analysis and robust synthesis under uncertainty conditions. Springer, London

Polyakov A, Efimov D, Fridman E, Perruquetti W (2016) On homogeneous distributed parameter systems. IEEE Trans Auto Control 61:3657–3662

Rosier L (1992) Homogeneous Lyapunov function for homogeneous conitnuous vector field. Syst Control Lett 19:467–473

Sastry S (1999) Nonlinear systems - analysis, stability, and control. Springer, New York

Sontag ED, Wang Y (1995) On characterizations of the input-to-state stability property. Syst Control Lett 24:351–359

Song Y, Wang Y, Holloway J, Krstic M (2017a) Time-varying feedback for regulation of normal-form nonlinear systems in prescribed finite time. Automatica 83:243–251

Song Y, Wang Y, Holloway J, Krstic M (2017b) Time-varying feedback for stabilization in prescribed finite time. Int J Robust Nonlinear Control 29(3):618–633

Walker JA (1980) Dynamical systems and evolution equations. Plenum Press, New York

Yoshizawa T (1963) Asymptotic behavior of solutions of a system of differential equations. Contrib Differ Equ 1:371–387

Zubov VI (1964) Methods of A.M. Lyapunov and their applications. Noordhoff, Leiden

Chapter 5
Control Lyapunov Functions

The present chapter is centered around the CLF paradigm (Sontag and Sussman, 1995) that underlies the principal feedback design methods such as Bellman dynamic programming in optimal control and nonlinear $\mathscr{H}_\infty$ approach in the robust synthesis. CLFs are first illustrated with quadratic forms, which result in a simple LMI criterion of the asymptotic stability of linear systems. Similar LMI-based conditions are then derived for homogeneous systems by using generalized polynomial forms. After that, the CLF approach is applied to describe the well-known Lyapunov minmax and speed gradient methods, and multiple Lyapunov functions are additionally constructed for VSS based on the Lyapunov gradient PDE. Finally, appropriate solutions of Hamilton–Jacobi PDIs are involved to specify CLFs for achieving a prescribed closed-loop performance of uncertain VSS with resets.

5.1 Lyapunov Algebraic Equation and Quadratic Forms

Origins of the CLF approach may be found in the Lyapunov algebraic equation

$$A^T P + PA = -Q, \tag{5.1}$$

the feasibility of which in the class of positive definite matrices $P, Q \in \mathbb{R}^{n\times n}$ is necessary and sufficient for the linear system

$$\dot{x} = Ax \tag{5.2}$$

to be exponentially stable in $\mathbb{R}^n$. It is actually clear that once the Lyapunov equation (5.1) holds true with positive definite matrices P and Q, the quadratic form

$$V(x) = x^T Px \tag{5.3}$$

Y. Orlov, *Nonsmooth Lyapunov Analysis in Finite and Infinite Dimensions*,
Communications and Control Engineering,
https://doi.org/10.1007/978-3-030-37625-3_5

constitutes a Lyapunov function of the linear system (5.2) such that its time derivative

$$\dot{V}(x(t)) = x^T PAx + x^T A^T Px = -x^T Qx \tag{5.4}$$

computed along the trajectories of the system, is negative definite. And vice versa. Given an arbitrary $Q > 0$, the matrix

$$P = \int_0^\infty e^{A^T t} Q e^{At} dt \tag{5.5}$$

is well-defined for an exponentially stable system (5.2). Moreover, it is positive definite, and by inspection, the Lyapunov equation (5.1) is satisfied with such $P > 0$ and $Q > 0$:

$$A^T P + PA = \int_0^\infty A^T e^{A^T t} Q e^{At} + e^{A^T t} Q e^{At} A dt = \int_0^\infty d\left(e^{A^T t} Q e^{At}\right) = -Q.$$

Then it becomes clear that for a nonlinear version

$$\dot{x} = A_0 x + Bu + f(x)x \tag{5.6}$$

of (5.2), controlled by the input $u \in \mathbb{R}^m$ and affected by the external disturbance $f(x)x$ where $f(x) : \mathbb{R}^n \to \mathbb{R}^n$ is uniformly bounded in x, the above quadratic form can be viewed as a CLF. In other words, the quadratic form (5.3) is capable of generating a linear state feedback

$$u = Kx \tag{5.7}$$

with a certain $K \in \mathbb{R}^{n \times m}$, resulting in the exponential stability of the closed-loop system.

To support the latter claim, suppose that $A_0 + BK$ is a Hurwitz matrix so that there exist $P > 0$ and $Q > 0$, satisfying the Lyapunov equation (5.1) with $A = A_0 + BK$. Then differentiating the quadratic form $V(x) = x^T Px$ along the closed-loop system (5.6), (5.7) yields

$$\dot{V}(x(t)) \leq -x^T(t)[Q - FI]x(t) < 0 \tag{5.8}$$

provided that the LMI $Q > FI$ is feasible where the constant $F > 0$ is an upper bound of the disturbance magnitude $\|f(x)\|$ and I is the identity matrix.

Thus, the quadratic CLF (5.1) allows one to derive the stabilizing linear feedback (5.7) such that the LMI $Q - FI > 0$ holds true with the Lyapunov pair (P, Q), solving the Lyapunov equation (5.1), specified with $A = A_0 + BK$. For instance, whenever the matrices A_0 and B are controllable it can be done by selecting K and Q such that $A + BK$ is Hurwitz and the minimal eigenvalue of Q is higher than F. Just in the case, the quadratic form (5.3) with P, given by (5.5) where $A = A_0 + BK$, is actually a Lyapunov function that ensures the asymptotic stability of the closed-loop system (5.6), (5.7).

5.2 Generalized Forms

Recently, quadratic and polynomial forms with respect to an augmented state vector, which along with the state components, captures their fractional degrees, have been utilized for the stability analysis of VSS. To illustrate an idea of constructing a Lyapunov function in such a specific generalized form (GF), twisting and supertwisting algorithms are chosen to anticipate the general exposition.

5.2.1 *Semiglobal Strict Lyapunov Functions of Twisting VSS*

The twisting algorithm was initially motivated to attenuate the chattering phenomenon while rejecting a low frequency disturbance $\phi(t)$ in the scalar system

$$\dot{x} = u(t) + \phi(t), \tag{5.9}$$

using a dynamic actuator

$$\dot{u} = v(t) \tag{5.10}$$

driven by its own, possibly discontinuous input $v(t)$. The so-called twisting input

$$v(t) = -a \text{ sign } x(t) - b \text{ sign } \dot{x}(t) \tag{5.11}$$

with $a > b > 0$ was proposed (see Levant 1993 and references therein) to robustly stabilize the overall system (5.9), (5.10) in finite time in the presence of a low frequency disturbance such that

$$ess \sup_t |\dot{\phi}(t)| < \min\{b, a - b\}. \tag{5.12}$$

The resulting closed-loop system

$$\begin{aligned} \dot{x} &= y, \\ \dot{y} &= -a \text{ sign } x - b \text{ sign } y + \dot{\phi}(t) \end{aligned} \tag{5.13}$$

written in terms of x and $y = \dot{x}$ was then shown (cf. Levant 1993, Theorem 2) to generate a FTS SOSM in the origin where the influence of uniformly bounded disturbances (5.12) was rejected. The finite time global stability was thus achieved in the closed-loop regardless of whichever disturbance (5.12) affects the system. The qualitative behavior of the twisting system (5.12), (5.13) is depicted in Fig. 5.1.

Remarkably, the discontinuous input (5.11) was applied to the actuator dynamics (5.10) rather than to the plant dynamics (5.9). Therefore, the original system (5.9) was enforced by the continuous signal $u(t)$, which is generated by the discontinuous signal (5.11), passing through an integrator.

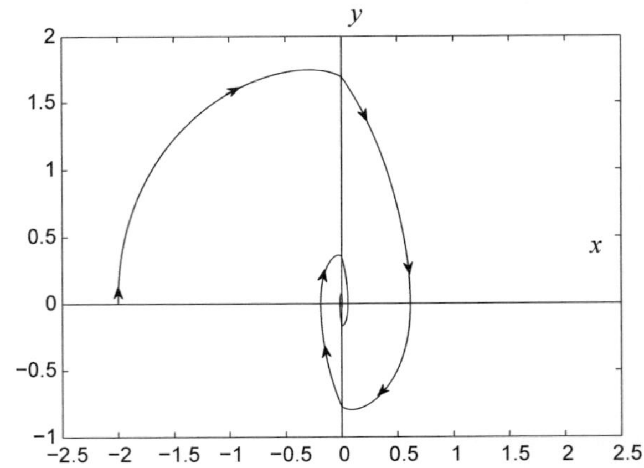

Fig. 5.1 Phase trajectory of the twisting system (5.12), (5.13)

Later on Alvarez et al. (2000), Orlov et al. (2003), the twisting algorithm was augmented with PD controller to globally stabilize uncertain mechanical systems with Coulomb friction forces. Following Orlov (2003), the design of semiglobal strict Lyapunov functions in quadratic GFs is illustrated for a simple frictional oscillator

$$\begin{aligned} \dot{x}_\omega &= y_\omega, \\ \dot{y}_\omega &= -a\text{sign } x_\omega - b\text{sign } y_\omega - hx_\omega - py_\omega + \omega(x_\omega, y_\omega, t) \end{aligned} \tag{5.14}$$

driven by a twisting+PD actuator. Such an actuated oscillator is manipulatable with the gains h, p, a, b, possibly adsorbing viscous and Coulomb friction coefficients, and it is assumed to be affected by a piecewise continuous nonlinear perturbation $\omega(x, y, t)$, uniformly bounded

$$|\omega(x, y, t)| \leq M \tag{5.15}$$

for all continuity points (x, y, t) and some $M > 0$.

If $a > b > 0$ and the bound M is small enough, namely,

$$0 < M < b < a - M, \tag{5.16}$$

whereas the linear gains are nonpositive, i.e.,

$$h \geq 0, \ p \geq 0, \tag{5.17}$$

the equiuniform stability of (5.14) can be verified with the non-strict Lyapunov function

$$\tilde{V}(x, y) = a|x| + \frac{1}{2}(y^2 + hx^2) \tag{5.18}$$

of the quadratic form with respect to the augmented state vector $(x, y, \sqrt{|x|})$.

1. *Equiuniform stability.* Differentiating (5.18) along the perturbed system (5.14), initialized outside the origin, yields the equiuniform estimate

$$\dot{\tilde{V}}(x_\omega(t), y_\omega(t)) = -b|y_\omega(t)| - py_\omega^2(t) + y_\omega(t)\omega(x_\omega(t), y_\omega(t), t) \leq -(b - M)|y_\omega(t)| \quad (5.19)$$

for almost all t regardless of whichever disturbance (5.16) affects the system. While deriving (5.19), it has been taken into account that just a sliding mode of the second order may occur in the origin $x = 0$ and $y = 0$, otherwise, no sliding modes occur along the vertical axis $x = 0$ where the Lyapunov function (5.18) is not differentiable.

It follows that the uncertain system (5.14)–(5.17) is equiuniformly stable. Indeed, the positive definite function (5.18) determines a norm in the Euclidean space $\mathbb{R}^2$ and due to (5.19), this norm does not increase along the solutions of (5.14)–(5.17). Thus, initialized in an arbitrarily small vicinity

$$D_\epsilon = \{(x, y) \in \mathbb{R}^2 : \tilde{V}(x, y) \leq \epsilon\} \quad (5.20)$$

of the origin, the uncertain system (5.14)–(5.17) cannot leave this vicinity, regardless of whichever admissible uncertainty ω affects the system.

2. *Global asymptotic stability.* Now let us prove that (5.14)–(5.17) is globally asymptotically stable, i.e., the limiting relations

$$\lim_{t\to\infty} y_\omega(t) = 0 \quad (5.21)$$

$$\lim_{t\to\infty} x_\omega(t) = 0 \quad (5.22)$$

hold for each solution $(x_\omega(t), y_\omega(t))$ individually. The idea behind this proof is inspired from the Extended Invariance Principle, Theorem 4.8, and it is based on the use of an auxiliary indefinite function, coupled to the non-strict Lyapunov function (5.18).

To begin with, let us note that by virtue of (5.19) all possible solutions of (5.14)–(5.17), initialized at $t_0 \in \mathbb{R}$ within a compact set

$$D_R = \{(x, y) \in \mathbb{R}^2 : \tilde{V}(x, y) \leq R\} \quad (5.23)$$

are a priori estimated by

$$\sup_{t\in[t_0,\infty)} \tilde{V}(x_\omega(t), y_\omega(t)) \leq R. \quad (5.24)$$

In order to justify (5.21) it remains to integrate inequality (5.19) on solutions of (5.14)–(5.17), initialized at $t_0 \in \mathbb{R}^1$ within the compact set (5.23) where R is arbitrarily large. Taking into account (5.15) and (5.24), one obtains

$$\int_{t_0}^{t} |y_\omega(t)|dt \leq (b - M)^{-1}[R - \tilde{V}(x_\omega(t), y_\omega(t))] \leq (b - M)^{-1}R \quad (5.25)$$

for all $t \geq t_0$ and for any solution $(x_\omega(t), y_\omega(t))$ of (5.14)–(5.17), initialized within D_0. Moreover, the integrand $y_\omega(t)$ is uniformly continuous on $[t_0, \infty)$ for these solutions because both $y_\omega(t)$ and $\dot{y}_\omega(t)$ are uniformly bounded in $t \geq t_0$. Thus, the validity of (5.21) is straightforwardly obtained by applying Barbalat's Lemma 2.2 to the integral inequality (5.25).

In turn, relation (5.22) is established by differentiating an auxiliary (indefinite!) function

$$U(x, y) = xy \tag{5.26}$$

along the trajectories of the uncertain system (5.14)–(5.17):

$$\begin{aligned}\dot{U}(x_\omega(t), y_\omega(t)) &= y_\omega^2(t) - a|x_\omega(t)| - bx_\omega(t) sign\ y_\omega(t) - hx_\omega^2(t) \\ &- px_\omega(t)y_\omega(t) + x_\omega(t)\omega(x_\omega(t), y_\omega(t), t) \\ &\leq y_\omega^2(t) - |x_\omega(t)|[a + bsign\ x_\omega(t) sign\ y_\omega(t) \\ &- \omega(x_\omega(t), y_\omega(t), t) sign\ x_\omega(t)] - px_\omega(t)y_\omega(t) \\ &\leq y_\omega^2(t) - (a - b - M)|x_\omega(t)| - px_\omega(t)y_\omega(t). \end{aligned} \tag{5.27}$$

Then integrating (5.27) on solutions of (5.14)–(5.17), initialized at $t_0 \in \mathbb{R}$ within the compact set (5.23) with R arbitrarily large, and employing (5.24)–(5.25), one obtains

$$\begin{aligned}(a - b - M)\int_{t_0}^{t} |x_\omega(\tau)|d\tau &\leq [U(x_\omega(t_0), y_\omega(t_0)) - U(x_\omega(t), y_\omega(t))] \\ + \int_{t_0}^{t} y_\omega(\tau)[y_\omega(\tau) - px_\omega(\tau)]d\tau &\leq [x_\omega(t_0)y_\omega(t_0) - x_\omega(t)y_\omega(t)] \\ &+\{\max_{\tau\in[t_0,t]} |y_\omega(\tau)| + p \max_{\tau\in[t_0,t]} |x_\omega(\tau)|\} \int_{t_0}^{t} |y_\omega(\tau)|d\tau \\ &\leq \frac{1}{2}[x_\omega^2(t_0) + y_\omega^2(t_0) + x_\omega^2(t) + y_\omega^2(t)] \\ + \left(\sqrt{2R} + \frac{1}{a}R\right) \int_{t_0}^{t} |y_\omega(\tau)|d\tau &\leq 2R + \frac{1}{a^2}R^2 \\ &+(b - M)^{-1}R\left(\sqrt{2R} + \frac{1}{a}R\right), \end{aligned} \tag{5.28}$$

where due to (5.24) the integrand $x_\omega(t)$ is uniformly continuous on $[t_0, \infty)$ (actually, both $x_\omega(t)$ and its time derivative $\dot{x}_\omega(t) = y_\omega(t)$ are uniformly bounded in $t \geq t_0$ for the solutions of (5.14), initialized within (5.23)).

Thus, Barbalat's Lemma 2.2 is applicable to (5.28). By applying this lemma, the limiting relation (5.22) is obtained, and the proof of the global asymptotic stability of the uncertain system (5.14)–(5.17) is completed.

3. *Semiglobal strict Lyapunov functions.* To strictify the above Lyapunov function (5.18) let us parameterize it

$$V_R(x, y) = \tilde{V}(x, y) + \kappa_R U(x, y) = a|x| + \frac{1}{2}(y^2 + hx^2) + \kappa_R xy \tag{5.29}$$

with $R > 0$ and with appropriately chosen positive weight parameters

$$\kappa_R < \min\left\{1, \frac{2a^2}{R}, \frac{a(b-M)}{a\sqrt{2R} + pR}\right\} \tag{5.30}$$

such that each GF (5.29) is positive definite on the corresponding compact set

$$D_R = \{(x_1, x_2) \in \mathbb{R}^2 : \tilde{V}(x_1, x_2) \le R\}. \tag{5.31}$$

Thus, the higher value of the parameter R, featuring the domain of the GF representative V_R, the smaller value of the cross-term weight κ.

The function $V_R(x, y)$, thus constructed, is indeed positive definite on compacta (5.31) because (5.31) implies that $|x| \le \frac{R}{a}$, $|y| \le \sqrt{2R}$, and therefore

$$\begin{aligned} a|x| + \frac{1}{2}(y^2 + hx^2) + \kappa_R xy &\ge a|x| + \frac{1}{2}(y^2 + hx^2) - \frac{1}{2}\kappa_R x^2 - \frac{1}{2}\kappa_R y^2 \\ &\ge \left(a - \frac{R}{2a}\kappa_R\right)|x| + \frac{1}{2}(1-\kappa_R)y^2 + \frac{1}{2}hx^2 > 0 \end{aligned} \tag{5.32}$$

for all $(x, y) \in D_R \setminus \{(0, 0)\}$ and $\kappa_R > 0$, satisfying (5.30).

Moreover, the time derivative of (5.29), computed along the trajectories of the uncertain system (5.14)–(5.17), initialized within D_R, is equiuniformly negative definite. Indeed, by employing (5.19) and (5.27) one has

$$\begin{aligned} \dot{V}_R(x_\omega(t), y_\omega(t)) \le -(b-M)|y_\omega(t)| + \\ \kappa_R y_\omega^2(t) - \kappa_R(a-b-M)|x_\omega(t)| - \kappa_R p x_\omega(t) y_\omega(t). \end{aligned} \tag{5.33}$$

Since it has been shown at the first step that the trajectories of (5.14)–(5.17), starting in the region D_R, cannot leave this region, by utilizing (5.24) it follows

$$\begin{aligned} \dot{V}_R(x_\omega(t), y_\omega(t)) \le -\left[b - M - \kappa_R\left(\sqrt{2R} + \frac{pR}{a}\right)\right]|y_\omega(t)| \\ -\kappa_R(a-b-M)|x_\omega(t)| \le -c_R[|y_\omega(t)| + |x_\omega(t)|], \end{aligned} \tag{5.34}$$

where $c_R = \min\{b - M - \kappa_R(\sqrt{2R} + \frac{pR}{a}), \kappa_R(a-b-M)\} > 0$ by virtue of (5.30). To this end, (5.34) results in

$$\dot{V}_R(x_\omega(t), y_\omega(t)) \le -K_R V_R(x_\omega(t), y_\omega(t)), \tag{5.35}$$

where

$$K_R = \frac{2ac_R}{\max\{2a^2 + hR, a\sqrt{2R} + 2\kappa_R R\}} > 0$$

and the upper estimate

$$V_R(x, y) \leq \frac{2a^2 + hR}{2a}|x| + \frac{a\sqrt{2R} + 2\kappa_R R}{2a}|y|$$

of the Lyapunov function (5.29) on compacta (5.23) has been used. Thus, the desired equiuniform negative definiteness

$$\dot{V}_R(x, y) \leq -K_R V_R(x, y) \tag{5.36}$$

is concluded for all $(x, y) \in D_R$.

Hence, the family (5.29) does represent semiglobal strict Lyapunov functions of the perturbed oscillator (5.14). Certainly, the exponential decay (5.36) on compacta (5.23) does not lead to the global exponential stability because the decay rate $K_R \to 0$ as $R \to \infty$, but it does guarantee the global equiuniform asymptotic stability of (5.14).

4. *Global equiuniform asymptotic stability.* Since the differential inequality (5.35) holds on the solutions of the uncertain system (5.14)–(5.17), initialized within the compact set (5.23), the function $V_R(x_\omega(t), y_\omega(t))$ exponentially decays

$$V_R(x_\omega(t), y_\omega(t)) \leq V_R(x_\omega(t_0), y_\omega(t_0))e^{-K_R(t-t_0)} \tag{5.37}$$

on these solutions with the decay rate K_R, independent of the uncertainty ω. While viewed on compacta (5.23), the functions $V_R(x, y)$ and $\tilde{V}(x, y)$ are equivalent in the sense that

$$L_R\tilde{V}(x, y) \leq V_R(x, y) \leq M_R\tilde{V}(x, y) \tag{5.38}$$

for all $(x, y) \in D_R$ and positive constants L_R, M_R, satisfying

$$L_R < \min\left\{\frac{2a^2 - R\kappa_R}{2a^2}, 1 - \kappa_R\right\}, \ M_R > \max\left\{\frac{2a^2 + R\kappa_R}{2a^2}, 1 + \kappa_R\right\}. \tag{5.39}$$

The above relations (5.37) and (5.38), coupled together, ensure that the function $\tilde{V}(x, y)$ exponentially decays

$$\begin{aligned}\tilde{V}(x_\omega(t), y_\omega(t)) &\leq L_R^{-1}M_R\tilde{V}(x_\omega(t_0), y_\omega(t_0))e^{-K_R(t-t_0)} \\ &\leq L_R^{-1}M_R Re^{-K_R(t-t_0)}\end{aligned} \tag{5.40}$$

on the solutions of (5.14)–(5.17), equiuniformly in the uncertainty ω and the initial data, located within an arbitrarily large set (5.23). Clearly, this proves that the uncertain system (5.14)–(5.17) is globally equiuniformly asymptotically stable.

5. *Global equiuniform finite time stability.* Finally, the equiuniform finite time convergence is established based on the quasihomogeneity principle of Theorem 3.5. Taking into account that the PD+disturbance term $-hx_1 - px_2 + \omega(x_1, x_2, t)$ remains locally uniformly bounded for admissible disturbances (5.15), (5.16), system (5.14) is embedded into the differential inclusion framework (3.192) to ensure by inspection of the homogeneity definition of Sect. 3.6.5 that it is indeed locally (but not globally) quasihomogeneous of the same degree $q = -1$ and dilation $r = (2, 1)$ as that of the nominal twisting system (5.14) with $h, p = 0$ and $\omega(x, y, t) \equiv 0$.

Thus, the PD-augmented perturbed twisting system (5.14) is globally equiuniformly asymptotically stable and it is locally quasihomogeneous of degree $q < 0$. By applying Theorem 3.5, the global equiuniform finite time stability of (5.14) is established. Summarizing, the following result (Orlov 2003, Theorem 4) has been proven.

Theorem 5.1 *Let Conditions* (5.15)–(5.17) *be satisfied. Then the uncertain discontinuous system* (5.14) *is globally equiuniformly finite time stable in the origin.*

Before passing to the construction of strict GF Lyapunov functions for the super-twisting algorithm, a result, similar to Theorem 5.1, is obtained for a constrained double integrator.

5.2.1.1 Case Study Under Unilateral Constraint

Suppose now that the twisting algorithm

$$u(x_1, x_2) = -a\,\mathrm{sign}(x_1) - b\,\mathrm{sign}(x_2) \tag{5.41}$$

enforces the perturbed double integrator

$$\dot{x}_1 = x_2 \tag{5.42a}$$
$$\dot{x}_2 = u(x_1, x_2) + \omega(t) \tag{5.42b}$$
$$x_1 \geq 0 \tag{5.42c}$$
$$x_2(t_k^+) = -\bar{e}x_2(t_k^-) \quad \text{if} \quad x_2(t_k^-) < 0 x_1(t_k) = 0, \tag{5.42d}$$

operating under a unilateral position constraint.

As before, x_1, x_2 are the position and the velocity respectively, the twisting control input (5.41) comes with positive gains $a > b > 0$, the disturbance $\omega(t)$ is piecewise continuous, whose upper bound (5.15) respects (5.16), t_k is the kth jump time instant where the velocity undergoes a reset or jump, $\bar{e}$ represents the loss of energy. The equalities (5.42a) and (5.42b) represent the continuous dynamics without jumps in velocity. The inequality (5.42c) represents the unilateral constraint on the position x_1

which evolves in a domain with a boundary. It is assumed that the jump event occurs instantaneously within an infinitesimally small time and hence mathematically can be represented by *Newton's restitution rule*, given by (5.42d) where $\bar{e} \in (0, 1)$. It is clear that the twisting control law (5.41) undergoes a jump whenever the state x_2 undergoes a jump (this is similar to the existing literature Grizzle et al. 2001) and such an impact VSS has been addressed in Sect. 1.2.

Following Oza et al. (2014), the present investigation aims to prove the finite time stability of the closed-loop system (5.41), (5.42) in the presence of velocity jumps without the need to analyze the Lyapunov function at the jump instants. The stability analysis is motivated by the destabilizing effect of the jump when the double integrator is not actuated. Such a destabilization is contrary, for example, to the self-stabilizing nature of a bouncing ball, considered in Sect. 1.2.2, where impact with the ground stabilizes the motion with loss of energy at the time of impact.

The nonsmooth Zhuravlev–Ivanov transformation

$$x_1 = |s|, x_2 = R\, v \operatorname{sign}(s), \;\; R = 1 - k \operatorname{sign}(s\, v), \;\; k = \tfrac{1-\bar{e}}{1+\bar{e}} \tag{5.43}$$

(see Sect. 1.2.1) is subsequently invoked to transform the closed-loop impact system (5.41), (5.42) into a jump-free VSS

$$\begin{aligned} \dot{s} &= R\, v \\ \dot{v} &= R^{-1} \operatorname{sign}(s)\, (u(|s|, Rv\operatorname{sign}(s)) + \omega(|s|, Rv\operatorname{sign}(s), t))\,. \end{aligned} \tag{5.44}$$

Since the resulting transformed system is jump-free, its solutions are well-posed in the sense of Filippov, an attribute absent in the case of the original jump system. As an immediate consequence, the resulting transformed system turns out to be a valid candidate, analyzable by means of semiglobal nonsmooth Lyapunov functions, identified in Sect. 5.2.1 for the twisting system in the constraint-free environment.

To perform the nonsmooth Lyapunov analysis of Sect. 5.2.1 let us combine (5.41) and (5.43) to represent the twisting controller (5.41) in terms of the transformed coordinates:

$$u(|s|, R\, v \operatorname{sign}(s)) = -b \operatorname{sign}(sv) - a \tag{5.45}$$

By substituting (5.45) into (5.44), the closed-loop system in the new coordinate frame takes the form

$$\begin{aligned} \dot{s} &= R\, v \\ \dot{v} &= -b\, R^{-1} \operatorname{sign}(v) - a\, R^{-1} \operatorname{sign}(s) + R^{-1} \operatorname{sign}(s) \omega(t). \end{aligned} \tag{5.46}$$

It should be noted from the transformation definition (5.43) that the origin $s = v = 0$ of the transformed system (5.46) corresponds to the origin $x_1 = x_2 = 0$ of the underlying system (5.41), (5.42). Although transformation (5.43) is not invertible, it is important to note that while starting from the transformed system (5.46), the original dynamics can be recovered via (5.43). Moreover, due to one-to-one correspondence between the two coordinate systems at the origin, the stability analysis can be performed in the transformed coordinates. The present nonsmooth transformation approach is confined to constraint surfaces of co-dimension one only (Artstein 1983). Its advantage is that the solutions of (5.46) are well-defined in the sense of Filippov. Furthermore, such a formulation admits both friction and jump phenomena, while guaranteeing the existence of a solution. Formulation (5.46) also captures the Zeno behavior with infinite rebounds once the system stabilizes to the origin on the constraint surface.

Finite time stability results can now be presented for the transformed system (5.46).

Lemma 5.1 *Consider the dynamic system* (5.46) *where R is given by* (5.43) *with $\bar{e} \in (0, 1)$. Then*

$$\mathrm{sign}(sv)\,\mathrm{sign}(R - R^{-1}) = -1. \tag{5.47}$$

Proof The parameter R is defined in (5.43). For the case when $\mathrm{sign}(sv) = -1$, R can be computed as $R = 1 - k\mathrm{sign}(sv) = 1 + k = \frac{2}{1+\bar{e}}$. Hence $R - R^{-1} = \frac{(\bar{e}+3)(1-\bar{e})}{2(1+\bar{e})}$. Noting that $\bar{e} \in (0, 1)$, it is indeed clear that $\mathrm{sign}(R - R^{-1}) = 1$. Hence this results in $\mathrm{sign}(sv)\,\mathrm{sign}(R - R^{-1}) = -1$.

For the case when $\mathrm{sign}(sv) = 1$, R can be computed as $R = 1 - k\mathrm{sign}(sv) = 1 - k = \frac{2\bar{e}}{1+\bar{e}}$. Hence $R - R^{-1} = \frac{(3\bar{e}+1)(\bar{e}-1)}{2\bar{e}(1+\bar{e})}$. Noting that $\bar{e} \in (0, 1)$, it follows that $\mathrm{sign}(R - R^{-1}) = -1$, and hence $\mathrm{sign}(sv)\,\mathrm{sign}(R - R^{-1}) = -1$. ■

The transformed system (5.46) is not a standard double integrator system and it does not fall into existing SOSM algorithms, introduced in Levant (1993), because R causes discontinuity in the right-hand side of the first equation of (5.46). It is also of interest to note that the discontinuity in (5.46) is caused by the fact that R switches between two positive values on the sets {(s, v):s=0}, {(s, v):v=0} of Lebesgue measure zero. The two equivalent values

$$R = \begin{cases} R_1 = \frac{2}{1+\bar{e}}, & \text{if } \mathrm{sign}(sv) = -1, \\ R_2 = \frac{2\bar{e}}{1+\bar{e}}, & \text{if } \mathrm{sign}(sv) = 1 \end{cases} \tag{5.48}$$

of R are then computed according to the Filippov Definition 2.1. Taking this into account, it is trivial to note that given $\bar{e} \in (0, 1)$, the relations

$$\begin{gathered} R_1 > R_2 > 0\,,\ R_1^{-1} < R_2^{-1}, |R_1 - R_1^{-1}| < |R_2 - R_2^{-1}| \\ |R_1 - R_1^{-1}| = \tfrac{3+\bar{e}}{2}|k|\,,\ |R_2 - R_2^{-1}| = \tfrac{3\bar{e}+1}{2\bar{e}}|k| \end{gathered} \tag{5.49}$$

are concluded from the computations in Lemma 5.1. The following result is thus in force in the absence of external disturbances.

Theorem 5.2 *Given $M = 0$, the impact system* (5.41), (5.42) *and its transformed version* (5.44), (5.45) *are globally finite time stable.*

Proof Lyapunov stability analysis can be performed in the transformed coordinates since both the set of expressions (5.41), (5.42) and (5.44), (5.45) represent the same system. Let a Lyapunov function candidate be given as follows:

$$V(s, v) = a\,|s| + \tfrac{1}{2}\,v^2 \tag{5.50}$$

By computing the temporal derivative of this function along the system trajectories in (5.44), (5.45) with $M = 0$, it is obtained that,

$$\dot{V} \leq a|v|\,|R - R^{-1}|sign(sv)\,\text{sign}(R - R^{-1}) - bR^{-1}|v|. \tag{5.51}$$

From Lemma 5.1, Eq. (5.51) can be simplified as

$$\dot{V} \leq -a|v|\,|R - R^{-1}| - b\,R^{-1}\,|v| \tag{5.52}$$

It can be verified that $R^{-1} > 0$ for either sign of $sgn(s\,v)$ since $\bar{e} \in (0, 1)$. Since the equilibrium point $s = v = 0$ is the only trajectory of (5.46) on the invariance manifold $v = 0$ where $\dot{V}(s, v) = 0$, the differential inclusion (5.44), (5.45) is globally uniformly asymptotically stable by applying the invariance principle of Theorem 4.35 to the underlying VSS (see also Alvarez et al. 2000; Shevitz and Paden 1994 for the invariance principle applicability to autonomous discontinuous systems). Moreover, the system described in (5.44), (5.45) is globally homogeneous of the negative degree $q = -1$ with respect to dilation $r = (2, 1)$ and is globally uniformly finite time stable according to the homogeneity principle of Theorem 3.4. ■

The disturbance-free transformed system (5.46), corresponding to the disturbance upper bound $M = 0$, is a globally homogeneous system. Consider next the case when M takes a nonzero value. As in the unconstrained case, the twisting controller (5.45) is capable of rejecting any disturbance ω with a uniform upper bound (5.15).

Theorem 5.3 *The closed-loop impact system* (5.41), (5.42) *and its transformed version* (5.44), (5.45) *are globally finite time stable, regardless of whichever disturbance* ω*, satisfying conditions* (5.15), (5.79)*, affects the system.*

Proof The proof is constructed in several steps.
1. Global Asymptotic Stability. Under the conditions (5.15), (5.16) of this theorem, the time derivative of the Lyapunov function (5.50), computed along the trajectories of (5.44), (5.45), is estimated as follows:

$$\begin{aligned}\dot{V} &= a|v|\,|R - R^{-1}|\text{sign}(sv)\,\text{sign}(R - R^{-1}) - b\,R^{-1}\,|v| + R^{-1}\,|v|\text{sign}(sv)\,\omega \\ &\leq -a|v|\,|R - R^{-1}| - (b - M)R^{-1}|v|.\end{aligned} \tag{5.53}$$

The first term in the last inequality follows from Lemma 5.1. Since $M < b$ by conditions (5.15), (5.16) of this theorem, the global asymptotic stability of (5.44), (5.45) is then established by applying the invariance principle of Theorem 4.35.
2. Semiglobal Strong Lyapunov Functions. The goal of this step is to show the existence of a parameterized family of local Lyapunov functions $V_{\tilde{R}}(s, v)$, $\tilde{R} > 0$ such that each $V_{\tilde{R}}(s, v)$ is well-posed on the corresponding compact set

$$D_{\tilde{R}} = \{(s, v) \in \mathbf{R}^2 : V(s, v) \leq \tilde{R}\}. \tag{5.54}$$

In other words, $V_{\tilde{R}}(s, v)$ is to be positive definite on $D_{\tilde{R}}$ and its derivative, computed along the trajectories of the uncertain system (5.44), (5.45) with initial conditions within $D_{\tilde{R}}$, is to be negative definite in the sense that

$$\dot{V}_{\tilde{R}}(s, v) \leq -W_{\tilde{R}}(s, v) \tag{5.55}$$

for all $(s, v) \in D_{\tilde{R}}$ and some $W_{\tilde{R}}(s, v)$, positive definite on $D_{\tilde{R}}$. A parameterized family of Lyapunov functions $V_{\tilde{R}}(s, v)$, $\tilde{R} > 0$, with the properties defined above are constructed by combining the Lyapunov function V of (5.50), whose time derivative along the system motion is only negative semidefinite, with the indefinite function $U(s, v) = s\,v$. The resulting family is given by

$$V_{\tilde{R}}(s, v) = V(s, v) + \kappa_{\tilde{R}} U(s, v) = a\,|s| + \tfrac{1}{2} v^2 + \kappa_{\tilde{R}}\, s\, v, \tag{5.56}$$

where the weight parameter $\kappa_{\tilde{R}}$ is chosen small enough, namely,

$$\kappa_{\tilde{R}} < min \left\{ 1,\ \tfrac{2a^2}{\tilde{R}},\ \tfrac{a|R_1 - R_1^{-1}| + R_1^{-1}(b-M)}{R_1\sqrt{2\tilde{R}}} \right\}, \tag{5.57}$$

and R_1 is defined in (5.48). It can be noted from (5.54) that the following inequalities

$$|s| \leq \tfrac{\tilde{R}}{a}, \quad |v| \leq \sqrt{2\tilde{R}} \tag{5.58}$$

hold true. Hence, the Lyapunov function (5.56) is concluded to be positive definite on compact set (5.54) for all $(s, v) \in D_{\tilde{R}} \backslash \{0, 0\}$ and $\kappa_{\tilde{R}} > 0$, satisfying (5.57). Indeed,

$$\begin{aligned} V_{\tilde{R}}(s, v) &= a\,|s| + \tfrac{1}{2}\,v^2 + \kappa_{\tilde{R}}\, s\, v \geq a|s| + \tfrac{1}{2}\,v^2 - \tfrac{1}{2}\kappa_{\tilde{R}} s^2 - \tfrac{1}{2}\kappa_{\tilde{R}}\, v^2 \\ &\geq \left(a - \tfrac{\kappa_{\tilde{R}}\tilde{R}}{2a}\right)|s| + \tfrac{1}{2}(1 - \kappa_{\tilde{R}})\, v^2 \ > 0. \end{aligned} \tag{5.59}$$

The time derivative of the indefinite function $U(s, v)$ along the trajectories of the uncertain system (5.44), (5.45) is obtained as follows

$$\begin{aligned} \dot{U}(s, v) &= R\, v^2 + s\left(-b\, R^{-1}\,\mathrm{sign}(v) - a\, R^{-1}\ \mathrm{sign}(s) + R^{-1}\, sign(s)\omega\right) \\ &= R\, v^2 - bR^{-1}|s|\mathrm{sign}(sv) - a\, R^{-1}\,|s| + R^{-1}|s|\,\omega \\ &\leq R\, v^2 - R^{-1}|s|\,(a - b - M)\,. \end{aligned} \tag{5.60}$$

Then by combining (5.53) and (5.60), the time derivative of (5.56) is estimated by

$$\dot{V}_{\tilde{R}} \leq -a|v|\,|R - R^{-1}| - (b - M)\, R^{-1}\,|v| + \kappa_{\tilde{R}} R\, v^2 - \kappa_{\tilde{R}} R^{-1}|s|\,(a - b - M)\,. \tag{5.61}$$

The state function R in (5.61) keeps switching between the two values as shown in Lemma 5.1. This corresponds to the fact that the rate of decay of the Lyapunov function (5.59) switches depending on R. Considering the slowest decay, a conservative estimate

$$\dot{V}_{\tilde{R}} \leq -a|v|\,|R_1 - R_1^{-1}| - (b-M)R_1^{-1}|v| + \kappa_{\tilde{R}}R_1 v^2 - \kappa_{\tilde{R}}R_1^{-1}|s|\,(a-b-M) \tag{5.62}$$

of the upper bound (5.61) is readily obtained, using Lemma 5.1 and relation (5.49). Noting that, due to (5.53), all possible solutions of the uncertain system (5.44), (5.45), initialized at $t_0 \in \mathbf{R}$ within the compact set (5.54), are a priori estimated by

$$\sup_{t\in[t_0,\infty)} V(s,v) \leq \tilde{R}, \tag{5.63}$$

and taking into account that (5.58) holds true within the compact set (5.54), estimate (5.62) can be manipulated to

$$\begin{aligned}\dot{V}_{\tilde{R}} \leq -\left(a\,|R_1 - R_1^{-1}| + (b-M)\,R_1^{-1} - \kappa_{\tilde{R}}R_1\sqrt{2\tilde{R}}\right)|v| \\ -\kappa_{\tilde{R}}R_1^{-1}(a-b-M)|s| \leq -c_{\tilde{R}}\,[|s|+|v|]\end{aligned} \tag{5.64}$$

where

$$c_{\tilde{R}} = \min\left\{\kappa_{\tilde{R}}R_1^{-1}(a-b-M),\ a|R_1 - R_1^{-1}| + (b-M)R_1^{-1} - \kappa_{\tilde{R}}R_1\sqrt{2\tilde{R}}\right\}. \tag{5.65}$$

It follows from (5.57) that $c_{\tilde{R}} > 0$. Hence (5.64) results in

$$\dot{V}_{\tilde{R}} \leq -K_{\tilde{R}}V_{\tilde{R}}(s,v) \tag{5.66}$$

where

$$K_{\tilde{R}} = c_{\tilde{R}}\left[\max\left\{\frac{2a^2+\kappa_{\tilde{R}}\tilde{R}}{2a}, \sqrt{\frac{\tilde{R}}{2}}(1+\kappa_{\tilde{R}})\right\}\right]^{-1} > 0,$$

and the upper estimate

$$V_{\tilde{R}} \leq \frac{2a^2+\kappa_{\tilde{R}}\tilde{R}}{2a}|s| + \sqrt{\frac{\tilde{R}}{2}}(1+\kappa_{\tilde{R}})|v|$$

of the Lyapunov function (5.59) on compact set (5.54) has been used. Hence the desired uniform negative definiteness (5.55) is obtained with $W_{\tilde{R}}(s,v) = K_{\tilde{R}}V_{\tilde{R}}(s,v)$.
3. Global Uniform Asymptotic Stability. Since the inequality (5.66) holds on the solutions of the uncertain system (5.44), (5.45), initialized within the compact set (5.54), the function $V_{\tilde{R}}(s,v)$ decays exponentially

$$V_{\tilde{R}}(s(t),v(t)) \leq V_{\tilde{R}}(s(t_0),v(t_0))e^{-K_{\tilde{R}}(t-t_0)} \tag{5.67}$$

on these solutions with the decay rate $K_{\tilde{R}}$ which depends on the gain parameters a, b, the disturbance magnitude bound M, and the system characteristic R_1. Due to (5.59), the following inequality

$$L_{\tilde{R}} V(s, v) \leq V_{\tilde{R}}(s, v) \leq M_{\tilde{R}} V(s, v) \tag{5.68}$$

holds on the compact set (5.54) for all $(s, v) \in D_{\tilde{R}}$ and positive constants $L_{\tilde{R}}$, $M_{\tilde{R}}$, satisfying

$$L_{\tilde{R}} < \min\left\{\frac{2a^2 - \tilde{R}\kappa_{\tilde{R}}}{2a^2}, 1 - \kappa_{\tilde{R}}\right\}, \quad M_{\tilde{R}} > \max\left\{\frac{2a^2 + \tilde{R}\kappa_{\tilde{R}}}{2a^2}, 1 + \kappa_{\tilde{R}}\right\}. \tag{5.69}$$

The above inequalities (5.67) and (5.68) ensure that the function $V(s, v)$ decays exponentially

$$V(s(t), v(t)) \leq L_{\tilde{R}}^{-1} M_{\tilde{R}} V(s(t_0), v(t_0)) e^{-K_{\tilde{R}}(t-t_0)} \leq L_{\tilde{R}}^{-1} M_{\tilde{R}} \tilde{R}\, e^{-K_{\tilde{R}}(t-t_0)} \tag{5.70}$$

on the solutions of (5.44), (5.45) uniformly in ω and the initial data, located within an arbitrarily large set (5.54). This proves that the uncertain system (5.44), (5.45) is globally uniformly asymptotically stable around the origin $(s, v) = (0, 0)$.
4. Global Uniform Finite Time Stability. By virtue of (5.16), the piecewise continuous uncertainty $R_1^{-1}\omega(t)sign(s)$ in the right-hand side of system (5.44), (5.45) is locally uniformly bounded by $R_1^{-1}M$ whereas the remaining part of the feedback is globally homogeneous with homogeneity degree $q = -1$ with respect to dilation $r = (r_1, r_2) = (2, 1)$. Noting that $q + r_2 \leq 0$, the globally uniformly asymptotically stable system (5.44), (5.45) and in turn the original impact system (5.41), (5.42) are globally finite time stable according to the quasihomogeneity principle of Theorem 3.5. This completes the proof. ■

Theorem 5.3 is extendible to second order feedback linearizable nonlinear systems of relative degree 2. Indeed, consider the nonlinear system

$$\dot{x} = f(x) + g(x)u \tag{5.71a}$$

$$h(x) = x_1 \tag{5.71b}$$

$$\dot{x}(t_k^+) = \bar{e}\dot{x}(t_k^-) \qquad \text{if } \dot{x} < 0, x = 0, \tag{5.71c}$$

where $u \in \mathrm{R}$ is the control input, $x \in \mathrm{R}^2$ is the state vector, $h(x)$ is the output, and the Lie derivative $\mathscr{L}_g \mathscr{L}_f^{-1} h(x)$ takes nonzero values due to the feedback linearisability (Byrnes and Isidori 1991). Such systems may occur, for example, in the field of robotics where $\mathscr{L}_g \mathscr{L}_f^{-1} h(x)$ can take nonzero values. Also, the full-state feedback control synthesis often becomes feasible as it is realistic to assume the availability of both position and velocity for a broad class of mechanical systems.

It is well-known (Byrnes and Isidori 1991) that the control law

$$u(x) = \frac{1}{\mathscr{L}_g \mathscr{L}_f^{-1} h(x)} \left(-\mathscr{L}_f{}^2 h(x) + v(y)\right) \tag{5.72}$$

transforms the continuous part ($t \notin \{t_k\}$) of the system dynamics (5.71) into the following double integrator:

$$\dot{y}_1 = y_2, \quad \dot{y}_2 = \nu(y_1, y_2). \tag{5.73}$$

It remains to check how the impact map (5.71c) is affected by the transformation. It should be noted that the new coordinates are given as $y_1 = h(x) = x$ and $y_2 = \mathscr{L}_f h(x) = \dot{x}$. Hence, the same impact map as that given by (5.71c) holds true and the control law (5.72) with $\nu(y) = -b\text{sign}(y_2) - a\text{sign}(y_1)$ results in the feedback-linearized closed-loop system

$$\begin{aligned} &\dot{y}_1 = y_2 && \\ &\dot{y}_2 = -b\text{sign}(y_2) - a\text{sign}(y_1) && \text{if } y_1 \neq 0 \\ &\dot{y}_2(t_k^+) = \bar{e}\dot{y}_2(t_k^-) && \text{if } y_2 < 0,\, y_1 = 0, \end{aligned} \tag{5.74}$$

which is the same as that described by (5.41), (5.42). By Theorem 5.3, the finite time stability of system (5.74) is concluded regardless of whichever external disturbance of a sufficiently small magnitude affects the system, thereby ensuring the same for a class of nonlinear systems (5.71) in the presence of jumps in the velocity.

5.2.2 *Strict Lyapunov Functions of Supertwisting VSS*

The supertwisting position feedback

$$u = v - \alpha\sqrt{|x|}, \; \dot{v} = -\beta \text{ sign } x \tag{5.75}$$

with positive gains α and β was developed as an alternative to the twisting state feedback (5.11) to globally stabilize the scalar system (5.9) in finite time regardless of whichever low frequency external disturbance ϕ affected the system (see Levant 1993, Sect. 4 and references, quoted therein). While being driven by the position feedback controller (5.75), system (5.9), rewritten in terms of $x_1 = x$ and $x_2 = v + \phi$, and $\omega = \dot{\phi}$, takes the form

$$\begin{aligned} \dot{x}_1 &= x_2 - \alpha\sqrt{|x_1|}, \\ \dot{x}_2 &= -\beta \text{ sign } x_1 + \omega(t), \end{aligned} \tag{5.76}$$

written in terms of $x_1 = x$ and $x_2 = v + \phi$, and $\omega = \dot{\phi}$. The resulting closed-loop system (5.76) proves to be capable of generating a SOSM in the origin despite of the presence of external disturbances $\omega(t)$ of low magnitudes. The desired robustness and FTS properties are thus imposed on the underlying system (5.9) (Levant 1993, Theorem 5) in the absence of the velocity measurement. The price to pay for that is to deal with undesired high frequency state oscillations around the origin, generated by the non-Lipschitz input component $\sqrt{|x|}$ of the supertwisting controller (5.75), the state derivative $-\frac{\text{sign } |x|}{2\sqrt{|x|}}$ of which escapes to infinity as $x \to 0$.

For the supertwisting system (5.76), a quadratic GF Lyapunov function

$$V(x_1, x_2) = \zeta^T \frac{1}{2} \begin{bmatrix} 4\beta + \alpha^2 & -\alpha \\ -\alpha & 2 \end{bmatrix} \zeta \tag{5.77}$$

was specified in Moreno and Osorio (2008, 2012) in terms of the modified state vector

$$\zeta^T = \left[\sqrt{|x_1|} \text{sign}\, x_1,\ x_2 \right]. \tag{5.78}$$

Taking into account that beyond the equilibrium $x_1 = x_2 = 0$, system (5.76) generates no sliding modes on the discontinuity surface $x_1 = 0$ where the Lyapunov function (5.77) is not differentiable, the time derivative $\dot{V}(x(t))$ is therefore estimated in the standard manner (without invoking Dini gradients and contingent derivatives) along the supertwisting dynamics $x(t) = (x_1(t), x_2(t))^T$ of (5.76) with the nontrivial component $x_1 \neq 0$. Provided that

$$ess \sup_t |\omega(t)| \leq M \tag{5.79}$$

for some constant $M > 0$ and

$$\beta > 3M + 2M^2\alpha^{-2}, \tag{5.80}$$

an upper estimate

$$\dot{V}(x(t)) \leq -q\sqrt{V(x(t))} \tag{5.81}$$

of the time derivative of the Lyapunov function (5.77) is obtained with

$$q = \lambda_{min}^{1/2}(P)\lambda_{max}^{-1}(P)\lambda_{min}(Q),$$

where $\lambda_{max}(P)$, $\lambda_{min}(P)$, $\lambda_{min}(Q)$ stand for the maximal and, respectively, minimal eigenvalues of the positive definite matrices

$$P = \frac{1}{2} \begin{bmatrix} 4\beta + \alpha^2 & -\alpha \\ -\alpha & 2 \end{bmatrix}, \quad Q = \frac{1}{2} \begin{bmatrix} 2\beta + \alpha^2 - M & -\alpha - 2M\alpha^{-1} \\ -\alpha - 2M\alpha^{-1} & 1 \end{bmatrix}, \tag{5.82}$$

which satisfy the algebraic Lyapunov equation (5.1). By applying Lemma 4.1 to the differential inequality (5.81), the supertwisting system (5.76) is concluded to be globally equiuniformly FTS. The above relations (5.80)–(5.82) are clearly found in a computable form and details of their derivation are postponed for the general LMI-based exposition of the next section.

In the rest of the section, using quadratic Lyapunov GFs is additionally investigated by following a more delicate route (Orlov et al. 2011) to derive a less conservative subordination of the supertwisting gains (5.80). The supertwisting generalization

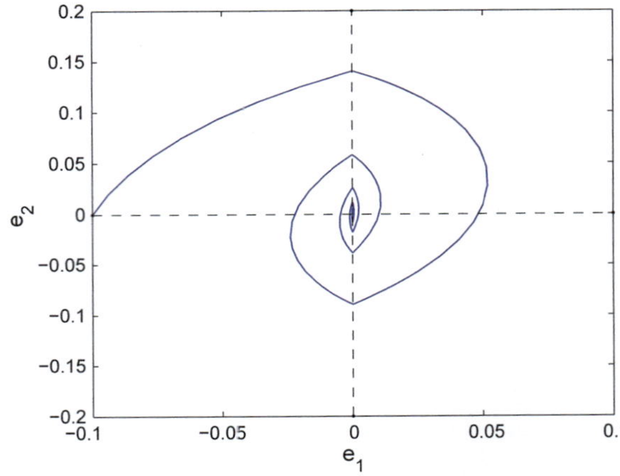

Fig. 5.2 Phase trajectory of the supertwisting generalization (5.83)

$$\begin{aligned} \dot{e}_1 &= e_2 - k_1\sqrt{|e_1|}sign\ e_1 - k_2e_1, \\ \dot{e}_2 &= -k_3 sign\ e_1 - k_4e_1 + \omega(t) \end{aligned} \tag{5.83}$$

affected by a low magnitude disturbance ω and augmented with linear gains $k_1, k_2 \geq 0$, is subsequently analyzed with the GF

$$V_1(e_1, e_2) = 2k_3|e_1| + k_4e_1^2 + \frac{1}{2}s^2(e_1.e_2) + \frac{1}{2}e_2^2, \tag{5.84}$$

specified with the term

$$s(e_1, e_2) = e_2 - k_1\sqrt{|e_1|}sign\ e_1 - k_2e_1, \tag{5.85}$$

matching the right-hand side of the first state equation (5.83). The qualitative behavior of such a generalization is depicted in Fig. 5.2.

Consider the supertwisting generalization (5.83), affected by a uniformly bounded disturbance (5.79). Letting

$$\begin{aligned} &k_1, k_3 > 0, k_2, k_4 \geq 0, \text{ if } k_4 = 0 \text{ then } k_2 = 0, \\ &M < \min\left\{\frac{k_1}{2}, \frac{k_1k_3}{1+k_1}, k_3\right\}, \end{aligned} \tag{5.86}$$

and setting

$$\gamma = \begin{cases} \min\left\{\frac{2(k_1k_3-M-Mk_1}{4k_3+3k_1^2}, \frac{k_1-2M}{4}\right\} & \text{if } k_2 = k_4 = 0 \\ \min\left\{\frac{2(k_1k_3-M-Mk_1}{4k_3+3k_1^2}, \frac{k_1-2M}{4}, \frac{2k_1k_4}{2k_4+3k_2^2}\right\} & \text{otherwise} \end{cases}, \tag{5.87}$$

the time derivative of the GF (5.84), computed along the system dynamics (5.83), is shown to respect the upper estimate (5.81) with $q = \sqrt{2k_3}\gamma$ and γ, given by (5.87). Moreover, the following result (Orlov et al. 2011, Theorem 5) is in force.

Theorem 5.4 *Suppose the gains k_i, $i = 1, \ldots, 4$ of system* (5.83) *are subordinated according to relations* (5.86) *with respect to the upper bound* (5.79) *on the disturbance magnitude. Then system* (5.83) *is globally equiuniformly FTS and an upper bound of its settling time function $T(e_1(0), e_2(0))$ is given by*

$$T(e_1(0), e_2(0)) \le \gamma^{-1}\sqrt{\frac{2V_1(e_1(0), e_2(0))}{k_3}}. \tag{5.88}$$

Proof To begin with, let us note that the origin $e_1 = e_2 = 0$ is an equilibrium of the discontinuous system (5.83) in the sense of Filippov, whereas no other sliding modes occur on the discontinuity manifold $e_1 = 0$. Indeed, it follows from the first equation of (5.83) that $e_2 = 0$ whenever the first component e_1 stays in the vertical axis $e_1 = 0$. Thus, the GF (5.84) proves to be differentiable along the solutions of (5.83) everywhere outside the origin.

With this in mind, the time derivative of the GF (5.84) trajectories of (5.83) is well-defined beyond the origin. Furthermore, it proves to be negative semidefinite for the disturbance-free system (5.83) with $\omega(t) \equiv 0$. To conclude this, denote

$$\begin{aligned} W(e_1, s) &= k_2 s^2 |e_1|^{\frac{1}{2}} + \frac{1}{2}k_1 s^2 + k_1 k_4 e_1^2 \\ &\quad + k_2 k_3 |e_1|^{\frac{3}{2}} + k_2 k_4 |e_1|^{\frac{5}{2}} + k_1 k_3 |e_1|. \end{aligned} \tag{5.89}$$

Then taking into account that

$$|e_1| \le \frac{V_1}{2k_3} \quad \Rightarrow \quad -|e_1|^{-\frac{1}{2}} \le -\left(\frac{V_1}{2k_3}\right)^{-\frac{1}{2}}, \tag{5.90}$$

one arrives at

$$\begin{aligned} \dot{V}_1 &= 2sk_3 \text{sign } e_1 + 2sk_4 e_1 - e_2(k_3 \text{sign } e_1 + k_4 e_1) \\ &\quad -s\left[\frac{1}{2}k_1 s|e_1|^{-\frac{1}{2}} + k_3 \text{sign } e_1 + k_4 e_1 + k_2 s\right] \\ &= -k_2 s^2 - \frac{1}{2}k_1 s^2 |e_1|^{-\frac{1}{2}} - k_1 k_4 |e_1|^{\frac{3}{2}} - k_2 k_3 |e_1| \\ &\quad -k_2 k_4 e_1^2 - k_1 k_3 |e_1|^{\frac{1}{2}} = -|e_1|^{-\frac{1}{2}}[k_2 s^2 |e_1|^{\frac{1}{2}} \\ &\quad +\frac{1}{2}k_1 s^2 + k_1 k_4 e_1^2 + k_2 k_3 |e_1|^{\frac{3}{2}} + k_2 k_4 |e_1|^{\frac{5}{2}} \\ &\quad +k_1 k_3 |e_1|] = -|e_1|^{-\frac{1}{2}} W(e_1, s) \le -\left(\frac{V_1}{2k_3}\right)^{-\frac{1}{2}} W(e_1, s). \end{aligned} \tag{5.91}$$

For the perturbed system (5.83) with $\omega(t) \neq 0$, the GF time derivative estimation (5.91) is readily modified to

$$\dot{V}_1 = -|e_1|^{-\frac{1}{2}} W(e_1, s) + 2sw + k_1 w |e_1|^{\frac{1}{2}} sign\, e_1 + k_2 w e_1, \tag{5.92}$$

and under conditions (5.79), (5.86) of the theorem, it is further specified to

$$\begin{aligned} \dot{V} &\leq -|e_1|^{-\frac{1}{2}} W(e_1, s) + 2M|s| + Mk_1|e_1|^{\frac{1}{2}} + Mk_2|e_1| \\ &= -|e_1|^{-\frac{1}{2}}\{W - M[2|s||e_1|^{\frac{1}{2}} - k_1|e_1| - k_2|e_1|^{\frac{3}{2}}]\} \leq -|e_1|^{-\frac{1}{2}} W_0(e_1, s), \end{aligned} \tag{5.93}$$

where the well-known inequality $2|s||e_1|^{\frac{1}{2}} \leq s^2 + |e_1|$ is taken into account and

$$\begin{aligned} W_0(e_1, s) = k_2 s^2 |e_1|^{\frac{1}{2}} + \left(\frac{k_1}{2} - M\right) s^2 + k_1 k_4 e_1^2 + (k_2 k_3 - M k_2)|e_1|^{\frac{3}{2}} \\ + (k_1 k_3 - M - M k_1)|e_1| + k_2 k_4 |e_1|^{\frac{5}{2}}. \end{aligned} \tag{5.94}$$

Since

$$V_1(e_1, e_2) \leq \left(2k_3 + \frac{3}{2}k_1^2\right)|e_1| + \left(k_4 + +\frac{3}{2}k_2^2\right) e_1^2 + 2s^2 \tag{5.95}$$

it follows that

$$W_0(e_1, s(e_1, e_2)) \geq \gamma V_1(e_1, e_2) \tag{5.96}$$

for all $e_1, e_2 \in \mathbb{R}$ and for $s(e_1, e_2)$ and γ, given by (5.85) and (5.87), respectively. Then relations (5.90), (5.93), and (5.95), coupled together ensure the upper estimate

$$\dot{V}(x(t)) \leq -\gamma \sqrt{2k_3} \sqrt{V(x(t))}. \tag{5.97}$$

In order to complete the proof it remains to apply Lemma 4.1 to the differential inequality (5.97), thus validating both the global equiuniform FTS of the modified supertwisting system (5.83) and its settling time estimate (5.88). ■

In a particular case of $k_2 = k_4 = 0$ and $M \to \infty$, condition (5.86) of Theorem 5.4 is simplified to $k_1 > 2M$, $k_3 > M$ whereas an excessive gain choice $k_3 > 3M$ would be required with the straightforward GF approach (5.80), inherited from Moreno and Osorio (2008), Moreno and Osorio (2012).

5.2.3 GF Lyapunov Functions of Homogeneous Systems and Their LMI-Based Construction

The capability of seeking a Lyapunov function in a quadratic GF has been illustrated in Sect. 5.2.2 with the supertwisting system. As proposed in Sanchez and Moreno

(2016), this idea is further developed towards homogeneous dynamic systems in a polynomial GF with respect to an augmented state vector that along with the state components captures their fractional degrees.

Consider a homogeneous system

$$\dot{x} = f(x) \tag{5.98}$$

of degree q and dilation $r = (r_1, \dots, r_n) \in \mathbb{R}^n_{>0}$ with the state vector $x \in \mathbb{R}^n$ and the vector field $f(x) = [f_1(x), \dots, f_n(x)]^T$, each component of which is in a GF

$$f_i(x) = \sum_{j=1}^{n} \alpha_j |x_j|^{k_j} \operatorname{sign} x_j + \sum_{i=1}^{s} \beta_i \Pi_{j=1}^{n} |x_j|^{k_{ij}} \operatorname{sign} x_j, \tag{5.99}$$

where s is a natural number, α_j, β_i are reals, and k_j, k_{ij} are rationals for all $j = 1, \dots, n,\ i = 1, \dots, s$.

Note that a GF can be viewed as a polynomial of certain fractional degrees of the state components. Generally speaking, such a polynomial is not differentiable in the origin. Moreover, it admits discontinuities at $x_i = 0$ whenever $k_i \Pi_{j=1}^{n} k_{ij} = 0$ for some $i = 1, \dots, n$. Just in the case, the meaning of (5.99) is adopted in the Filippov sense.

An effective tool of constructing Lyapunov functions for homogeneous systems in the GF (5.99) is based on the observation that a Lyapunov function candidate, chosen in a GF, yields the time derivative along the GF system trajectories to be in a GF as well. The resulting construction procedure, inherited from Sanchez and Moreno (2016), is as follows.

Step 1. A homogeneous Lyapunov function candidate

$$V(x) = \sum_{j=1}^{n} \eta_j |x_j|^{\frac{m}{r_j}} + \sum_{i=1}^{p} \theta_i \Pi_{j=1}^{n} |x_j|^{\kappa_{ij}} \operatorname{sign} x_j \tag{5.100}$$

of degree m is involved with some natural p, some rational κ_{ij}, and some real η_j and θ_i, $j = 1, \dots, n,\ i = 1, \dots, p$. Relation (5.100) determines a positive definite function $V(x)$ provided that $\eta_j > 0,\ j = 1, \dots, n$ whereas $\theta_i\ i = 1, \dots, p$ are sufficiently small. In turn, the homogeneity of the function $V(x)$ is ensured by the parameter relationship

$$\sum_{j=1}^{n} r_j \kappa_{ij} = m \tag{5.101}$$

to be imposed on (5.100).

Step 2. Function (5.100) is additionally guaranteed to be differentiable if

$$m \geq \max_i \{r_i\} \text{ and } \kappa_{ij} \geq 1,\ j = 1, \dots, n,\ i = 1, \dots, p. \tag{5.102}$$

Step 3. Differentiating (5.100) along the trajectories of (5.98) yields

$$\dot{V}(x) = \frac{\partial V(x)}{\partial x} f(x) = -W(x) \tag{5.103}$$

with some GF $W(x)$.

Step 4. More restrictions are imposed on the parameters $\kappa_{ij}, \theta_i,\ j = 1, \dots, n,\ i = 1, \dots, p$ to ensure that the GF $W(x)$ is positive definite.

Step 5. The parameter restrictions, deduced at Steps 1–4, are summarized in the form of inequalities which are linear in the parameters of the Lyapunov function candidate (5.100) and are also linear in the plant parameters of (5.99). Although the resulting inequalities are in general bilinear in the overall plant and Lyapunov GF parameters, it might happen that they are representable in the form of LMIs, similar to the LMI-based analysis of linear time-invariant systems. Once the LMIs, thus obtained, turns out to be feasible, their solution determine a Lyapunov function, guaranteeing the asymptotic stability of the homogeneous GF system (5.98), (5.99).

The interested reader may refer to Sanchez and Moreno (2018) for the procedure verification details. If the procedure is capable of validating the Lyapunov time derivative (5.103) to be not simply negative definite, but also satisfying the stronger differential inequality (4.17) of Lemma 4.1 the finite time stability of the system in question is additionally guaranteed. This way has successfully been used in Moreno (2012) to construct a global FTS Lyapunov function in the GF

$$V_{tw}(x, y) = \left[(\mu + 2M + \frac{\sqrt{8}}{3}\nu)|x| + y^2 \right]^{\frac{3}{2}} + \nu xy, \quad \nu, \mu > 0 \tag{5.104}$$

for the twisting system

$$\begin{aligned} \dot{x} &= y, \\ \dot{y} &= -a \operatorname{sign} x - b \operatorname{sign} y + \psi(t), \end{aligned} \tag{5.105}$$

whose parameters a, b and external disturbance $\psi(t)$ are such that

$$b + \mu > a - M > b > M + \frac{\sqrt{8}}{3}\nu \tag{5.106}$$

where $M > ess \sup_t |\psi(t)|$ is an upperbound on the disturbance magnitude. A more general procedure of the construction of FTS Lyapunov functions, which relies on solving a PDE counterpart of the Lyapunov algebraic equation (5.1), is illustrated next.

5.3 Construction of Multiple FTS Lyapunov Functions via Solving Lyapunov Gradient Equation

The construction methodology of multiple FTS Lyapunov functions, proposed in Polyakov and Poznyak (2009), is based on the state space of the dynamic system

$$\dot{x} = \phi(x, t) \tag{5.107}$$

with a piecewise continuous right-hand side $\phi(\cdot) : \mathbb{R}^{n+1} \to \mathbb{R}^n$ to be partitioned into subdomains, in which a local Lyapunov function $V(x)$ is designed analytically. The methodology relies on specific positive definite solutions of the first order PDE

$$\nabla^T V(x)\tilde{\phi}(x) = -qV^\rho \tag{5.108}$$

with some $\rho \in [0, 1)$ and the left-hand side, written in the gradient form, where the time-invariant function $\tilde{\phi}(x)$ properly estimates $\phi(x, t)$ under potential uncertainties of the right-hand side of (5.107) to ensure that $\nabla^T V(x)\phi(x, t) \le -qV^\rho$.

Since along with $\phi(x, t)$, its estimate $\tilde{\phi}(x)$ is also admitted to be piecewise continuous rather than continuous, the above PDE is locally solved in each continuity subdomain of $\tilde{\phi}(x)$. The globally defined function $V(x)$, which is composed of such local solutions and which is generally speaking multi-valued on the subdomain boundaries, determines a multiple Lyapunov function of the underlying system (5.107). Once the resulting function $V(x)$ is everywhere continuous, it might be viewed as a common Lyapunov function of (5.107). Thus, (5.107) is guaranteed to be FTS with no extra condition on the multiple Lyapunov function (cf. Theorem 4.2). The PDE (5.108) is further referred to as a *Lyapunov gradient equation*.

To exemplify the methodology, let us consider the second order control system

$$\begin{aligned} &\dot{x}_1(t) = x_2(t), \qquad t > 0, \\ &\dot{x}_2(t) = a(x_1(t), x_2(t), t) + b(x_1(t), x_2(t), t)u(x_1(t), x_2(t)), \\ &|a| < M, \quad 0 < b_{\min} \le b \le b_{\max} \end{aligned} \tag{5.109}$$

with scalar state components x_1, x_2, a scalar relay feedback u, switching between u_0 and $-u_0$, and scalar functions a, b, depending on both the state and time variables. Then solving the corresponding PDE (5.108) for system (5.109) with $\rho = 0$ results in the unified *Lyapunov control function candidate*

$$\begin{aligned} V_u(x_1, x_2) = q\left(\omega\sqrt{\left|x_1 - \frac{x_2^2}{2(\gamma+\mu u(x_1,x_2))}\right|} - \frac{x_2}{\gamma+\mu u(x_1,x_2)}\right), \\ q > 0, \quad \omega > 0, \quad \mu \in \mathbb{R} \text{ and } \gamma \in \mathbb{R}. \end{aligned} \tag{5.110}$$

If given a relay feedback law $u(x_1, x_2)$, the parameters q, ω, μ and γ can be chosen to make the function V_u continuous and positive definite then (5.110) determines a FTS Lyapunov function of the relay system (5.109).

In Polyakov and Poznyak (2012), Lyapunov functions of the form (5.110) and high precision (sharp) settling time estimates are found by quadrants for the earlier introduced Fuller system (1.8)–(1.9) as well as for SOSM control algorithms. In a particular case of the predetermined parameters $a \equiv 0$ and $b \equiv 1$, the Lyapunov control function V_u, given by (5.110) and specified with $q = \mu = 1, \gamma = 0$, determines, by means of Lemma 4.1, the settling time function of the system (5.109) with a relay stabilizing feedback u. For the disturbance-free twisting system (5.105) with $\psi(t) \equiv 0$, the Lyapunov function

$$\begin{aligned} V_{u_{tw}}(x_1, x_2) &= \omega\sqrt{|x_1| + \frac{x_2^2}{2(a+b\text{sign}(x_1x_2))}} + \frac{|x_2|\text{sign}(x_1x_2)}{a+b\text{sign}(x_1x_2)}, \\ \omega &= \frac{2\sqrt{2}a}{a^2-b^2}\left(\frac{1}{\sqrt{a-b}} - \frac{1}{\sqrt{a+b}}\right)^{-1}, \quad a > b > 0 \end{aligned} \tag{5.111}$$

is, for instance, obtained by substituting the twisting control law $u_{tw}(x_1, x_2) = -a\text{sign}\, x_1 - b\text{sign}\, x_2$ into (5.110).

5.4 Lyapunov Minmax Approach and Speed Gradient Method

The Lyapunov minmax approach aims to synthesize a stabilizing feedback in such a manner that the time derivative of a Lyapunov control function candidate, e.g., selected for a nominal system, is minimized against the maximal plant perturbation provided that magnitude bounds are known a priori both for admissible control actions and parametric/external disturbances. The approach, pioneered in Gutman (1979), Gutman and Leitmann (1976), was then applied for the SM control design, resulting in a discontinuous *unit control input* (Orlov and Utkin 1998) of the unit norm everywhere except for the discontinuity manifold, and it further anticipated the popular linear $\mathscr{H}_\infty$ synthesis (Doyle et al. 1989) and its nonlinear counterpart (Isidori and Astolfi 1992).

In analogy to standard component-wise SM control signals, the unit control with sufficiently high magnitude can enforce an asymptotically stable sliding mode, which is robust against matched disturbances. An important point is that the trajectories of the closed-loop system never pass through the discontinuity manifold. The system stability is thus analyzed beyond the manifold. Once the trajectory is on the discontinuity manifold, smooth dynamics restore, and the standard Lyapunov theory is in force.

The Lyapunov minmax approach was initially developed for affine systems

$$\dot{x}(t) = f(x(t), t) + B(x(t), t)u(t) + h(x, t), \quad x \in \mathbb{R}^n, \ u \in \mathbb{R}^m \tag{5.112}$$

with matched disturbances such that $h(x, t) = B(x, t)\gamma(x, t)$ and $\|\gamma(x, t)\| \leq \gamma_0$ for some constant γ_0. Let the equation

$$\dot{x}(t) = f(x(t), t) \tag{5.113}$$

represent an open-loop nominal system. For simplicity, the nominal system (5.113) is assumed to be asymptotically stable with some a priori known positive definite continuously differentiable Lyapunov function V, such that its time derivative, computed along the trajectories of (5.113), is negative definite, i.e.,

$$W_0(x, t) = \nabla^T V(x) f(x, t) \leq -W_1(x), \quad \forall\, (x, t) \in \mathbb{R}^{n+1}, \tag{5.114}$$

where $W_1(x)$ is a continuous, positive definite function. Then the time derivative of V on the trajectories of the perturbed system (5.112) takes the form

$$W = \frac{dV}{dt} = W_0 + \nabla^T V \cdot B(u + \gamma). \tag{5.115}$$

Let system (5.112) be driven by the control input

$$u = -\nu U(s(x, t)), \qquad U = \frac{s}{\|s\|} \tag{5.116}$$

with a gain ν such that

$$\nu > \gamma_0, \tag{5.117}$$

and m-vector-function

$$s(x, t) = B^T(x, t) \nabla V(x). \tag{5.118}$$

The control feedback (5.116) is called the unit control law since $\|U\| = 1$ everywhere beyond $s = 0$ where it undergoes a discontinuity.

Thus due to (5.117), the time derivative of the Lyapunov function $V(x)$, computed on the trajectories of the closed-loop system (5.112), is negative definite:

$$W = W_0 - \nu \parallel B^T \nabla V \parallel + \nabla^T V \cdot B\gamma \leq -W_1 - \parallel B^T \nabla V \parallel \left[\nu - \gamma_0\right] \leq -W_1. \tag{5.119}$$

It means that the closed-loop system is globally asymptotically stable.

It is of interest to note that in contrast to the conventional sliding mode control signals, which undergo discontinuities whenever a component of the sliding manifold changes its sign, the unit control action is a continuous state function until the manifold $s = 0$ is reached. Due to this difference, the unit control method is an appropriate tool of discontinuous control design not only in a finite-dimensional state space, but also in an infinite-dimensional state-space where control inputs are not (or even cannot be) represented in a component-wise form.

An appropriate generalization of the unit control design for dynamic systems in a Hilbert space is as follows. For infinite-dimensional dynamics

$$\dot{x} = Ax + Bu + h(x, t) \tag{5.120}$$

evolving in a Hilbert space and affected by an unknown disturbance, the state and control variables are elements of Hilbert spaces $x \in X$, $u \in U$, $h : \mathbb{R} \times X \rightarrow X$ is a disturbance, $A : D(A) \subset X \rightarrow X$ is a closed linear operator with the domain $D(A)$ dense in X, and $B : U \rightarrow X$ is a linear bounded operator. The unknown disturbance $h(x, t)$ is assumed to be bounded with an a priori known norm bound such that the open-loop system (5.120) possesses a strong solution for an arbitrary initial condition $x(0) \in D(A)$ (e.g., h is locally Lipschitz continuous). Certainly, such a system can robustly be stabilized if the matching condition

$$h(x, t) = B\gamma(x, t) \tag{5.121}$$

holds for some $\gamma : \mathbb{R} \times X \rightarrow U$ of uniformly bounded magnitude $\|\gamma(x, t)\| \leq \gamma_0$ and $\gamma_0 \in \mathbb{R}_{>0}$.

To facilitate the exposition, let us assume that the nominal system $\dot{x} = Ax$ is exponentially stable with a continuous positive definite Lyapunov functional $V : X \rightarrow \mathbb{R}$, available to the designer. In other words, let there exist a continuous positive definite functional $V(x)$ such that $\dot{V} = V_x Ax < 0$ for all $x \in D(A)$ where $V_x = \nabla_x V : X \rightarrow \mathbb{R}$ is a linear functional (that might be viewed as a counterpart to the finite-dimensional operator $\nabla^T V : \mathbb{R}^n \rightarrow \mathbb{R}$) with the property

$$\lim_{\|\Delta x\| \rightarrow 0} \frac{V(x + \Delta x) - V(x) - V_x \Delta x}{\|\Delta x\|} = 0, \quad \Delta x \in X. \tag{5.122}$$

Calculating the time derivative of $V(x)$ on the trajectories of the original systems yields

$$\dot{V} = V_x Ax + V_x B(u + \gamma). \tag{5.123}$$

Let $(V_x B)^* \in U$ be an adjoint operator (Rudin 1991) to $V_x B$, ensuring that $V_x B\, (\nabla_x V B)^* = \|\nabla_x V B\|^2$. Thus, for the unit control control law

$$u = -\nu \frac{(V_x B)^*}{\|V_x B\|}, \quad \nu > \gamma_0 \tag{5.124}$$

it is derived that

$$\dot{V} \leq V_x Ax + \|V_x B\|(-\nu + \gamma_0) < 0. \tag{5.125}$$

Hence the origin $x = 0$ proves to be globally asymptotically stable.

To complete this section the speed gradient method, extending the Lyapunov minmax approach to a set stabilization non-affine control systems, is briefly reviewed in the finite-dimensional flavor. More details on the method can be found in Fradkov (2007), Fradkov and Pogromsky (1998).

Let the control system be modeled as

$$\dot{x} = F(x, u), \tag{5.126}$$

where $x \in R^n$ is the state and $u \in R^m$ is the input (controlling signal). Let the control goal be expressed as the limiting relation

$$Q(x(t)) \to 0 \text{ as } t \to \infty \tag{5.127}$$

with a performance index $Q(\cdot) \in R$, specified along the system trajectories. Note that in general, $Q(x) = 0$ is not a singleton, but it specifies a desired destination set. The speed gradient algorithm, aiming to achieve (5.127), is designed in the form

$$u = -\Psi(\nabla_u \xi(x, u)), \tag{5.128}$$

where by the chain rule, $\xi(x, u) = Q_x F(x, u)$ is the speed of changing $Q(x(t))$ along the trajectories of (5.126) and vector $\Psi(z)$ forms an acute angle with the vector z, i.e., $\Psi(z)^T z > 0$ when $z \neq 0$.

The first step of the speed gradient procedure is to compute the speed $\dot{Q}$. The second step consists in evaluating the gradient $\nabla_u \dot{Q}(x, u)$ with respect to the control input u. As a final step, the vector-function $\Psi(z)$ should be chosen to meet the acute angle condition. For instance, the choice $\Psi(z) = Kz$ with $K > 0$ yields the *proportional* (with respect to speed gradient) feedback

$$u = -\gamma \nabla_u \xi(x, u), \tag{5.129}$$

whereas the choice $\Psi(z) = \gamma \text{sign}\, z$ yields the *relay* algorithm

$$u = -\gamma \text{ sign } \nabla_u \xi(x, u). \tag{5.130}$$

The *integral* form of the speed gradient algorithm

$$\frac{du}{dt} = -\gamma\ \nabla_u \xi(x, u) \tag{5.131}$$

is also appropriate as well as combined (e.g., proportional–integral) forms are.

The motivating idea behind the speed gradient choice (5.129) is that moving along the antigradient of the speed $\dot{Q}$ results in the decay of $\dot{Q}$. It may eventually lead to achieving the primary goal (5.127). However, to guarantee (5.127) some additional assumptions are needed (Fradkov and Pogromsky 1998).

5.5 Construction of Lyapunov Functions Using Proximal Solutions of Hamilton–Jacobi PDI

As shown in the previous section, the Lyapunov minmax approach is perfectly suited to reject matched disturbances whereas in the general case of mismatched disturbances only their attenuation becomes available. The nonlinear $\mathscr{H}_\infty$ approach of Isidori and Astolfi (1992) is adopted to subsequently address the mismatched disturbance attenuation for nonlinear dynamics, evolving under a scalar unilateral constraint. Given such a constraint $F(x,t) : \mathbb{R}^{n+1} \to \mathbb{R}_{\geq 0}$, the underlying continuous dynamics are governed by

$$\dot{x} = f(x,t) + g(x,t)w, \quad z = h(x,t) \tag{5.132}$$

beyond the surface $F(x,t) = 0$ when the constraint inactive, and the dynamics are subject to the restitution rule

$$x(t_i+) = \mu(x(t_i-), t_i) + \Omega(x(t_i-), t_i)w_i^d, \quad z_i^d = x(t_i+), \quad i = 1, 2, \ldots \tag{5.133}$$

at a priori unknown collision time instants $t = t_i,\ i = 1, 2, \ldots,$ when the system trajectory hits the surface $F(x,t) = 0$.

Hereinafter, $x \in \mathbb{R}^n$ represents the state vector; $w \in \mathbb{R}^l$ and $w_i^d \in \mathbb{R}^q$ collect exogenous signals affecting the motion of the system (external forces, including impulsive ones, as well as model imperfections); the output variables $z \in \mathbb{R}^m$ and $z_i^d \in \mathbb{R}^n$ are responsible for the performance of the system; the matrix functions f, g, h, μ, Ω are of appropriate dimensions.

Throughout, the following *assumptions* are imposed on the hybrid system in question:

(A1) The functions F, g, h, μ, Ω are piecewise continuous and uniformly bounded in t and locally Lipschitz continuous in x;

(A2) The function

$$f(x,t) = \begin{cases} f^+(x,t) & \text{if } s(x,t) > 0 \\ f^-(x,t) & \text{if } s(x,t) < 0\,. \end{cases} \tag{5.134}$$

undergoes discontinuities on a time-varying surface

$$S(t) = \{x \in \mathbb{R}^n : s(x,t) = 0\}, \tag{5.135}$$

determined by a smooth scalar function $s(x,t)$, and $f(x,t)$ is piecewise continuous and uniformly bounded in t and locally Lipschitz continuous in x outside the discontinuity surface (5.135);

(A3) The origin is an equilibrium point of the unforced system

$$\dot{x} = f(x,t) \tag{5.136}$$

and $h(0) = 0$, and $\mu(0) = 0$.

The $\mathcal{L}_2$-gain concept, given below, is a straightforward extension of the $\mathcal{L}_2$-gain Definition 3.16 for the underlying hybrid system (5.132), (5.133).

Definition 5.1 Given a real number $\gamma > 0$, referred to as a disturbance attenuation level, it is said that the hybrid system (5.132), (5.133) locally possesses *$\mathcal{L}_2$-gain less than γ* if the inequality

$$\int_0^T \|z(t)\|^2 \mathrm{d}t + \sum_{i=1}^{N_T} \|z_i^d\|^2 < \gamma^2 \left[\int_0^T \|w(t)\|^2 \mathrm{d}t + \sum_{i=1}^{N_T} \|w_i^d\|^2 \right] + \beta_0(x(0)) + \sum_{k=1}^{N_T} \beta_k(x(t_k^-)) \qquad (5.137)$$

holds for some positive definite functions $\beta_k(x)$, $k = 0, \ldots, N_T$, for all segments $T > 0$ and a natural N_T such that $t_{N_T} \leq T < t_{N_T+1}$, for all piecewise continuous disturbances $w(t)$ and discrete ones w_i^d, $i = 1, 2, \ldots$, for which the state trajectory of the underlying system starting from an initial point $x(t_0) = x_0 \in \mathcal{U}$ remains in some neighborhood $\mathcal{U} \in \mathbb{R}^n$ of the origin for all $t \in [0, T]$.

It is worth noticing that the above $\mathcal{L}_2$-gain definition is consistent with the notion of dissipativity, introduced in Willems (1972) and Hill and Moylan (1980), and with iISS notion (Hespanha et al. 2008), and it represents a natural extension to hybrid systems (see, e.g., the works Baras and James 1993; Lin and Byrnes 1996; Nešić et al. 2008; Yuliar et al. 1998). In order to facilitate the exposition the underlying system, chosen for treatment, has been prespecified with the postimpact velocity value $x(t)$ in the discrete output z_i^d of (5.133). The general case of a certain function $\kappa(x(t))$ of the postimpact velocity value in the discrete output (5.133) can be treated in a similar manner because the $\mathcal{L}_2$-gain inequality (5.137) is flexible in the choice of positive definite functions $\beta_k(x)$, $k = 0, \ldots, N_T$.

The present objective is to locally construct a so-called storage Lyapunov function to ensure not only the asymptotic stability of the disturbance-free system (5.132), (5.133) but also disturbance attenuation in the presence of sufficiently small disturbances. An appropriate construction, developed in Osuna et al. (2016), relies on positive definite proximal solutions of the Hamilton–Jacobi PDI

$$\frac{\partial V}{\partial t} + \frac{\partial V}{\partial x} f(x, t) + \frac{1}{4\gamma^2} \frac{\partial V}{\partial x} g(x, t) g^T(x, t) \left(\frac{\partial V}{\partial x} \right)^T + h^T(x, t) h(x, t) < 0 \qquad (5.138)$$

with some positive γ (see Definition 3.5 for the proximal solution concept). Such a Lyapunov function construction becomes feasible under the hypotheses, specified below in a domain $B_\delta^n = \{x \in \mathbb{R}^n : \|x\| \leq \delta\}$ of interest:

(H1) The norm of the matrix function Ω is uniformly upper bounded by $\frac{\sqrt{2}}{2}\gamma$, i.e.,

$$\|\Omega(x,t)\| \le \frac{\sqrt{2}}{2}\gamma \quad \text{for all } t \in \mathbb{R}; \tag{5.139}$$

(H2) The Hamilton–Jacobi inequality (5.138) possesses a local positive definite proximal solution $V(x,t)$ under some positive γ;

(H3) Hypothesis (H2) is satisfied with the function $V(x,t)$ which decreases along the direction μ, i.e., the inequality

$$V(x,t) \ge V(\mu(x,t),t) \tag{5.140}$$

holds for all $x \in B_\delta^n$ and all t.

The resulting construction is summarized from Osuna et al. (2016, Theorem 5).

Theorem 5.5 *Consider the hybrid system* (5.132), (5.133) *with Assumptions (A1)–(A3). Given $\gamma > 0$, suppose that Hypotheses (H1) and (H2) are satisfied for the system in a domain B_δ^n with a function $V(x,t)$. Then, the hybrid system* (5.132), (5.133) *locally possesses $\mathcal{L}_2$-gain less than γ. Once Hypothesis (H3) is satisfied as well, a proximal solution $V(x,t)$ of the Hamilton–Jacobi inequality* (5.138) *proves to be a strict Lyapunov function of the disturbance-free version of the hybrid system* (5.132), (5.133), *thereby guaranteeing the asymptotic stability of the internal dynamics* (5.132), (5.133) *under $w \equiv 0$, $w_i^d \equiv 0$.*

Proof The proof is rather technical and it follows the standard arguments of the nonlinear $\mathcal{L}_2$-gain analysis of Isidori and Astolfi (1992) and Van der Schaft (1992), recently extended in Osuna et al. (2016) to discontinuous (Filippov) vector fields under unilateral constraints. It is clear that Lemma 3.5 is applicable to a proximal solution $V(x,t)$ of the Hamilton–Jacobi inequality (5.138), regardless of whether it is viewed on the solutions $x(t)$ of the disturbance-free system (5.136) beyond the discontinuity manifold (5.135) or it is viewed along the discontinuity manifold (5.135). Then relations (3.141), (3.142), (5.138), coupled together, result in

$$\frac{\mathrm{d}}{\mathrm{d}t}V(x,t) = DV(x,t;\dot{x},1) \le \frac{\partial V}{\partial t} + \frac{\partial V}{\partial x} f(x,t) < 0. \tag{5.141}$$

With (5.141) in mind, Hypotheses (H2) and (H3) ensure that Theorem 4.9 is applicable to the disturbance-free version (5.136) of the hybrid system (5.132), (5.133), which is thus established to be asymptotically stable.

It remains to demonstrate that the disturbed system (5.132), (5.133) locally possesses $\mathcal{L}_2$-gain less than γ. For this purpose, let us first focus on the system dynamics beyond the discontinuity manifold (5.135) and let us introduce the multi-valued function

$$H(x,w,t) = \frac{\partial V(x,t)}{\partial x}[f(x,t) + g(x,t)w] + h^T(x,t)h(x,t) - \gamma^2 w^T w, \tag{5.142}$$

where $\left(\frac{\partial V}{\partial x}, \frac{\partial V}{\partial t}\right)^T \in \partial^P V(x,t)$. Clearly, the multi-valued function (5.142) is quadratic in w. Then

$$\left.\frac{\partial H(x,w,t)}{\partial w}\right|_{w=\alpha(x,t)} = \frac{\partial V(x,t)}{\partial x} g(x,t) - 2\gamma^2 \alpha^T(x,t) = 0 \tag{5.143}$$

for $\left(\frac{\partial V}{\partial x}, \frac{\partial V}{\partial t}\right)^T \in \partial^P V(x,t)$ and

$$\alpha^T(x,t) = \frac{1}{2\gamma^2} \frac{\partial V(x,t)}{\partial x} g(x,t). \tag{5.144}$$

Expanding the quadratic function $H(x,w,t)$ in Taylor series, we derive that

$$H(x,w,t) = H\left(x, \alpha(x,t), t\right) - \gamma^2 \|w - \alpha(x,t)\|^2 \tag{5.145}$$

where $H\left(x, \alpha(x,t)\right) < 0$ due to (5.138). Hence,

$$H(x,w,t) < -\gamma^2 \|w - \alpha(x,t)\|^2 \tag{5.146}$$

and employing (5.142) and (5.145) we arrive at

$$\frac{\partial V(x,t)}{\partial x} [f(x,t) + g(x,t)w] < -\gamma^2 \|w - \alpha(x,t)\|^2 - \|h(x,t)\|^2 + \gamma^2 \|w\|^2. \tag{5.147}$$

By applying Lemma 3.5 and taking (5.147) into account, a time derivative estimate of the solution $V(x,t)$ of the Hamilton–Jacobi inequality (5.138) on the trajectories of (5.132) between the collision time instants is given as

$$\frac{\mathrm{d}}{\mathrm{d}t} V(x,t) < -\gamma^2 \|w - \alpha(x,t)\|^2 - \|z\|^2 + \gamma^2 \|w\|^2. \tag{5.148}$$

Following the same line of reasoning, estimates (5.147) and (5.148) are additionally verified for the system dynamics along the discontinuity surface (5.135) to hold true with $f(x,t) = f_0(x,t)$, $g(x,t) = g_0(x,t)$, given by (3.137), (3.138), and with $\alpha(x,t)$, now specified by (5.144) with $g(x,t) = g_0(x,t)$.

Thus, between any successive collision time instants, estimate (5.148) remains in force regardless of whether the trajectory evolves beyond the discontinuity surface (5.135) or SMs occur along it. By integrating (5.148) between the collision time instants t_{i-1} and t_i and by summing the resulting inequalities over $i = 1, \ldots, N_T$, it follows that

$$\sum_{i=1}^{N_T}\int_{t_{i-1}}^{t_i}[\gamma^2\|w\|^2-\|z(t)\|^2]\mathrm{d}t > \sum_{i=1}^{N_T}\int_{t_{i-1}}^{t_i}\frac{\mathrm{d}V(x(t))}{\mathrm{d}t}\mathrm{d}t \\ +\sum_{i=1}^{N_T}\gamma^2\int_{t_{i-1}}^{t_i}\|w(t)-\alpha(x(t),t)\|^2\mathrm{d}t > 0. \tag{5.149}$$

Skipping positive terms in the right-hand side of (5.149) yields

$$\int_0^T(\gamma^2\|w\|^2-\|z(t)\|^2)\mathrm{d}t > V(x(t_{N_T}+))+\sum_{i=0}^{N_T}[V(x(t_i^-))-V(x(t_i^+))]-V(x(t_0-)). \tag{5.150}$$

Since the function V is Lipschitz continuous by Hypotheses (H2) the following relation:

$$|V(x(t_i^-))-V(x(t_i^+))|\leq L_\delta\|x(t_i^-)-x(t_i^+)\| < L_\delta[\|x(t_i^-)\|+\|x(t_i^+)\|] \tag{5.151}$$

holds true with $L_\delta > 0$ being Lipschitz constant of V in the domain $B_\delta^n \in \mathbb{R}^n$. Relations (5.150) and (5.151), coupled together, ensure the inequality

$$\int_{t_0}^T(\gamma^2\|w\|^2-\|z(t)\|^2)\mathrm{d}t > -2L_\delta\sum_{i=1}^{N_T}\|x(t_i^-)\|-V(x(t_0)) \tag{5.152}$$

in the domain $B_\delta^{2n}\in\mathbb{R}^{2n}$. Apart from this, inequality

$$\sum_{i=1}^{N_T}\|z_i^d\|^2=\sum_{i=1}^{N_T}\|x(t_i^+)\|^2\leq 2\sum_{i=1}^{N_T}\|\mu(x(t_i^-),t_i)\|^2+2\sum_{i=1}^{N_T}\|\Omega(x(t_i^-),t)w_i^d\|^2\leq \\ 2\sum_{i=1}^{N_T}\|\mu(x(t_i^-),t_i)\|^2+\gamma^2\sum_{i=1}^{N_T}\|w_i^d\|^2 \tag{5.153}$$

is ensured by relation (5.139) of Hypothesis (H1). Thus, combining (5.152) and (5.153), one derives

$$\int_{t_0}^T\|z(t)\|^2\mathrm{d}t+\sum_{i=1}^{N_T}\|z_i^d\|^2 < V(x(t_0))+\gamma^2\left[\int_{t_0}^T\|w(t)\|^2\mathrm{d}t+\sum_{i=1}^{N_T}\|w_i^d\|^2\right]+ \\ 2\sum_{i=1}^{N_T}\|\mu(x(t_i^-),t_i)\|^2+2L_\delta\sum_{i=1}^{N_T}\|x(t_i^-)\|, \tag{5.154}$$

i.e., the disturbance attenuation inequality (5.137) is established with the positive definite functions

$$\beta_0(x)=V(x,t_0),\quad \beta_i(x)=2L_\delta\|x\|+2\|\mu(x,t_i)\|^2,\ i=1,\ldots,N. \tag{5.155}$$

Theorem 5.5 is thus proved. ■

A remarkable consequence of Theorem 5.5, given below, is for a particular case where no unilateral constraints are imposed on the underlying system.

Theorem 5.6 *Consider system* (5.132) *with Assumptions (A1)–(A3) and with no state restitution* (5.133). *Given* $\gamma > 0$, *suppose that Hypotheses (H2) is satisfied for such a constraint-free system in a domain* B_δ^n *with a function* $V(x, t)$. *Then, system* (5.132) *locally possesses* $\mathscr{L}_2$*-gain less than* γ, *and a proximal solution* $V(x, t)$ *of the Hamilton–Jacobi inequality* (5.138) *is a strict Lyapunov function of the disturbance-free version of system* (5.132) *that guarantees the asymptotic stability of the internal dynamics* (5.132) *under* $w \equiv 0$.

Proof Since Hypotheses (H1) and (H3) are irrelevant for the constraint-free system (5.132), Theorem 5.6 represents a straightforward consequence of Theorem 5.5. ■

5.6 Concluding Remarks

Nonsmooth Lyapunov functions and their distinct constructions are proposed for a wide class of dynamic systems and Filippov vector fields, affected by external disturbances and state resets. Although the development focuses on the finite-dimensional setting, however, its infinite-dimensional extension is actually possible. This claim is further supported by applications, which apart from the proposed CLF constructions, utilize their counterparts to deal with distributed parameter and time-delay systems.

References

Alvarez J, Orlov Y, Acho L (2000) An invariance principle for discontinuous dynamic systems with applications to a Coulomb friction oscillator. J Dyn Syst Meas Control 74:190–198

Artstein Z (1983) Stabilization with relaxed controls. Nonlinear Anal Theory Methods Appl 7:1163–1173

Baras J, James M (1993) Robust output feedback control for discrete-time nonlinear systems: the finite-time case. In: IEEE conference on decision and control, pp 51–55

Byrnes C, Isidori A (1991) Asymptotic stabilization of minimum phase nonlinear systems. IEEE Trans Auto Control 36(10):1122–1137

Doyle JC, Glover K, Khargonekar P, Francis BA (1989) State-space solutions to standard $\mathscr{H}_2$ and $\mathscr{H}_\infty$ control problems. IEEE Trans Auto Control 34:831–847

Fradkov AL (2007) Cybernetical physics: from control of chaos to quantum control. Springer, Berlin

Fradkov AL, Pogromsky AY (1998) Introduction to control of oscillations and chaos. World Scientific Pub. Co., Singapore

Grizzle J, Abba G, Plestan F (2001) Asymptotically stable walking for biped robots: analysis via systems with impulse effects. IEEE Trans Auto Control 46:51–64

Gutman S (1979) Uncertain dynamic systems - a Lyapunov min-max approach. IEEE Trans Auto Control AC-24:437–449

Gutman S, Leitmann G (1976) Stabilizing feedback control for dynamic systems with bounded uncertainties. In: IEEE conference on decision and control, pp 94–99

Hespanha JP, Liberzon D, Teel AR (2008) Lyapunov conditions for input-to-state stability of impulsive systems. Automatica 44(11):2735–2744

Hill D, Moylan P (1980) Connections between finite-gain and asymptotic stability. IEEE Trans Auto Control 25(5):931–936

Isidori A, Astolfi A (1992) Disturbance attenuation and $\mathscr{H}_\infty$-control via measurement feedback in nonlinear systems. IEEE Trans Auto Control 37:1283–1293

Levant A (1993) Sliding order and sliding accuracy in sliding mode control. Int J Control 58:1247–1263

Lin W, Byrnes C (1996) $\mathscr{H}_\infty$-control of discrete-time nonlinear systems. IEEE Trans Auto Control 41(4):494–510

Moreno J (2012) A lyapunov approach to output feedback control using second-order sliding modes. IMA J Math Control Inf 29:291–308

Moreno J, Osorio M (2008) A Lyapunov approach to second-order sliding mode controllers and observers. In: Proceedings 47th conference on decision and control, pp 2856–2861

Moreno J, Osorio M (2012) Strict lyapunov functions for the supertwisting algorithm. IEEE Trans Auto Control 57:1035–1040

Nešić D, Zaccarian L, Teel AR (2008) Stability properties of reset systems. Automatica 44(8):2019–2026

Orlov Y (2003) Finite time stability of homogeneous switched systems. In: Proceedings of the 42nd conference on decision and control, pp 4271–4276

Orlov Y, Alvarez J, Acho L, Aguilar L (2003) Global position regulation of friction manipulators via switched chattering control. Int J Control 76:1446–1452

Orlov Y, Austin Y, Chevallereau C (2011) Finite time stabilization of a perturbed double integrator Part I: continuous sliding mode-based output feedback synthesis. IEEE Trans Auto Control 56:614–618

Orlov Y, Utkin V (1998) Unit sliding mode control in infinite-dimensional systems. J Appl Math Comput Sci 8:7–20

Osuna T, Montano O, Orlov Y (2016) Nonlinear $\mathscr{L}_2$-gain analysis of hybrid systems in the presence of sliding modes and impacts. Math Probl Eng, Article ID 9074096, 10 p

Oza HB, Orlov YV, Spurgeon SK (2014) Finite time stabilization of a perturbed double integrator with unilateral constraints. Math Comput Simul 95:200–212

Polyakov A, Poznyak A (2009) Reaching time estimation for super-twisting second order sliding mode controller via lyapunov function designing. IEEE Trans Auto Control 54:1951–1955

Polyakov A, Poznyak A (2012) Unified Lyapunov function for a finite-time stability analysis of relay second-order sliding mode control systems. IMA J Math Control Inf 29:529550

Rudin W (1991) Functional analysis, 2nd edn. McGraw-Hill, New York

Sanchez T, Moreno J (2016) A Constructive Lyapunov function design method for a class of homogeneous systems. In: Proceedings of the 53rd conference on decision and control, pp 5500–5505

Sanchez T, Moreno J (2018) Design of Lyapunov functions for a class of homogeneous systems: generalized forms approach. Int J Robust Nonlinear Control. https://doi.org/10.1002/rnc.4274

Shevitz D, Paden B (1994) Lyapunov stability theory of nonsmooth systems. IEEE Trans Auto Control 39:1910–1914

Sontag ED, Sussman HJ (1995) Nonsmooth control-Lyapunov functions. In: Proceedings of the 34th IEEE decision control conference, pp 2799–2805

Van der Schaft A (1992) $\mathscr{L}_2$-gain analysis of nonlinear systems and nonlinear state feedback control. IEEE Trans Auto Control 37:770–784

Willems J (1972) Dissipative dynamical systems Part I: general theory. Arch Rat Mech Anal 45(5):321–351

Yuliar S, James M, Helton J (1998) Dissipative control systems synthesis with full state feedback. Math Control Signals Syst 11(4):335–356

Part III
Lyapunov Redesign

Constructive Lyapunov functions, developed so far, are not only powerful for the robust stability analysis but they are also expected to be useful for the control synthesis, accompanied by gain tuning to meet the desired closed-loop performance regardless of whichever mismatched disturbance affects the system. Attaining a certain disturbance attenuation level is subsequently illustrated for SM controllers of the first and second orders by tuning their gains to ensure the feasibility of appropriate proximal solutions of the corresponding Hamilton–Jacobi PDI. A similar procedure is then developed for a parabolic PDE in the presence of external disturbances.

Based on quadratic Lyapunov functionals, which are always available for linear systems, adaptive plant identification is additionally presented in the linear PDE and time delay settings. The persistent excitation concept, inspired from the finite-dimensional treatment, is properly generalized for linear time delay and distributed parameter systems to ensure their identifiability. Once such a system is persistently excited its identifiability is guaranteed, and an adaptive identifier, asymptotically estimating unknown plant parameters, becomes available from the Lyapunov redesign that ensures a strict quadratic Lyapunov functional to exist. The adaptive identifier construction is worked out side by side for one-dimensional diffusion–reaction and string PDEs as well as for linear-time delay systems.

Specific CLF applications to robust orbital stabilization of a biped gait under a unilateral constraint and to energy control of nonlinear sine-Gordon PDE model of continuum of oscillators make the presentation appropriately complete.

Chapter 6
Lyapunov-Based Tuning

The primary concern of the present chapter is to test, by means of nonsmooth Lyapunov functions, the capability of SM control algorithms to attenuate mismatched disturbances with a priori unknown bounds on their magnitudes. While SM control algorithms are well-recognized to be capable of rejecting matched disturbances of bounded magnitude, the relevant works (see the milestone monograph Utkin 1992 and references therein), assume typically that only such disturbances affect the underlying system. The objective of the subsequent investigation is to demonstrate that SM algorithms (standard FOSM and popular SOSM algorithms, including twisting and supertwisting ones) are additionally capable of attenuating unbounded mismatched disturbances. The investigation, made within the Hamilton–Jacobi PDE-based paradigm, developed in Sect. 5.5 for the nonsmooth $\mathscr{L}_2$-gain analysis, focuses on feedback linearizable VSS in a Euclidean state space, and it follows the line of reasoning, recently proposed in Osuna et al. (2016, 2018a, b). Apart from the finite-dimensional treatment, the attenuation of mismatched disturbances is extended to DPS, governed by a parabolic PDE.

6.1 $\mathscr{L}_2$-Gain Tuning of First-Order Sliding Modes

A SISO system

$$\begin{aligned} \dot{x}_1 &= x_2 + d_1(x,t)w_1(t) \\ &\vdots \\ \dot{x}_{n-1} &= x_n + d_{n-1}(x,t)w_{n-1}(t) \\ \dot{x}_n &= \phi(x,t) + b(x,t)u + d_n(x,t)w_n(t) \end{aligned} \tag{6.1}$$

Y. Orlov, *Nonsmooth Lyapunov Analysis in Finite and Infinite Dimensions*, Communications and Control Engineering,
https://doi.org/10.1007/978-3-030-37625-3_6

chosen for the investigation, is of relative degree n and it is given in the normal form of Byrnes and Isidori (1991) where $x = (x_1, \ldots, x_n)^T \in \mathbb{R}^n$ is the state vector, $w = (w_1, \ldots, w_n)^T \in \mathbb{R}^n$ is the disturbance vector, $u \in \mathbb{R}^1$ is the control input, and the scalar functions $\phi(x,t), b(x,t), d_i(x,t),\ i = 1, \ldots, n$ are continuous on the domain

$$B_R = \{x \in \mathbb{R}^n : \Sigma_{i=1}^n x_i^2 \le R^2\} \tag{6.2}$$

of a radius $R > 0$ for all $t \in \mathbb{R}$ and sufficiently smooth in both arguments. It is assumed throughout that

$$b(x,t) \ge b_0, \tag{6.3}$$

$$|\phi(x,t)| \le L_\phi,\ \ |d_i(x,t)| \le L_i,\ i = 1, \ldots, n \tag{6.4}$$

for all $x \in B_R, t \in \mathbb{R}^1$ and some bounds $b_0, L_\phi, L_i > 0$, known a priori.

The closed-loop system (6.1) is throughout accompanied by a FOSM controller

$$u = -M\text{sign}(s), \tag{6.5}$$

switching on a linear surface $s(x) = 0$, governed by

$$s = x_n + \sum_{k=1}^{n-1} c_k x_k. \tag{6.6}$$

The objective is to demonstrate that such a closed-loop system is capable of not only rejecting matched uniformly bounded disturbances, but also attenuating locally square- integrable (possibly unbounded) disturbances (including mismatched ones) provided that the switching magnitude M and parameters $c_i,\ i = 1, \ldots, n$ of the linear switching surface $s(x) = 0$ are properly tuned. The performance output of interest is specified to $z = [x_1, \ldots, x_{n-1}, s(x)]^T$ in order to bring the sliding variable s into play. The plant dynamics (6.1) are thus represented in the generic form

$$\dot{x} = f(x,t) + g(x,t)w(t), \tag{6.7}$$

$$z = h(x) \tag{6.8}$$

with

$$x = \left[x_1, \ldots, x_n\right]^T, \tag{6.9}$$

$$w = \left[w_1, \ldots, w_n\right]^T, \tag{6.10}$$

$$g(x,t) = diag[d_1(x,t), \ldots, d_n(x,t)], \tag{6.11}$$

$$f(x,t) = \left[x_2, \ldots, x_n, \phi(x,t) - Mb(x,t)\text{sign}\left(x_n + \sum_{k=1}^{n-1} c_k x_k\right)\right]^T, \tag{6.12}$$

$$h(x) = [x_1, \ldots, x_{n-1}, s(x)]^T. \tag{6.13}$$

The parameters of the SM controller (6.5)–(6.6) are subsequently tuned to attenuate external disturbances according to Theorem 5.6.

6.1.1 Tuning Under Full-State Information

For an appropriate parameter tuning, the nonsmooth positive definite function

$$V = \tilde{x}^T P \tilde{x} + |s| \tag{6.14}$$

specified with s, given by (6.6), with $\tilde{x}=[x_1, \ldots, x_{n-1}]^T$, and with some $P = P^T > 0$ to be designed, is utilized to validate Hypothesis (H2) of Theorem 5.6. More specifically, the Hamilton–Jacobi inequality

$$\frac{\partial V}{\partial t} + \frac{\partial V}{\partial x} f(x,t) + \frac{1}{4\gamma^2}\frac{\partial V}{\partial x} g(x,t) g^T(x,t) \left(\frac{\partial V}{\partial x}\right)^T + h^T(x)h(x) < 0 \tag{6.15}$$

is verified with the above function (6.14) side by side out of and, respectively, along the switching surface $s = 0$. Clearly, this yields the complete analysis of the SM system (6.7)–(6.13) to possess the $\mathscr{L}_2$-gain less than a certain disturbance attenuation level $\gamma > 0$.

It is worth noticing that while verifying the Hamilton–Jacobi inequality (6.15) along SMs on the switching surface $s = 0$, it should be specified according to Definition 3.5 to its counterpart

$$\frac{\partial V}{\partial t} + \frac{\partial V}{\partial x} f_0(x,t) + \frac{1}{4\gamma^2}\frac{\partial V}{\partial x} g_0(x,t) g_0^T(x,t) \left(\frac{\partial V}{\partial x}\right)^T + h^T(x)h(x) < 0 \tag{6.16}$$

with the Filippov vector fields $f_0(x,t)$, $g_0(x,t)$, which determine the SMs of the system and which are thus given by relations (3.137), (3.138), (6.11), (6.12). In turn,

$$V\big|_{s=0} = \tilde{x}^T P \tilde{x} \tag{6.17}$$

represents the restriction of (6.14) on the surface $s = 0$.

6.1.1.1 Verification of the Hamilton–Jacobi Inequality Outside the Switching Surface

Substituting (6.14) into the left-hand side of (6.15), specified with (6.11)–(6.13), one has

$$\begin{aligned}
\mathcal{H} &= 2\sum_{k=1}^{n-1}\left[\left(\sum_{j=1}^{n-1} P_{kj}x_j\right)x_{k+1}\right] + \sum_{k=1}^{n-1} c_k x_{k+1}\operatorname{sign}(s) + s^2 \\
&- \left[Mb(x,t) - \phi(x,t)\operatorname{sign}(s) - \frac{1}{4\gamma^2}\left(\sum_{k=1}^{n-1} c_k^2 d_k^2(x,t) + d_n^2(x,t)\right)\right] + x_1^2 + \cdots \\
&+ x_{n-1}^2 + \frac{1}{\gamma^2}\sum_{k=1}^{n-1}\left[d_k^2(x,t)\left(\sum_{j=1}^{n-1} P_{kj}x_j\right)^2 + c_k d_k^2(x,t)\operatorname{sign}(s)\sum_{j=1}^{n-1} P_{kj}x_j\right] \\
&\le 2\sum_{k=1}^{n-1}\left[\left(\sum_{j=1}^{n-1} P_{kj}x_j\right)x_{k+1}\right] + \sum_{k=1}^{n-1} c_k|x_{k+1}| + s^2 \\
&- \left[Mb_0 - |\phi(x,t)| - \frac{1}{4\gamma^2}\left(\sum_{k=1}^{n-1} c_k^2 d_k^2(x,t) - d_n^2(x,t)\right)\right] + x_1^2 + \cdots + x_{n-1}^2 \\
&+ \frac{1}{\gamma^2}\sum_{k=1}^{n-1}\left[d_k^2(x,t)\left(\sum_{j=1}^{n-1} P_{kj}x_j\right)^2 + c_k d_k^2(x,t)\left|\sum_{j=1}^{n-1} P_{kj}x_j\right|\right],
\end{aligned} \tag{6.18}$$

where the Hamiltonian $\mathcal{H}$ stands for the left-hand side of the Hamilton–Jacobi inequality (6.15), thus obtained. Within the domain B_R, given by (6.2), inequality (6.18) is simplified to

$$\begin{aligned}
\mathcal{H} &\le -Mb_0 + L_\phi + \frac{1}{4\gamma^2}\left(\sum_{k=1}^{n-1} c_k^2 L_k^2 + L_n^2\right) \\
&+R^2\left(2\|P\| + \frac{1}{\gamma^2}L_d^2\|P\|^2 + 2c + 1\right) + R\sqrt{n}\left(c + \frac{1}{\gamma^2}cL_d^2\|P\|\right),
\end{aligned} \tag{6.19}$$

where

$$c = \max_k c_k,\ L_d = \max_k L_k, \tag{6.20}$$

$L_\phi, L_i,\ i = 1, \ldots, n$ have been defined in (6.4), and the well-known inequality

$$\left(\sum_{i=1}^{n} a_i\right)^2 \le n\sum_{i=1}^{n} a_i^2 \tag{6.21}$$

has been taken into account. It follows that the Hamiltonian $\mathcal{H}$ is negative definite within the ball B_R provided that the controller gain M is chosen according to

$$M(R) > b_0^{-1}\left[\frac{1}{4\gamma^2}\left(\sum_{k=1}^{n-1} c_k^2 L_k^2 + L_n^2\right) + R^2\left(2\|P\| + \frac{1}{\gamma^2}d^2\|P\|^2 + 2c + 1\right)\right.$$
$$\left. + R\sqrt{n}\left(c + \frac{1}{\gamma^2}cL_d^2\|P\|\right)\right]. \tag{6.22}$$

Thus, under condition (6.22), the corresponding Hamilton–Jacobi inequality (6.15) is shown to locally (within B_R) hold outside the switching surface (6.6).

6.1.1.2 Verification of the Hamilton–Jacobi Inequality Along the Switching Surface

The sliding mode equation, governing the system dynamics on the switching surface (6.6), is obtained by applying the equivalent control method (Utkin 1992), earlier described in Sect. 2.2.2. Thus, if confined to the switching manifold (6.6), the system (6.1) is reduced to the system

$$\begin{aligned} \dot{x}_1 &= x_2 + d_1(x,t)w_1(t) \\ &\vdots \\ \dot{x}_{n-1} &= -\sum_{k=1}^{n-1} c_k x_k + d_{n-1}(x,t)w_{n-1}(t) \end{aligned} \tag{6.23}$$

and its output (6.13) is then specified to

$$z = h(x,t) = [x_1, \ldots, x_{n-1}, 0]^T. \tag{6.24}$$

In turn, the positive definite function (6.14) on the switching surface $s(x) = 0$ takes the form (6.17). Let the parameters $c_1, \ldots, c_{n-1}$ be chosen according to Utkin (1992) to ensure that the internal dynamics of the sliding modes (6.23) (when $w_1 = \cdots = w_{n-1} = 0$) are exponentially stable with an arbitrarily large decay rate $\kappa > 0$. In other words, a positive definite matrix P is designed in such a manner that the time derivative of the Lyapunov function (6.17) along the solution of the disturbance-free system (6.23) satisfies

$$\dot{V}\big|_{s=0} \le -\kappa\|\tilde{x}\|^2 \tag{6.25}$$

with $\kappa > 0$ chosen arbitrarily large. Let us now demonstrate that the Hamilton–Jacobi inequality (6.16), while being specified for the sliding mode equation (6.23), is satisfied with the positive definite function (6.17). By substituting (6.17) into the Hamilton–Jacobi inequality (6.16), specified for (6.23), and taking into account that

$$h^T(x,t)h(x,t) = \sum_{i=1}^{n-1} x_i^2 = \|\tilde{x}\|^2,$$

one derives

$$\mathscr{H} \leq -\kappa\|\tilde{x}\|^2 + \frac{1}{2\gamma^2}L_d^2\|P\|^2\|\tilde{x}\|^2 + \|\tilde{x}\|^2 < 0, \tag{6.26}$$

provided the parameter κ is chosen according to

$$\kappa > \frac{1}{2\gamma^2}L_d^2\|P\|^2 + 1. \tag{6.27}$$

The validity of the Hamilton–Jacobi inequality (6.16) is thus straightforwardly verified on the switching surface (6.6) subject to the parameter choice (6.27).

6.1.1.3 Mismatched Disturbance Attenuation

Since Theorem 5.6 remains applicable to system (6.7)–(6.13), its application in the present circumstances reproduces the result, established in Osuna et al. (2018b).

Theorem 6.1 *Given arbitrary $\gamma > 0$ and radius $R > 0$, let the switching magnitude $M > 0$ be chosen to meet inequality* (6.22) *whereas the surface parameters $c_1, \ldots, c_n$ be such that* (6.25) *holds true with* (6.17) *and κ, satisfying* (6.27). *Then the disturbance-free system* (6.1)–(6.6) *with $w = 0$ is asymptotically stable and its perturbed version possesses $\mathscr{L}_2$-gain less than γ with respect to output* (6.13) *locally within the ball B_R of radius R.*

Proof Being successively specified for system (6.7)–(6.13) outside and along the switching surface $s = 0$, the Hamilton–Jacobi inequality (6.15) and its sliding mode counterpart (6.16) have, respectively, been validated in Sects. 6.1.1.1 and 6.1.1.2. Hypothesis (H2) of Theorem 5.6 is thus shown to be in force for the system in question. Hence, Theorem 5.6 is applicable to system (6.7)–(6.13). The proof is then completed by an appropriate application of Theorem 5.6. ■

6.1.2 Tuning of SM Estimator Gains

In what follows, the state $x_1(t)$ is accepted to be the only available measurement of system (6.1) which is corrupted by a disturbance $w_0(t) \in \mathbb{R}$ of class C^1, i.e., the scalar measurement output is given by

$$y = x_1 + w_0(t). \tag{6.28}$$

The estimator design to be developed is made under extra assumptions that the functions $\phi(x, t)$ and $b(x, t)$ are of class C^1 in the state variable x and on domain (6.2), the magnitudes of their spatial derivatives $\phi_x(x, t)$ and $b_x(x, t)$ possess upper estimates $K_\phi, K_b > 0$ which, generally speaking, depend on the domain radius R and which are known a priori. It is thus ensured that

$$\phi(x,t) - \phi(\hat{x},t) = \Phi(x,\hat{x},t)(x-\hat{x}), \quad b(x,t) - b(\hat{x},t) = B(x,\hat{x},t)(x-\hat{x}) \tag{6.29}$$

where $\Phi = (\Phi_1, \dots, \Phi_n)$, $B = (B_1, \dots, B_n)$, and

$$\Phi_i(x,\hat{x},t) \in [-K_\phi, K_\phi], \quad B_i(x,\hat{x},t) \in [-K_b, K_b], \; i = 1, \dots, n \tag{6.30}$$

for all $x, \hat{x} \in B_R$ and $t \in \mathbb{R}$.

By differentiating (6.28), let us represent the underlying system (6.1) in terms of its output dynamics

$$\begin{aligned} \dot{y} &= x_2 + d_1(x,t)w_1(t) + \dot{w}_0(t) \\ \dot{x}_2 &= x_3 + d_2(x,t)w_2(t) \\ &\vdots \\ \dot{x}_{n-1} &= x_n + d_{n-1}(x,t)w_{n-1}(t) \\ \dot{x}_n &= \phi(x,t) + b(x,t)u + d_n(x,t)w_n(t). \end{aligned} \tag{6.31}$$

The following switched observer

$$\begin{aligned} \dot{\hat{x}}_1 &= \hat{x}_2 + \beta_1 \text{sign}(y - \hat{x}_1) \\ \dot{\hat{x}}_2 &= \hat{x}_3 + \beta_2 \text{sign}(y - \hat{x}_1) \\ &\vdots \\ \dot{\hat{x}}_{n-1} &= \hat{x}_n + \beta_{n-1} \text{sign}(y - \hat{x}_1) \\ \dot{\hat{x}}_n &= \phi(\hat{x},t) + b(\hat{x},t)u + \beta_n \text{sign}(y - \hat{x}_1), \end{aligned} \tag{6.32}$$

is well-recognized (Utkin 1992) to estimate the state $x(t) \in \mathbb{R}^n$ of system (6.1) under matching disturbances.

It is intended to tune the observer parameters $\beta_1, \dots, \beta_n$ to arrive at the estimation errors dynamics of the $\mathscr{L}_2$-gain less than a certain disturbance attenuation level γ even in the presence of mismatched disturbances. Certainly, the above system (6.32) may be viewed as an observer of (6.1) just in the absence of mismatched disturbances provided that the magnitude of the switched observer injection exceeds that of the matched disturbance term $\phi(x,t) + d_n(x,t)w(t)$ to be known a priori; otherwise it should be viewed as a state estimator imposing the desired disturbance attenuation features on the estimation error dynamics.

For the purpose of the $\mathscr{L}_2$-gain tuning, let us rewrite the state estimator (6.32) in terms of the estimation errors

$$\begin{aligned} e_1 &= y - \hat{x}_1 \\ e_2 &= x_2 - \hat{x}_2 \\ &\vdots \\ e_n &= x_n - \hat{x}_n, \end{aligned} \tag{6.33}$$

whose dynamics are clearly governed by

$$\begin{aligned}
\dot{e}_1 &= e_2 - \beta_1 \mathrm{sign}(e_1) + d_1(x,t)w_1(t) + \dot{w}_0(t) \\
\dot{e}_2 &= e_3 - \beta_2 \mathrm{sign}(e_1) + d_2(x,t)w_2(t) \\
&\vdots \\
\dot{e}_{n-1} &= e_n - \beta_{n-1} \mathrm{sign}(e_1) + d_{n-1}(x,t)w_{n-1}(t) \\
\dot{e}_n &= \Sigma_{i=1}^{n}(\Phi_i + uB_i)e_i - \beta_n \mathrm{sign}(e_1) + d_n(x,t)w_n(t) - (\Phi_1 + uB_1)\omega_0(t),
\end{aligned} \tag{6.34}$$

where the nonlinearity representation (6.29) has been utilized. By evaluating the performance of the error dynamics with the error output

$$z_e = h(e,t) = [e_1, \ldots, e_n]^T, \tag{6.35}$$

the estimation error system (6.34)–(6.35) is readily represented in the generic form (6.7), (6.8) where the state vector, the disturbance vector, and the plant matrices are, respectively, given by

$$e = [e_1, \ldots, e_n]^T \tag{6.36}$$

$$w_e = [w_0, w_1, \ldots, w_n, \dot{w}_0]^T \tag{6.37}$$

$$g(e,x,\hat{x},t) = \begin{bmatrix} 0 & d_1 \ 0 \ \ldots \ 0 \ 1 \\ 0 & 0 \ d_2 \ 0 \ \ldots 0 \\ \vdots & 0 \ 0 \ \ddots \ldots 0 \\ -(\Phi_1 + uB_1) & 0 \ \ldots \ 0 \ d_n \ 0 \end{bmatrix} \tag{6.38}$$

$$f(e,x,\hat{x},t) = \begin{bmatrix} e_2 - \beta_1 \mathrm{sign}(e_1) \\ e_3 - \beta_2 \mathrm{sign}(e_1) \\ \vdots \\ e_n - \beta_{n-1} \mathrm{sign}(e_1) \\ \Sigma_{i=1}^{n}(\Phi_i + uB_i)e_i - \beta_n \mathrm{sign}(e_1) \end{bmatrix}. \tag{6.39}$$

In order to apply Theorem 5.6 to the $\mathscr{L}_2$-gain analysis of the estimation error dynamics (6.34) Hypothesis (H2) of Theorem 5.6 is subsequently validated with the nonsmooth positive definite function

$$V_e = |e_1| + \tilde{e}^T Q \tilde{e} \tag{6.40}$$

where $\tilde{e} = [e_2, e_3, \ldots, e_n]^T$, and a symmetric positive definite matrix $Q \in \mathbb{R}^{(n-1)\times(n-1)}$ is to be designed.

6.1.2.1 Verification of the Hamilton–Jacobi Inequality Outside the Estimator Switching Surface

Substituting (6.40) into the left-hand side of (6.15), specified with (6.35)–(6.39), one derives

$$
\begin{aligned}
\mathcal{H}_e &= e_2 \mathrm{sign}(e_1) - \mathrm{sign}^2(e_1)\Big[\beta_1 - \frac{1}{4\gamma^2}\left(d_1^2+1\right)\Big] + e_1^2 + \cdots + e_n^2 \\
&+ 2\sum_{k=2}^{n-1}\sum_{i=2}^{n} Q_{k-1,i-1} e_i e_{k+1} - 2\sum_{k=2}^{n-1}\sum_{i=2}^{n} Q_{k-1,i-1} e_i \beta_k \mathrm{sign}(e_1) \\
&- 2\sum_{i=2}^{n} Q_{n-1,i-1} e_i \Big[\beta_n \mathrm{sign}(e_1) - \Sigma_{i=1}^{n}(\Phi_i + uB_i)e_i\Big] \\
&+ \frac{1}{4\gamma^2}\sum_{k=2}^{n}\Big[2\sum_{i=2}^{n} Q_{k-1,i-1} e_i (d_k - \delta_{kn}(\Phi_1 + uB_1))\Big]^2,
\end{aligned}
\tag{6.41}
$$

where the Hamiltonian $\mathcal{H}_e$ stands for the left-hand side of the Hamilton–Jacobi inequality (6.15), thus obtained, and the Kronecker symbol $\delta_{kn} = 0$ if $k \neq n$ and $\delta_{nn} = 1$. It follows that

$$
\begin{aligned}
\mathcal{H}_e &\le |e_2| - \Big[\beta_1 - \frac{1}{4\gamma^2}\left(L_d^2+1\right)\Big] + 2\sum_{k=2}^{n-1}\sum_{i=2}^{n} \left|Q_{k-1,i-1} e_i e_{k+1}\right| \\
&+2\sum_{k=2}^{n-1}\sum_{i=2}^{n} \left|Q_{k-1,i-1} e_i\right| |\beta_k| + 2\sum_{i=2}^{n}\left|Q_{n-1,i-1} e_i\right| \Big[|\beta_n| + \Sigma_{j=1}^{n}|\Phi_j + uB_j||e_j|\Big] \\
&+\frac{(L_d + |\Phi_1 + uB_1|)^2}{4\gamma^2}\sum_{k=2}^{n}\left(2\sum_{i=2}^{n}\left|Q_{k-1,i-1} e_i\right|\right)^2 + e_1^2 + \cdots + e_n^2,
\end{aligned}
\tag{6.42}
$$

where the upperbounds b^0, L_ϕ, L_d are inherited from (6.3), (6.4). Taking into account (6.21) and applying relations (6.5), (6.29), one concludes that

$$2\sum_{k=2}^{n-1}\sum_{i=2}^{n}\left|Q_{k-1,i-1} e_i e_{k+1}\right| \le \|Q\| R^2(2n-3), \tag{6.43}$$

$$2\sum_{k=2}^{n-1}\sum_{i=2}^{n}\left|Q_{k-1,i-1} e_i\right| |\beta_k| \le 2\beta \|Q\| \sqrt{n}(n-2)R, \tag{6.44}$$

$$2\sum_{i=2}^{n}\left|Q_{n-1,i-1} e_i\right| |\beta_n| \le 2\beta \|Q\| \sqrt{n} R, \tag{6.45}$$

$$2\sum_{i=2}^{n}\left|Q_{n-1,i-1}e_i\right|\left(\Sigma_{j=1}^{n}|\Phi_j+uB_j||e_j|\right)\leq 2(K_\phi+MK_b)\|Q\|nR^2, \tag{6.46}$$

$$\frac{(L_d+|\Phi_1+uB_1|)^2}{4\gamma^2}\sum_{k=2}^{n}\left(2\sum_{i=2}^{n}\left|Q_{k-1,i-1}e_i\right|\right)^2\leq\frac{(L_d+K_\phi+MK_b)^2 n(n-2)}{\gamma^2}\|Q\|^2R^2 \tag{6.47}$$

within the ball $B_R=\left\{e=(e_1,\ldots,e_n)^T\in\mathbb{R}^n: e_1^2+\cdots+e_n^2\leq R^2\right\}$ of radius R, provided that

$$|\beta_k|\leq\beta,\ k=2,\ldots,n. \tag{6.48}$$

By employing the above relations, inequality (6.42) is then reduced to

$$\begin{aligned}\mathscr{H}_e\leq&-\beta_1+\frac{1}{4\gamma^2}\left(L_d^2+1\right)+R\left[2\beta\|Q\|\sqrt{n}+2\beta\|Q\|\sqrt{n}(n-2)+1\right]\\&+R^2\frac{(L_d+K_\phi+MK_b)^2n(n-2)}{\gamma^2}\|Q\|^2\\&+R^2\left[\|Q\|(2n-3)+2(K_\phi+MK_b)\|Q\|n+1\right].\end{aligned} \tag{6.49}$$

Once the estimator parameter $\beta_1=\beta_1(R)$ is chosen large enough, namely,

$$\begin{aligned}\beta_1(R)>&\frac{1}{4\gamma^2}\left(L_d^2+1\right)+R\left[2\beta\|Q\|\sqrt{n}(n-1)+1\right]\\&+R^2\frac{(L_d+K_\phi+MK_b)^2n(n-2)}{\gamma^2}\|Q\|^2\\&+R^2\left[\|Q\|(2K_\phi+2MK_b+2n-3)+1\right]\end{aligned} \tag{6.50}$$

to particularly suppress the upper estimate (6.48) of the other parameters, the Hamiltonian $\mathscr{H}_e$, evaluated by (6.49), proves to be negative definite within the ball B_R. Thus, under the parameter subordination (6.50), the corresponding Hamilton–Jacobi inequality (6.15) is shown to locally hold outside the switching surface $e_1=0$ within the ball B_R.

6.1.2.2 Verification of the Hamilton–Jacobi Inequality on the Estimator Switching Surface

The sliding mode equation, governing the estimator dynamics on the switching surface $e_1=0$, is obtained by applying the equivalent control method, earlier described in Sect. 2.2.2. Thus, if confined to the switching surface $e_1=0$, the estimation error dynamics (6.34) are reduced to

$$\begin{aligned}
\dot{e}_2 &= e_3 + d_2(x,t)w_2(t) - \frac{\beta_2}{\beta_1}\left[e_2 + d_1(x,t)w_1(t) + \dot{w}_0(t)\right] \\
\dot{e}_3 &= e_4 + d_3(x,t)w_3(t) - \frac{\beta_3}{\beta_1}\left[e_2 + d_1(x,t)w_1(t) + \dot{w}_0(t)\right] \\
&\vdots \\
\dot{e}_{n-1} &= e_n + d_{n-1}(x,t)w_{n-1}(t) - \frac{\beta_{n-1}}{\beta_1}\left[e_2 + d_1(x,t)w_1(t) + \dot{w}_0(t)\right] \\
\dot{e}_n &= \Sigma_{i=2}^{n}(\Phi_i + uB_i)e_i + d_n(x,t)w_n(t) - (\Phi_1 + uB_1)w_0(t) \\
&\quad - \frac{\beta_n}{\beta_1}\left[e_2 + d_1(x,t)w_1(t) + \dot{w}_0(t)\right],
\end{aligned} \tag{6.51}$$

and its output (6.35) is, respectively, specified to

$$z_e^0 = h(e) = [0, e_2, \ldots, e_n]^T . \tag{6.52}$$

The resulting system (6.51)–(6.52) is then readily represented in the generic form (6.7), (6.8) with the disturbance vector (6.37), and the state vector and the plant matrices, given by

$$\tilde{e} = \left[e_2, \ldots, e_n\right]^T, \quad f(\tilde{e}, x, \hat{x}, t) = F(x, \hat{x}, t)\tilde{e}, \tag{6.53}$$

$$F(x,\hat{x},t) = \begin{bmatrix} -\frac{\beta_2}{\beta_1} & 1 & 0 & 0 & \cdots & 0 \\ -\frac{\beta_3}{\beta_1} & 0 & 1 & 0 & \cdots & 0 \\ \vdots & & & & & \\ -\frac{\beta_{n-1}}{\beta_1} & 0 & 0 & \cdots & 0 & 1 \\ \Phi_2 + uB_2 - \frac{\beta_n}{\beta_1} & \Phi_3 + uB_3 & \Phi_4 + uB_4 & \cdots & \Phi_{n-1} + uB_{n-1} & \Phi_n + uB_n \end{bmatrix} \tag{6.54}$$

$$g(\tilde{e},x,\hat{x},t) = \begin{bmatrix} 0 & -\frac{\beta_2}{\beta_1}d_1 & d_2 & 0 & 0 & \ldots & 0 & -\frac{\beta_2}{\beta_1} \\ 0 & -\frac{\beta_3}{\beta_1}d_1 & 0 & d_3 & 0 & \ldots & 0 & -\frac{\beta_3}{\beta_1} \\ \vdots & & & & & & & \\ 0 & -\frac{\beta_{n-1}}{\beta_1}d_1 & 0 & \ldots & 0 & d_{n-1} & 0 & -\frac{\beta_{n-1}}{\beta_1} \\ -\Phi_1 - uB_1 & -\frac{\beta_n}{\beta_1}d_1 & 0 & 0 & \ldots & 0 & d_n & -\frac{\beta_n}{\beta_1} \end{bmatrix} \tag{6.55}$$

whereas the positive definite function (6.40) is simplified to

$$V_{\tilde{e}} = \tilde{e}^T Q\tilde{e}. \tag{6.56}$$

By substituting (6.52)–(6.56) into the left-hand side of the Hamilton–Jacobi inequality (6.15), the Hamiltonian of the estimation error system along the switching surface $e_1 = 0$ is derived in the form

$$\mathscr{H}_{\tilde{e}} = \tilde{e}^T \left(F^T Q + QF + \frac{1}{\gamma^2} Qgg^T Q + I \right) \tilde{e} \tag{6.57}$$

It is clear that the Hamiltonian $H_{\tilde{e}}$ is negative definite iff

$$\beta_1^2 \left(F^T Q + QF + \frac{1}{\gamma^2} QGG^T Q + I \right) < 0 \tag{6.58}$$

or, equivalently, iff

$$\beta_1(\Gamma^T Q + Q\Gamma) + \frac{1}{\gamma^2} QGG^T Q + \beta_1^2 I < 0 \tag{6.59}$$

where $\Gamma = \beta_1 F,\ G = \beta_1 g$, and

$$\Gamma = \begin{bmatrix} -\beta_2 & \beta_1 & 0 & 0 & \cdots & 0 \\ -\beta_3 & 0 & \beta_1 & 0 & \cdots & 0 \\ \vdots & & & & & \\ -\beta_{n-1} & 0 & 0 & \cdots & 0 & \beta_1 \\ \beta_1(\Phi_2 + uB_2) - \beta_n & \beta_1(\Phi_3 + uB_3) & \beta_1(\Phi_4 + uB_4) & \cdots & \beta_1(\Phi_{n-1} + uB_{n-1}) & \beta_1(\Phi_n + uB_n) \end{bmatrix},$$

$$G = \begin{bmatrix} 0 & -\beta_2 d_1 & \beta_1 d_2 & 0 & 0 & \cdots & 0 & -\beta_2 \\ 0 & -\beta_3 d_1 & 0 & \beta_1 d_3 & 0 & \cdots & 0 & -\beta_3 \\ \vdots & & & & & & & \\ 0 & -\beta_{n-1} d_1 & 0 & \cdots & 0 & \beta_1 d_{n-1} & 0 & -\beta_{n-1} \\ -\beta_1(\Phi_1 + uB_1) & -\beta_n d_1 & 0 & 0 & \cdots & 0 & \beta_1 d_n & -\beta_n \end{bmatrix}.$$

Thus, the Hamilton–Jacobi inequality (6.15), viewed along the estimator switching surface $e_1 = 0$ and specified with (6.52)–(6.56), holds true provided that there exist a symmetric positive definite matrix $Q > 0$ and estimator gains $\beta_2, \ldots, \beta_n$ such that (6.59) is negative definite with a parameter β_1, prespecified according to (6.50).

Since by virtue of the above representation of the matrices Γ and G, inequality (6.59) proves to be quadratic in the parameters $\beta_1, \ldots, \beta_n$ to be identified the feasibility of (6.59) is readily reduced by Schur lemma to the feasibility of the corresponding LMIs. Similar to Boyd et al. (1994), the feasibility of the latter LMIs should be viewed only in the vertices (6.30) of the functions $\Phi(x, \hat{x}, t)$ and $B(x, \hat{x}, t)$ under the maximal control magnitude $u = M$, i.e., the extremal values $\pm K_\phi$ and $\pm K_b$ should be substituted for $\Phi_i,\ B_i,\ i = 1, \ldots, n$ into the LMIs, thus obtained, and the resulting LMIs should be coupled to (6.50) to be solved in $\beta_1, \ldots, \beta_n$ for some $\gamma > 0$. This observation allows one to apply the standard MATLAB package to tune the estimator gains for attenuating the mismatched disturbances lower than a prespecified level γ.

6.1.2.3 Estimator Robustness Against Mismatched Disturbances

The desired robustness features of the SM estimator are summarized from Osuna et al. (2018b).

Theorem 6.2 *Within the domain B_R, consider system* (6.1)*, driven by* (6.5)*. A switched estimator* (6.32) *is then designed, based on the system measurement* (6.28)*. Given γ, let there exist estimator gains $\beta_1, \ldots, \beta_n$ and positive definite matrix Q such that the parameter subordination* (6.50) *holds true and the quadratic inequality* (6.59) *is satisfied in the vertices* (6.30) *of the functions $\Phi(x, \hat{x}, t)$ and $B(x, \hat{x}, t)$ under the extremal values $\pm K_\phi$, $\pm K_b$ and the maximal control magnitude M, being substituted for Φ_i, B_i, $i = 1, \ldots, n$ and for u, respectively. Then the estimation error system* (6.34) *is internally asymptotically stable and it locally possesses $\mathscr{L}_2$-gain less than γ with respect to the error output* (6.35).

Proof The Hamilton–Jacobi inequality (6.15) and its sliding mode counterpart (6.16), specified for the estimation error system (6.34) outside and, respectively, along the switching surface $e_1 = 0$, have been validated in Sects. 6.1.2.1 and 6.1.2.2 with the positive definite function (6.40). Hypothesis (H2) of Theorem 5.6 is thus shown to be in force for the system in question. The applicability of Theorem 5.6 to the estimation error system (6.34) is then straightforwardly concluded. By applying Theorem 5.6 to (6.34) the proof is completed. ■

6.1.3 Tuning Under Incomplete State Information

In order to synthesize the output controller, fed by the only available system measurement (6.28), the state feedback law (6.5) is modified to the following one:

$$u = -M\text{sign}(\hat{s}) \tag{6.60}$$

where in contrast to (6.6), the switching surface

$$\hat{s}(x, e) = (x_n - e_n) + \sum_{k=1}^{n-1} c_k(x_k - e_k) = \hat{x}_n + \sum_{k=1}^{n-1} c_k\hat{x}_k \tag{6.61}$$

relies on the output $\hat{x} = (\hat{x}_1, \ldots, \hat{x}_n)^T$ of the proposed switched estimator (6.32), thereby utilizing the measured output (6.28) only. The system performance (6.13) is then updated to

$$z = [y, x_2, \ldots, x_{n-1}, \hat{s}, e_1, \ldots, e_n]^T. \tag{6.62}$$

The resulting SM output feedback synthesis couples derivations of Sects. 6.1.1 and 6.1.2 together to formalize it in the form, reminiscent of a separation principle.

Theorem 6.3 *Consider system* (6.1)*, driven by the switched output feedback controller* (6.60), (6.61)*, which is fed by the state estimator* (6.32)*, running in parallel. Given an arbitrary* $\gamma > 0$ *and the domain* B_R *of radius* $R > 0$*, suppose that the controller magnitude* $M > 0$ *is selected to meet inequality* (6.22) *whereas the surface parameters* $c_1, \dots, c_n$ *are such that* (6.25) *holds true with* (6.17) *and* κ*, satisfying* (6.27)*. Moreover, let a positive definite matrix* Q *and estimator gains* $\beta_1, \dots, \beta_n$ *be chosen to ensure that the parameter subordination* (6.50) *holds true and the quadratic inequality* (6.59) *is satisfied in the vertices* (6.30) *of the functions* $\Phi(x, \hat{x}, t)$ *and* $B(x, \hat{x}, t)$ *under the extremal values* $\pm K_\phi$, $\pm K_b$ *and the maximal control magnitude* M*, being substituted for* Φ_i, B_i, $i = 1, \dots, n$ *and for* u*, respectively. Then the closed-loop system* (6.1), (6.32), (6.60), (6.61) *is internally asymptotically stable and it possesses* $\mathcal{L}_2$*-gain less than* γ *with respect to outputs* (6.35), (6.62)*, locally within the ball* B_R *of radius* R.

Proof is rather technical and it is based on the validating the corresponding Hamilton–Jacobi inequalities (6.15), (6.16) outside and along the switching surfaces $\hat{s} = 0$ and/or $e_1 = 0$ with the positive definite function

$$V_0 = \tilde{x}^T P \tilde{x} + |\hat{s}| + |e_1| + \tilde{e}^T Q \tilde{e} \tag{6.63}$$

viewed over the closed-loop system (6.1), (6.32), (6.60), (6.61). The proof follows the same line of reasoning that has been used for Theorems 6.1 and 6.2, and it is, therefore, omitted. The reader may refer to Osuna et al. (2018b, Appendix A) for the detailed proof.

The effectiveness of the tuning procedure, resulting from Theorem 6.3, may be found in Osuna et al. (2018b) where the closed-loop performance is tested for a laboratory-inverted pendulum.

6.2 $\mathcal{L}_2$-Gain Tuning of Second-Order Sliding Modes

The $\mathcal{L}_2$-gain tuning is subsequently tested for SOSM feedback synthesis where the twisting state feedback controller is used along with the supertwisting algorithm which is implemented to design a velocity observer, running in parallel to feed the twisting controller. The tests to be conducted should demonstrate that SOSMs are also capable along with rejecting matched bounded disturbances to attenuate mismatched (possibly unbounded) disturbances of class L_2.

6.2.1 *Tuning of Twisting Controller*

To begin with, the $\mathcal{L}_2$-gain analysis is developed for the twisting controller

$$u = -k_1 \text{sign}(x_1) - k_2 \text{sign}(x_2) \tag{6.64}$$

applied to the double integrator

$$\dot{x}_1 = x_2, \quad \dot{x}_2 = u + \omega_u \tag{6.65}$$

only affected by a piecewise continuous matched disturbance $\omega_u \in \mathbb{R}$ what is actually the case of fully actuated electromechanical plants. The gains $k_1, k_2 \in \mathbb{R}^+$ of the state feedback twisting controller (6.64) should be selected to ensure a certain disturbance attenuation level. By Theorem 5.1, the twisting system (6.64), (6.65) is globally asymptotically stable if

$$\sup_{t \geq 0} |\omega_u(t)| < k_2 < k_1 - k_2.$$

Setting $x = \left[x_1 \; x_2\right]^T$, the following state-space representation of (6.64), (6.65)

$$\dot{x} = \begin{bmatrix} x_2 \\ -k_1 \text{sign}(x_1) - k_2 \text{sign}(x_2) + \omega_u \end{bmatrix} \tag{6.66}$$

is obtained and coupled to the preselected output

$$z = x \tag{6.67}$$

to arrive at the generic representation (6.7), (6.8) specified with

$$\varphi(x) = \begin{bmatrix} x_2 \\ -k_1 \text{sign}(x_1) - k_2 \text{sign}(x_2) \end{bmatrix}, \tag{6.68}$$

$$\psi(x) = \left[0 \; 1\right]^T, \tag{6.69}$$

$$h(x) = \left[x_1 \; x_2\right]^T. \tag{6.70}$$

To meet Hypothesis (H2) of Theorem 5.6 the Hamilton–Jacobi inequality (3.130) is verified with the following positive definite function:

$$V = \frac{1}{2}(x_1 + x_2)^2 + k_1 |x_1|. \tag{6.71}$$

Recall that such a kind of a Lyapunov function has been used in the proof of Theorem 5.1 on the FTS of the twisting algorithm.

Remark 6.1 It is clear that the closed-loop system (6.66) possesses no sliding modes on the discontinuity surface $x_1 = 0$ but the origin whereas on the discontinuity surface $x_2 = 0$, sliding modes, if any, turn out to be disturbance free because only matched disturbances are admitted. Thus, the $\mathscr{L}_2$-gain inequality (3.157) is trivially satisfied along sliding modes so that no need to verify the Hamilton–Jacobi inequality (6.16) on the discontinuity surfaces.

Substituting (6.68)–(6.71) in the left-hand side of (3.130) yields

$$\mathscr{H} = x_1 x_2 + x_2^2 - k_1|x_1| - k_2 x_1 \operatorname{sign}(x_2) - k_2|x_2| + \frac{1}{4\gamma^2}(x_1 + x_2)^2 + x_1^2 + x_2^2. \tag{6.72}$$

Then letting

$$k_1 > k_2, \tag{6.73}$$

and setting

$$\lambda_1 = \min\{k_1 - k_2, k_2\}, \qquad \lambda_2 = \frac{5}{2} + \frac{1}{2\gamma^2}, \tag{6.74}$$

one derives the inequality

$$\mathscr{H} \le -(\lambda_1 - \lambda_2\|x|)\,\|x\|, \tag{6.75}$$

which validates the negative definiteness of the Hamiltonian $\mathscr{H}$ locally within the ball

$$B_{R_c} = \left\{x \in \mathbb{R}^2 : \|x\| < R_c\right\} \tag{6.76}$$

of a radius R_c such that

$$\lambda_1 > \lambda_2 R_c. \tag{6.77}$$

Hypothesis (H2) is thus locally verified for all $x \in B_{R_c} \subset \mathbb{R}^2$ and by applying Theorem 5.6 to the twisting system (6.66), the following result (Osuna et al. 2018a) is then obtained.

Theorem 6.4 *Given arbitrary $\gamma > 0$ and radius $R_c > 0$, let the controller gains k_1 and k_2 be chosen to ensure that inequalities* (6.73), (6.77), *coupled to* (6.74), *hold true. Then the closed-loop system* (6.66) *is internally asymptotically stable and it possesses $\mathscr{L}_2$-gain less than γ with respect to the output* (6.67), *locally within the ball B_{R_c} of radius R_c.*

Proof Since Hypothesis (H2) of Theorem 5.6 has already been verified under conditions of the present theorem, the assertion of Theorem 6.4 is straightforwardly validated by Theorem 5.6, being applied to the twisting system (6.66). ■

6.2.2 Tuning of Supertwisting Estimator

The next investigation focuses on the $\mathscr{L}_2$-gain analysis of the supertwisting estimator

$$\dot{\tilde{x}}_1 = \tilde{x}_2 + m_1|y - \tilde{x}_1|^{\frac{1}{2}}\mathrm{sign}(y - \tilde{x}_1),$$
$$\dot{\tilde{x}}_2 = u + m_2\mathrm{sign}(y - \tilde{x}_1), \tag{6.78}$$

whose states $\tilde{x}_1$ and $\tilde{x}_2$ estimate the position x_1 of system (6.66) and its velocity x_2 based on the available measurement $y = x_1 + \omega_o$, corrupted by the measurement error $\omega_o \in \mathbb{R}$ whose second-order time derivative $\ddot{\omega}_o$ is assumed to be piecewise continuous. The estimator gains $m_1, m_2 \in \mathbb{R}^+$ are to be selected to ensure a certain disturbance attenuation level.

Setting the estimation error $e = \left[e_1\ e_2\right]^T$ as

$$e_1 = y - \tilde{x}_1, \quad e_2 = \dot{y} - \tilde{x}_2, \tag{6.79}$$

the following second-order estimation error system

$$\dot{e} = \begin{bmatrix} e_2 - m_1|e_1|^{\frac{1}{2}}\mathrm{sign}(e_1) \\ -m_2\mathrm{sign}(e_1) + \omega_u + \ddot{\omega}_o \end{bmatrix} \tag{6.80}$$

is obtained with the generic matched disturbance

$$\omega = \omega_u + \ddot{\omega}_o. \tag{6.81}$$

By Theorem 5.4, the supertwisting system (6.80) proves to be globally asymptotically stable provided that

$$\sup_{t\geq 0} |[\omega_u(t) + \ddot{\omega}_o(t)]| < \min\left\{\frac{1}{2}m_1, \frac{m_1 m_2}{1 + m_1}\right\}.$$

The subsequent $\mathscr{L}_2$-gain analysis of the estimation error system is coupled to the preselected output

$$z = e, \tag{6.82}$$

and system (6.80) is thus represented in the generic form (6.7), (6.8), specified with $x = e$ and

$$\varphi(e) = \begin{bmatrix} e_2 - m_1|e_1|^{\frac{1}{2}}\mathrm{sign}(e_1) \\ -m_2\mathrm{sign}(e_1) \end{bmatrix}, \tag{6.83}$$
$$\psi(e) = \left[0\ 1\right]^T, \tag{6.84}$$
$$h(e) = \left[e_1\ e_2\right]^T. \tag{6.85}$$

The Hamilton–Jacobi inequality (3.130), specified with (6.83)–(6.85), is now verified for the positive definite function

$$V = 2m_2|e_1| + \frac{1}{2}e_2^2 + \frac{1}{2}r^2(e_1, e_2) \tag{6.86}$$

where

$$r(e_1, e_2) = e_2 - m_1|e_1|^{\frac{1}{2}}\text{sign}(e_1). \tag{6.87}$$

Recall that such a Lyapunov function has been used in the proof of Theorem 5.4 on the FTS of the supertwisting algorithm.

It should be pointed out that Remark 6.1 applies here as well. With this in mind, substituting (6.83)–(6.85) and (6.86) into the left-hand side of the Hamilton–Jacobi inequality (3.130) yields

$$\begin{aligned}\mathscr{H} = -\frac{1}{2}m_1 r^2|e_1|^{-\frac{1}{2}} + m_2 r\text{sign}(e_1) - m_2 e_2\text{sign}(e_1)\\ + \frac{1}{4\gamma^2}(e_2 + r)^2 + e_1^2 + e_2^2.\end{aligned} \tag{6.88}$$

Then setting

$$\lambda_3 = 4\left(1 + \frac{1}{\gamma^2}\right) \tag{6.89}$$

and using (6.87), it follows that

$$\mathscr{H} \le -\left(m_1 m_2 - \frac{1}{4}\lambda_3 m_1^2|e_1|^{\frac{1}{2}} - |e_1|^{\frac{3}{2}}\right)|e_1|^{\frac{1}{2}} - \frac{1}{2}\left(m_1 - \lambda_3|e_1|^{\frac{1}{2}}\right)r^2(e_1, e_2)|e_1|^{-\frac{1}{2}}. \tag{6.90}$$

The latter inequality validates the negative definiteness of the Hamiltonian $\mathscr{H}$ within the strip

$$\mathscr{S}_{R_o} = \left\{e \in \mathbb{R}^2 : |e_1| < R_o^2\right\} \tag{6.91}$$

of some width $R_o > 0$ provided that

$$m_1 > \lambda_3 R_o, \tag{6.92}$$

$$m_2 > \frac{1}{4}\lambda_3 m_1 R_o + \frac{R_o^3}{m_1}. \tag{6.93}$$

Hypothesis (H2) of Theorem 5.6 is thus locally verified for all $e \in \mathscr{S}_{R_o} \subset \mathbb{R}^2$. By applying Theorem 5.6, the following result is reproduced from Osuna et al. (2018a).

Theorem 6.5 *Given arbitrary $\gamma > 0$ and radius $R_o > 0$, let the estimator gains m_1 and m_2 be chosen to ensure that inequalities* (6.92), (6.93), *coupled to* (6.89), *hold true. Then the estimation error system* (6.80) *is internally asymptotically stable and it possesses $\mathscr{L}_2$-gain less than γ with respect to the error output* (6.82), *locally within the strip $\mathscr{S}_{R_o}$ given by* (6.91).

Proof Hypothesis (H2) of Theorem 5.6 has already been verified to be in force for the estimation error system (6.80). Thus, Theorem 5.6 proves to be applicable to system (6.80), and the validity of Theorem 6.5 is then straightforwardly concluded from Theorem 5.6. ■

6.2.3 *Output Feedback Tuning*

Implementing the twisting controller (6.64), fed by the supertwisting velocity estimator (6.78), running in parallel, results in the closed-loop position feedback system (6.65), governed by

$$\dot{x}_1 = x_2, \quad \dot{x}_2 = -k_1 \text{sign}(y) - k_2 \text{sign}(\tilde{x}_2) + \omega_u. \tag{6.94}$$

The overall closed-loop system (6.80), (6.94) to be analyzed next admits the following state-space representation

$$\begin{aligned} \dot{y}_1 &= y_2, \\ \dot{y}_2 &= -k_1 \text{sign}(y_1) - k_2 \text{sign}(y_2 - e_2) + \omega, \\ \dot{e}_1 &= e_2 - m_1 |e_1|^{\frac{1}{2}} \text{sign}(e_1), \\ \dot{e}_2 &= -m_2 \text{sign}(e_1) + \omega, \end{aligned} \tag{6.95}$$

$$z = \begin{bmatrix} y_1 \ y_2 \ e_1 \ e_2 \end{bmatrix}^T, \tag{6.96}$$

in terms of $y_1 = y$, $y_2 = \dot{y}$, the estimation error (6.79), the performance output z, and the generic disturbance ω, given by (6.81). The generic representation (6.7), (6.8) is thus obtained with the state variable $x = z$ and

$$\varphi = \begin{bmatrix} y_2 \\ -k_1 \text{sign}(y_1) - k_2 \text{sign}(y_2 - e_2) \\ e_2 - m_1 |e_1|^{\frac{1}{2}} \text{sign}(e_1) \\ -m_2 \text{sign}(e_1) \end{bmatrix}, \tag{6.97}$$

$$\psi = \begin{bmatrix} 0 \ 1 \ 0 \ 1 \end{bmatrix}^T, \tag{6.98}$$

$$h = \begin{bmatrix} y_1 \ y_2 \ e_1 \ e_2 \end{bmatrix}^T. \tag{6.99}$$

The $\mathscr{L}_2$-gain analysis is successively applied beyond and on the discontinuity manifolds

$$S_c = \left\{(y_1, y_2, e_1, e_2) \in \mathbb{R}^4 : y_1 = y_2 = 0\right\}, \tag{6.100}$$

$$S_o = \left\{(y_1, y_2, e_1, e_2) \in \mathbb{R}^4 : e_1 = e_2 = 0\right\}. \tag{6.101}$$

It is worth noticing that sliding modes occur on the former manifold when the available measurement y and its time derivative reach the origin and remain there so that the $\mathcal{L}_2$-gain analysis is confined to that of the estimator. In turn, on the latter manifold, sliding modes occur when the estimation errors are nullified and the output feedback formally coincides with the state feedback.

6.2.3.1 $\mathcal{L}_2$-Gain Analysis Beyond the Discontinuity Manifolds

Conditions of Theorem 5.6 are first verified beyond the discontinuity manifolds (6.100) and (6.101). For this purpose, the positive definite function

$$V = \frac{1}{2}(y_1 + y_2)^2 + k_1|y_1| + 2m_2|e_1| + \frac{1}{2}e_2^2 + \frac{1}{2}r^2(e_1, e_2) \tag{6.102}$$

is introduced with $r(e_1, e_2)$ given by (6.87).

By substituting (6.97)–(6.99) and (6.102) into the Hamilton–Jacobi inequality (6.15), its left-hand side takes the form

$$\begin{aligned}
\mathcal{H} = {} & y_1 y_2 + y_2^2 + k_1 y_2 \text{sign}(y_1) - k_1|y_1| - k_2 y_1 \text{sign}(y_2 - e_2) \\
& -k_1 y_2 \text{sign}(y_1) - k_2 y_2 \text{sign}(y_2 - e_2) + 2m_2 r \text{sign}(e_1) \\
& -\frac{1}{2}m_1 r^2 |e_1|^{-\frac{1}{2}} - m_2 e_2 \text{sign}(e_1) - m_2 r \text{sign}(e_1) \\
& +\frac{1}{2\gamma^2}(y_1 + y_2 + e_2 + r)^2 + y_1^2 + y_2^2 + e_1^2 + e_2^2.
\end{aligned} \tag{6.103}$$

Employing the well-known inequalities $\left(\sum_{i=1}^n a_i\right)^2 \le 2\sum_{i=1}^n a_i^2$, $|a_1 - a_2| \ge |a_1| - |a_2|$, $a_1 a_2 \le \frac{1}{2}a_1^2 + \frac{1}{2}a_2^2$ and substituting $e_2 = r(e_1, e_2) + m_1|e_1|^{\frac{1}{2}}\text{sign}(e_1)$, resulting from (6.87), it follows that

$$\begin{aligned}
\mathcal{H} \le {} & \left(\frac{3}{2} + \frac{1}{\gamma^2}\right)y_1^2 - (k_1 - k_2)|y_1| + \left(\frac{5}{2} + \frac{1}{\gamma^2}\right)y_2^2 - k_2|y_2| \\
& - \frac{1}{2}(m_1 - 2k_2)\, r^2(e_1, e_2)|e_1|^{-\frac{1}{2}} + \left(2 + \frac{4}{\gamma^2}\right)r^2(e_1, e_2) + \left(2m_1^2 + \frac{1}{\gamma^2}m_1^2\right)|e_1| \\
& + e_1^2 - (m_1 m_2 - 2k_2 m_1 - k_2)\,|e_1|^{\frac{1}{2}}.
\end{aligned} \tag{6.104}$$

Taking into account relations (6.74) and (6.89), determining the parameters λ_1, λ_2, and λ_3, one then derives the inequality

$$\mathscr{H} \le -\left(\lambda_1 - \lambda_2\sqrt{y_1^2 + y_2^2}\right)\sqrt{y_1^2 + y_2^2} - \left(\frac{1}{2}m_1 - 2k_2 - \frac{1}{2}\lambda_3|e_1|^{\frac{1}{2}}\right) \times$$
$$r^2(e_1, e_2)|e_1|^{-\frac{1}{2}} - \left(m_1 m_2 - 2k_2 m_1 - k_2 - \frac{1}{2}\lambda_3 m_1^2|e_1|^{\frac{1}{2}} - |e_1|^{\frac{3}{2}}\right)|e_1|^{\frac{1}{2}}, \quad (6.105)$$

whose right-hand side is negative definite locally in the region

$$\mathscr{R} = \left\{(y_1, y_2, e_1, e_2) \in \mathbb{R}^4 : y_1^2 + y_2^2 < R_c^2 \text{ and } |e_1| < R_o^2\right\} \quad (6.106)$$

provided that relations (6.73), (6.77) hold and

$$\begin{aligned} m_1 &> 4k_2 + \lambda_3 R_o \\ m_2 &> 2k_2 - \frac{k_2}{m_1} - \frac{1}{2}\lambda_3 m_1 R_o - \frac{R_o^3}{m_1}. \end{aligned} \quad (6.107)$$

The Hamilton–Jacobi inequality (3.130) is thus locally verified and Hypothesis (H2) is locally validated beyond the discontinuity manifolds (6.100) and (6.101) within the region $\mathscr{R} \subset \mathbb{R}^4$.

6.2.3.2 $\mathscr{L}_2$-Gain Analysis on the Discontinuity Manifolds S_c and S_o

Once the trajectories of (6.95) evolve in the sliding mode along the discontinuity manifold (6.100), the underlying system (6.95) is reduced to the estimation error system of the second order which is given by (6.80), and the $\mathscr{L}_2$-gain analysis of this resulting system has been summarized in Theorem 6.5.

Once the trajectories of (6.95) evolve in the sliding mode along the discontinuity manifold (6.101), the underlying system (6.95) is reduced to the one of the second order which is given by (6.66), driven by the twisting state feedback (6.64), and Theorem 6.4 has validated the $\mathscr{L}_2$-gain analysis of the resulting sliding mode system.

6.2.3.3 Overall $\mathscr{L}_2$-Gain Analysis

Coupling the above derivations of Sects. 6.2.3.1 and 6.2.3.2 to Theorems 6.4 and 6.5 summarizes the $\mathscr{L}_2$-gain analysis of the overall output feedback system (6.95), proposed in Osuna et al. (2018a).

Theorem 6.6 *Given arbitrary $\gamma > 0$, and $R_c > 0$, and $R_o > 0$, let the controller and estimator gains k_1, k_2 and m_1, m_2 be chosen to ensure that inequalities* (6.73), (6.77), (6.107), *coupled to* (6.74) *and* (6.89), *hold true. Then, the overall system* (6.95) *is internally asymptotically stable and it possesses $\mathscr{L}_2$-gain less than γ with respect to the performance output* (6.96) *locally within the region $\mathscr{R}$, governed by* (6.106).

Proof Being coupled to Theorems 6.4 and 6.5, the $\mathscr{L}_2$-gain analysis of Sects. 6.2.3.1 and 6.2.3.2 verifies that Hypothesis (H2) of Theorem 5.6 is in force for the state-space representation (6.95), (6.96) of the system in question. Thus, Theorem 5.6 proves to be applicable to the overall system (6.95), (6.96), and the validity of Theorem 6.6 is then straightforwardly concluded from Theorem 5.6. ■

Experimental evidences, which support capabilities of the tuning procedure, resulting from Theorem 6.6, are given in Osuna et al. (2018a) where the closed-loop performance is tested for an industrial DC motor.

6.3 Settling Time Tuning of Enforced Double Integrator

The task of finite time stabilization, accompanied by a requirement to achieve a desired settling time is far from being trivial unless a specific finite time stable Lyapunov function (similar to that of Lemma 4.1) is available for the closed-loop system. In general, achieving the desired settling time without having a finite time stable Lyapunov function is actually a challenging problem and it is illustrated for a simple double integrator. The underlying methodology, inherited from Oza et al. (2011), Perruquetti et al. (2003) for settling the double integrator at the origin in the desired finite time, is to initially utilize globally exponentially stable linear feedback until the closed-loop trajectories enter an arbitrarily prespecified domain of attraction and then switch it to the finite time stabilizing twisting controller. Tuning rules are then explicitly derived to achieve the desired settling time.

Let the open-loop system dynamics be given as follows:

$$\begin{aligned} \dot{x}_1 &= x_2, \\ \dot{x}_2 &= u(x_1, x_2) + \omega(t), \end{aligned} \tag{6.108}$$

where $x = (x_1, x_2)^T$ is the state of the system, $\omega(t)$ is a uniformly bounded disturbance, and u is a control input. It is assumed that (i) all the states are available for feedback and (ii) an upper bound $N > 0$ on the disturbance term $|\omega| \leq N$ is known a priori.

The control aim is threefold:

1. To synthesize a finite time stabilizing control law $u(x_1, x_2)$.
2. To establish an upper bound $\mathscr{T}_s$ on the settling time of the closed-loop system (6.108) such that $x_1(t) = x_2(t) = 0$ for all $t \geq \mathscr{T}_s$.
3. To establish tuning rules for the parameters of the controller to achieve the desired settling time.

6.3.1 Switched Control Synthesis

The following variable structure feedback control law is proposed:

$$u(x_1, x_2) = \begin{cases} Lx, & x \notin \Gamma_R \\ \Phi(x), & x \in \Gamma_R \end{cases} \tag{6.109}$$

where

$$\begin{aligned} L &= \begin{bmatrix} -l_1 & -l_2 \end{bmatrix}, \\ \Phi(x) &= -\mu_2 \ \mathrm{sign}(x_1) - \mu_1 \ \mathrm{sign}(x_2), \\ \Gamma_R &= \{x : V(x_1, x_2) \leq R\}, \end{aligned} \tag{6.110}$$

and

$$V(x_1, x_2) = \mu_2 |x_1| + \frac{1}{2} x_2^2. \tag{6.111}$$

The parameters $l_1, l_2, \mu_1, \mu_2, R$ are positive scalars. The condition

$$0 < N < \mu_1 < \mu_2 - N \tag{6.112}$$

is imposed on the twisting controller gains to enforce finite time stability of the closed-loop system to the origin where N is a positive scalar. Furthermore, $R > 0$ can be chosen arbitrarily small or large. The proportional feedback gains l_1, l_2 represent the traditional state feedback. Figure 6.1 graphically depicts the new switched control law and the finite time stability of the origin. The point O_1 shows the arbitrary initial condition.

The underlying philosophy of the above switched control synthesis is the successive application of the classical linear state feedback law and the second-order

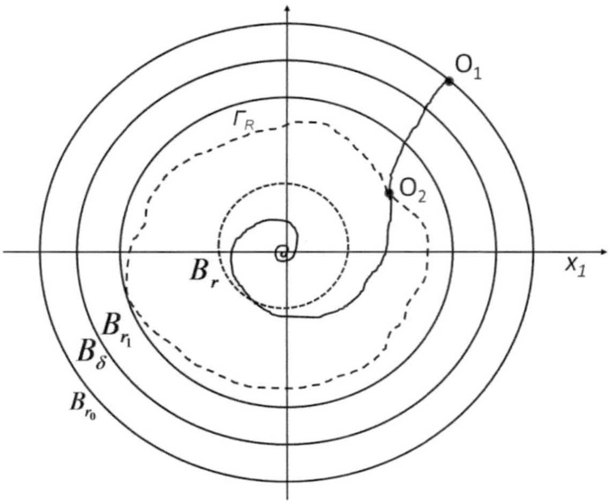

Fig. 6.1 Concept of finite settling time by utilizing step-by-step application of the linear controller and the twisting one, level set Γ_R, ball B_r, the outer ball B_{r_1} such that $\Gamma_R \subset B_{r_1}$ and the initial condition region B_{r_0}. © 2012 Oxford University Press. Reprinted from Oza et al. (2011) by permission of Oxford University Press

sliding mode controller. Following Oza et al. (2011), the resulting tuning rules are shown to guarantee the attainment of a specified settling time. Since the open-loop system (6.108) is controllable, there exist feedback gains l_1, l_2 such that the trajectories of the unperturbed closed-loop system exponentially decay to the origin with a desired rate. In turn, while being perturbed by an admissible uniformly bounded disturbance $\omega(t)$, the system states $x_1(t), x_2(t)$ are steered, as $t \to \infty$, to a ball $B_r = \{(x_1, x_2) : x_1^2 + x_2^2 \le r^2\}$ with an arbitrarily small radius r, which is determined by the gains l_1, l_2. It can also be deduced that the trajectories of the properly tuned closed-loop system enter the level set Γ_R of the function $V(x_1, x_2)$ in finite time (at some point O_2) with a desired exponential decay rate provided that B_r is small enough to be inside Γ_R so that the relation $B_r \cap \partial \Gamma_R = \emptyset$ holds true where $\partial \Gamma_R = \{(x_1, x_2) : \mu_2 |x_1| + \frac{1}{2} x_2^2 = R\}$ stands for the boundary of the level set Γ_R.

A switch from the linear state feedback control law to the twisting controller is then introduced at the time instant when the trajectory reaches the level set Γ_R. As established in Sect. 5.2.1, the function $V(x_1, x_2)$ is indeed a valid Lyapunov function when a twisting controller is utilized for feedback. Moreover, the trajectories of the closed-loop system can never leave the level set Γ_R of the Lyapunov function $V(x_1, x_2)$ once they are inside Γ_R. The fact that the temporal derivative of the Lyapunov function satisfies $\dot{V} \le 0 \, \forall x \in \Gamma_R$ along the trajectories of the closed-loop system is the principal argument behind choosing the level set Γ_R for switching from the linear to the twisting controller.

It should be noted that the settling time estimate and tuning algorithm, which are carried out in the sequel, remains valid even in the absence of the linear feedback. Such a pure twisting algorithm might be superior to that combined with the linear feedback, e.g., in the presence of small switching delays, which may provoke sliding modes on the switching set Γ_R, where a switch from the linear feedback to the twisting controller takes place. The price one should pay for that is slower convergence rate when the system is far from the origin, and in order to achieve the desired settling time higher twisting gains should be utilized.

6.3.2 Reaching Time Estimate of Linear Feedback

The well-known method of finding the explicit solution of the second-order linear time-invariant system in the canonical form is exploited and briefly outlined in this section. The unperturbed closed-loop system (6.108) with the feedback law $u(x_1, x_2) = Lx$ is given by

$$\begin{aligned} \dot{x}_1 &= x_2, \\ \dot{x}_2 &= -l_1 x_1 - l_2 x_2. \end{aligned} \tag{6.113}$$

Letting

$$l_1 = \alpha \lambda^2, \quad l_2 = (\alpha + 1) \lambda, \quad \alpha > 1, \lambda > 0$$

the eigenvalues

$$\lambda_1 = -\lambda, \quad \lambda_2 = -\alpha\,\lambda$$

are imposed on the closed-loop system. Then the general solution of (6.113) is represented in the form

$$\begin{aligned} x_1(t) &= \left[\alpha x_{1_0} + \frac{1}{\lambda} x_{2_0}\right] e^{-\lambda t} - \left[x_{1_0} + \frac{1}{\lambda} x_{2_0}\right] e^{-\alpha\,\lambda\,t}, \\ x_2(t) &= -\left[\alpha\lambda\, x_{1_0} + x_{2_0}\right] e^{-\lambda t} + \alpha\left[\lambda\, x_{1_0} + x_{2_0}\right] e^{-\alpha\,\lambda\,t}, \end{aligned} \tag{6.114}$$

where $x_1^0 = x_1(0),\ x_2^0 = x_2(0)$ are the initial conditions of the system (6.113), determining an initial condition ball $B_{r_0} = \sqrt{\left(x_1^0\right)^2 + \left(x_2^0\right)^2}$. Thus, estimating the reaching time $\mathscr{T}_1$ for the level set

$$\Gamma_{R_1} = \{(x_1, x_2) : V(x_1, x_2) \le R_1\}$$

of the Lyapunov function

$$V(x_1, x_2) = \mu_2 |x_1| + \frac{1}{2} x_2^2$$

viewed on the solutions (6.114), which are initialized within the ball B_{r_0}, yields

$$\begin{aligned} V(x_1, x_2) &\le \mu_2 \left|\alpha x_{1_0} + \frac{1}{\lambda} x_{2_0}\right| e^{-\lambda t} + \mu_2 \left|x_{1_0} + \frac{1}{\lambda} x_{2_0}\right| e^{-\alpha\,\lambda\,t} \\ &\quad + \frac{1}{2}\left[-(\alpha\lambda\, x_{1_0} + x_{2_0})\, e^{-\lambda t} + \alpha(\lambda\, x_{1_0} + x_{2_0})\, e^{-\alpha\,\lambda\,t}\right]^2 \\ &\le \mu_2 \left|\alpha x_{1_0} + \frac{1}{\lambda} x_{2_0}\right| e^{-\lambda t} + \mu_2 \left|x_{1_0} + \frac{1}{\lambda} x_{2_0}\right| e^{-\alpha\,\lambda\,t} \\ &\quad + (\alpha\lambda\, x_{1_0} + x_{2_0})^2 e^{-2\lambda t} + \frac{1}{2}\alpha^2(\lambda\, x_{1_0} + x_{2_0})^2 e^{-2\alpha\,\lambda\,t} \\ &\le \left[\mu_2(\alpha+1)|x_{1_0}| + \frac{2\mu_2}{\lambda}|x_{2_0}| + 2\alpha^2\lambda^2 x_{1_0}^2 + (\alpha^2+1)x_{2_0}^2\right] e^{-\lambda t} \le R_1. \end{aligned} \tag{6.115}$$

While deriving (6.115), the corresponding gain μ_2 of the twisting controller is viewed as a parameter and the well-known inequality $\frac{1}{2}(a+b)^2 \le a^2 + b^2$ is employed. The upper bound on the reaching time $\mathscr{T}_1$ is a solution of the transcendental inequality

$$\left[\mu_2(\alpha+1)|x_{1_0}| + \frac{2\mu_2}{\lambda}|x_{2_0}| + 2\alpha^2\lambda^2 x_{1_0}^2 + (\alpha^2+1)x_{2_0}^2\right] e^{-\lambda\,\mathscr{T}_1} \le R_1. \tag{6.116}$$

Taking into account $\lim_{\lambda \to \infty} \lambda^2 e^{-\lambda t} = 0$ for all $t > 0$, it follows that the function $\mathcal{T}_1(\lambda)$ of λ escapes to zero for all admissible parameters $x_{1_0}, x_{2_0}, \alpha, \mu_2, R_1$ as λ goes to infinity. The upper bound on the reaching time $\mathcal{T}_1(\lambda)$ is thus given by

$$\mathcal{T}_1 \leq \frac{1}{\lambda} \ln \left[\frac{\mu_2(\alpha+1)|x_{1_0}| + \frac{2\mu_2}{\lambda}|x_{2_0}| + 2\alpha^2\lambda^2 x_{1_0}^2 + (\alpha^2+1)x_{2_0}^2}{R_1} \right]. \tag{6.117}$$

It should be noted that (6.114) matches to the state transition matrix of (6.113) in the form

$$e^{At} = \begin{bmatrix} \alpha e^{-\lambda t} - e^{-\alpha \lambda t} & \frac{1}{\lambda}(e^{-\lambda t} - e^{-\alpha \lambda t}) \\ -\alpha \lambda e^{-\lambda t} + \alpha \lambda e^{-\alpha \lambda t} & -e^{-\lambda t} + \alpha e^{-\alpha \lambda t} \end{bmatrix}. \tag{6.118}$$

Consider now the perturbed version

$$\begin{aligned} \dot{x}_1 &= x_2 \\ \dot{x}_2 &= -l_1 x_1 - l_2 x_2 + \omega(t) \end{aligned} \tag{6.119}$$

of system (6.113) where the external disturbance is upper bounded

$$|\omega(t)| \leq N \tag{6.120}$$

for almost all $t \geq 0$ by some positive constant N. Then the solution of (6.119), initialized at the origin $x_1(0) = 0$, $x_2(0) = 0$, is given by

$$x(t) = \int_0^t e^{A(t-\tau)} \Omega(\tau) d\tau$$

where $\Omega = \begin{bmatrix} 0 & \omega \end{bmatrix}^T$. Thus,

$$\begin{aligned} x_1(t) &= \int_0^t \frac{1}{\lambda} \left[e^{-\lambda(t-\tau)} - e^{-\alpha\lambda(t-\tau)} \right] \Omega(\tau) d\tau, \\ x_2(t) &= \int_0^t \left[-e^{-\lambda(t-\tau)} + \alpha e^{-\alpha\lambda(t-\tau)} \right] \Omega(\tau) d\tau. \end{aligned} \tag{6.121}$$

The Lyapunov function

$$V(x_1, x_2) = \mu_2 |x_1| + \frac{1}{2} x_2^2$$

is now estimated on the solutions (6.121) as follows:

$$\begin{aligned} V(x_1, x_2) &\le \mu_2 N \int_0^t \frac{1}{\lambda}\left[e^{-\lambda(t-\tau)} + e^{-\alpha\lambda(t-\tau)}\right] d\tau + (\alpha+1)^2 N^2 \left(\int_0^t e^{-\lambda(t-\tau)} d\tau\right)^2 \\ &\le \frac{2\mu_2 N + (\alpha+1)^2 N^2}{\lambda^2}. \end{aligned} \tag{6.122}$$

It is readily established that the upper bound $\frac{2\mu_2 N+(\alpha+1)^2 N^2}{\lambda^2}$ in (6.122) escapes to zero as λ goes to infinity.

Clearly relations (6.115) and (6.122), coupled together, ensure that the perturbed system (6.119) with nontrivial initial conditions enters the level set $\Gamma_R = \{(x_1, x_2) : V(x_1, x_2) \le R\}$ with $R = R_1 + r$ and

$$r \ge \frac{2\mu_2 N + (\alpha+1)^2 N^2}{\lambda^2} \tag{6.123}$$

in the same reaching time $\mathscr{T}_1(\lambda)$ and $\mathscr{T}_1(\lambda) \to 0$ as $\lambda \to \infty$.

The next aim is to define the scalar $r_1 > 0$ such that the expression $\Gamma_R \subset B_{r_1}$ holds. In other words, the following implication

$$\frac{\mu_2 |x_1|}{R} + \frac{x_2^2}{2R} \le 1 \Rightarrow \left(\frac{x_1}{r_1}\right)^2 + \left(\frac{x_2}{r_1}\right)^2 \le 1 \tag{6.124}$$

is required. Let the following inequalities

$$\left(\frac{x_1}{r_1}\right)^2 \le \frac{\mu_2 |x_1|}{R}, \quad \left(\frac{x_2}{r_1}\right)^2 \le \frac{x_2^2}{2R} \tag{6.125}$$

be in force. Then the expression $(x_1, x_2) \in B_{r_1}$ holds true for every given point $(x_1, x_2) \in \Gamma_R$ in the state space. Note that the following inequality

$$|x_1| \le \frac{R}{\mu_2} \tag{6.126}$$

always holds true for all $x \in \Gamma_R$. The first inequality of (6.125) is simplified to

$$|x_1| \left(\frac{1}{r_1}\right)^2 \le \frac{\mu_2}{R} \tag{6.127}$$

Utilizing the relationship (6.126), the conservative requirement

$$|x_1| \left(\frac{1}{r_1}\right)^2 \le \frac{R}{\mu_2}\left(\frac{1}{r_1}\right)^2 \le \frac{\mu_2}{R} \tag{6.128}$$

is formulated to render sufficiently large value for the radius r_1. Hence, the upper bound

$$r_1 \geq \frac{R}{\mu_2}, \tag{6.129}$$

imposed on r_1, suffices to satisfy the first inequality of (6.125). Similarly, the second inequality of (6.125) leads to $r_1 \geq \sqrt{2R}$. Thus, the following estimate

$$r_1 = \max\left\{\frac{R}{\mu_2}, \sqrt{2R}\right\} \tag{6.130}$$

of the parameter r_1 is obtained.

Next, the definition of the scalar $r_2 > 0$ is to be obtained such that the expression $B_{r_2} \subset \Gamma_R$ holds. In other words, one has to require

$$\left(\frac{x_1}{r_2}\right)^2 + \left(\frac{x_2}{r_2}\right)^2 \leq 1 \Rightarrow \frac{\mu_2|x_1|}{R} + \frac{x_2^2}{2R} \leq 1. \tag{6.131}$$

It can be noted that for all $x \in B_{r_2}$, the implication

$$|x_1| \leq r_2 \Rightarrow \frac{\mu_2|x_1|}{R} \leq \frac{\mu_2 r_2}{R} \tag{6.132}$$

proves to be in force. Let the inequality

$$r_2 \leq \frac{\rho R}{\mu_2} \tag{6.133}$$

be satisfied with an arbitrary scalar $0 < \rho < 1$. Then it follows that

$$\frac{\mu_2|x_1|}{R} \leq \frac{\mu_2 r_2}{R} \leq \rho. \tag{6.134}$$

Furthermore, assuming the inequality

$$r_2 \leq \sqrt{2R(1-\rho)} \tag{6.135}$$

yields

$$\frac{x_2^2}{2R} \leq (1-\rho)\frac{x_2^2}{r_2^2}. \tag{6.136}$$

Hence, by combining (6.134) and (6.136), the true specification

$$r_2 = \min\left\{\frac{\rho R}{\mu_2}, \sqrt{2R(1-\rho)}\right\} \tag{6.137}$$

is concluded, thereby ensuring that

$$\frac{\mu_2|x_1|}{R} + \frac{x_2^2}{2R} \leq \rho + (1-\rho)\frac{x_2^2}{r_2^2} \leq 1 \tag{6.138}$$

for all $x \in B_{r_2}$. It should be noted that the inequality (6.138) is obtained from the fact that $\frac{x_2^2}{r_2^2} \leq 1 \quad \forall x \in B_{r_2}$. The aim $B_{r_2} \subset \Gamma_R$ is thus achieved.

It is assumed that the inequality $r_0 > r_1 > 0$ holds true. Otherwise, the twisting controller is used without application of the linear controller. Section 6.3.3 derives a finite settling time estimate to the origin starting from the radius r_1. Hence the settling time estimate derived holds true for any arbitrary initial condition.

Remark 6.2 It is possible to guarantee that the relation $r_2 < R$ holds true by imposing a tuning rule on the gain parameter μ_2. Subject to the condition

$$\mu_2 > \rho\sqrt{\tfrac{R}{2(1-\rho)}}, \tag{6.139}$$

relation (6.137) is simplified to

$$r_2 = \tfrac{\rho R}{\mu_2}. \tag{6.140}$$

Then $r_2 < R$ is guaranteed if the condition $\mu_2 > \rho$ is satisfied. Summarizing, the lower bound

$$\mu_2 > \rho \ \max\left\{\sqrt{\tfrac{R}{2(1-\rho)}}, \quad 1\right\} \tag{6.141}$$

is obtained as the first tuning rule on μ_2.

The reason why (6.128) is conservative can be seen in the outcome Eq. (6.129). Even small values of the state $|x_1|$ are replaced by the upper bound $\frac{R}{\mu_2}$ in the Eq. (6.128). Hence, the lower bound of the scalar r_1 is conservatively determined by the bound $\frac{R}{\mu_2}$. It should be noted that smaller values of r_1 may suffice if it was not for the conservatism of Eq. (6.128). The solution to reduce the conservatism is either to choose the switching boundary R small or the gain parameter μ_2 very high. Both approaches have advantages and disadvantages. Choosing R very small may cause the linear feedback gain to increase when dealing with fixed finite settling time requirement. This means that the switch from the linear to the twisting controller occurs very close to the origin, resulting in low-magnitude chattering. On the other hand, choosing μ_2 very high overcomes the disadvantage of the previous choice at the cost of high-magnitude chattering. It is recommended that the user strikes a judicious balance between linear and twisting controller gains to trade-off the above conflicting outcomes. This is always a plausible solution because the choice of switching parameter R can be arbitrarily made.

Fig. 6.2 Finite settling time for the twisting controller. © 2012 Oxford University Press. Reprinted from Oza et al. (2011) by permission of Oxford University Press

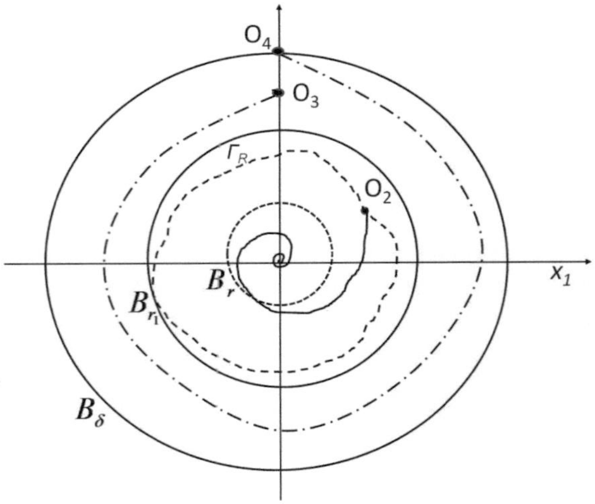

6.3.3 Settling Time Estimate of Twisting Controller

A finite upper bound on the settling time of the closed-loop system (6.108), (6.109) with the twisting controller applied is now presented. The time taken by the trajectories to travel from the point O_4 on the vertical positive semi-axis $e_1^+ = \{x \in \mathbf{R}^2 : x_1 = 0, x_2 > 0\}$ on the ball B_δ to the point O_3 on the e_1^+ axis is computed (see Fig. 6.2, reproduced from Oza et al. 2011). When the trajectories are initialized on the positive vertical axis at O_4, the factor by which it gets close to the origin after one revolution can be computed. The value of the intercept (point O_3) on positive vertical semi-axis after one iteration should be greater than the radius r_1 of the ball B_{r_1} containing the level set Γ_R. Such a value of δ will ensure that the settling time estimate will be more conservative than the one computed with the initialization on the level set. The motivation for such a choice of initialization of the trajectories on the ball B_δ stems from the fact that the trajectory, containing O_2 on the level set Γ_R, starting from any arbitrary point below O_4 on the e_1^+ axis cannot intersect the trajectory starting from the point O_4. The basis for this is the fact that different trajectories have no intersections because otherwise they would coincide with each other due to the uniqueness of the solution of the second-order linear double integrator system driven by the twisting controller.

The two-step settling time estimation is proposed. Firstly, a comparison system, the trajectory of which encompasses the actual system, is defined. Secondly, the comparison system is initialized on the positive vertical semi-axis e_1^+ with the coordinates $(0, \delta)$. Then the finite settling time is computed for the comparison system subject to the condition $\delta\Psi > r_1$ where Ψ is the factor by which the trajectory gets closer to the origin after one revolution at point O_3. Such an estimate is conservative as it encompasses the actual system trajectory.

Consider the closed-loop system,

$$\begin{aligned} \dot{x}_1 &= x_2, \\ \dot{x}_2 &= -\mu_1 \operatorname{sign}(x_2) - \mu_2 \operatorname{sign}(x_1) + \omega(t), \end{aligned} \tag{6.142}$$

where $\omega(t)$ is a bounded uncertainty which satisfies the upper bound (6.120). As a matter of fact, the meaning of the closed-loop system (6.142) with discontinuous right-hand side in the sense of Filippov. The method of considering a comparison system for such a second-order system has been detailed in Levant (1993), Orlov et al. (2003), Utkin (2015). The same method is inherited in the following and briefly outlined. The right-hand side of (6.142) is given by

$$\begin{aligned} \phi_1 &= x_2, \\ \phi_2 &= -\mu_1 \operatorname{sign}(x_2) - \mu_2 \operatorname{sign}(x_1) + \omega(t). \end{aligned} \tag{6.143}$$

Let a comparison system corresponding to (6.142) be as follows:

$$\begin{aligned} \dot{x}_1 &= x_2, \\ \dot{x}_2 &= -\left[\mu_1 - N\right] \operatorname{sign}(x_2) - \mu_2 \operatorname{sign}(x_1). \end{aligned} \tag{6.144}$$

In turn, the right-hand side of the comparison system (6.144) is specified by

$$\begin{aligned} \phi_1^c &= x_2, \\ \phi_2^c &= -\left[\mu_1 - N\right] \operatorname{sign}(x_2) - \mu_2 \operatorname{sign}(x_1). \end{aligned} \tag{6.145}$$

The comparison system (6.145) and the plant (6.143) are related by

$$\begin{aligned} \phi_1 &= \phi_1^c, \\ \phi_2 &= \phi_2^c + \Delta\phi, \end{aligned} \tag{6.146}$$

where

$$\Delta\phi = -N \operatorname{sign}(x_2) + \omega. \tag{6.147}$$

It is trivial to note that

$$\begin{aligned} \Delta\phi \le 0 &\quad \text{if } x \in (G_1 \cup G_4), \\ \Delta\phi \ge 0 &\quad \text{if } x \in (G_2 \cup G_3), \end{aligned} \tag{6.148}$$

where

$$\begin{aligned} G_1 &= \left\{x \in \mathbf{R^2} : x_1 > 0, x_2 > 0\right\}, \quad G_2 = \left\{x \in \mathbf{R^2} : x_1 > 0, x_2 < 0\right\}, \\ G_3 &= \left\{x \in \mathbf{R^2} : x_1 < 0, x_2 < 0\right\}, \quad G_4 = \left\{x \in \mathbf{R^2} : x_1 < 0, x_2 > 0\right\}. \end{aligned} \tag{6.149}$$

Hence, the inclusions (6.145) are readily seen to be more conservative dynamics than the original system (6.143) in the sense that the solutions $(x_1(t), x_2(t))$ of the original system (6.142) are bounded by the solutions $(x_1^c(t), x_2^c(t))$ of comparison system (6.144). For the purpose of estimating the finite settling time, it suffices to consider system (6.144), represented in the matrix form

$$\dot{x}(t) = A\,x(t) + B\,u(t), \tag{6.150}$$

where

$$x = \begin{bmatrix} x_1 \; x_2 \end{bmatrix}^T, \quad A = \begin{bmatrix} 0 & 1 \\ 0 & 0 \end{bmatrix}, \quad B = \begin{bmatrix} 0 \\ 1 \end{bmatrix} \\ u(t) = -(\mu_1 - N)\,\text{sign}(x_2) - \mu_2\,\text{sign}(x_1). \tag{6.151}$$

The motion in the state space is obtained, using the convolution integral:

$$x(t) = e^{At}\,x(0) + \int_0^t e^{A(t-\tau)} B\,u(\tau)d\tau. \tag{6.152}$$

Since the control switches on the the axes $x_1 = 0, x_2 = 0$, the integral (6.152) is required to be computed in each quadrant as follows:

$$\begin{aligned}
x(t) &= e^{At}\,x(0) + \int_0^t e^{A(t-\tau)} \begin{bmatrix} 0 \\ -\mu_1 - \mu_2 + N \end{bmatrix} d\tau \quad , x \in G_1; \\
x(t) &= e^{At}\,x(0) + \int_0^t e^{A(t-\tau)} \begin{bmatrix} 0 \\ \mu_1 - \mu_2 - N \end{bmatrix} d\tau \quad , x \in G_2; \\
x(t) &= e^{At}\,x(0) + \int_0^t e^{A(t-\tau)} \begin{bmatrix} 0 \\ \mu_1 + \mu_2 - N \end{bmatrix} d\tau \quad , x \in G_3; \\
x(t) &= e^{At}\,x(0) + \int_0^t e^{A(t-\tau)} \begin{bmatrix} 0 \\ -\mu_1 + \mu_2 + N \end{bmatrix} d\tau \quad , x \in G_4.
\end{aligned} \tag{6.153}$$

It is noted that using such integrals to define the solutions of the comparison system (6.144) is mathematically correct because as shown in Sect. 5.2.1, the twisting control law never generates a sliding mode on the switching lines $x_1 = 0$ and $x_2 = 0$. Hence, the solutions always cross the switching lines $x_1 = 0, x_2 = 0$ except at the origin $(x_1, x_2) = 0$.

Step 1: Time to travel from the point $O_4 \in B_\xi$ to point O_3 on the $e_1^+ = \{x \in \mathbf{R}^2 : x_1 = 0, x_2 > 0\}$ axis:

Case 1: $\text{sign}(x_1) = 1, \text{sign}(x_2) = 1$: Noting that the initial condition is assumed to lie in the first quadrant for a conservative estimate, Eq. (6.153) takes the form

$$x(t) = e^{At} \begin{bmatrix} 0 \\ \delta \end{bmatrix} + \left[\int_0^t e^{A(t-\tau)} d\tau \right] \begin{bmatrix} 0 \\ -\mu_1 - \mu_2 + N \end{bmatrix}, \tag{6.154}$$

where $\delta = x_2(t_0)$ represents the projection on the e_1^+ axis. The matrix exponential in (6.154) is computed as follows:

$$e^{At} = I + At + \frac{At^2}{2!} + \cdots. \tag{6.155}$$

Since $A^n = 0, \forall n \geq 2$, the Taylor representation (6.155) leads to

$$e^{At} = I + At = \begin{bmatrix} 1 & t \\ 0 & 1 \end{bmatrix}, \quad \int_0^t e^{-A\tau} d\tau = \begin{bmatrix} t & \frac{-t^2}{2} \\ 0 & t \end{bmatrix}. \tag{6.156}$$

Further simplification produces

$$\begin{bmatrix} x_1(t) \\ x_2(t) \end{bmatrix} = \begin{bmatrix} \delta t - \frac{1}{2}t^2(\mu_1 + \mu_2 - N) \\ \delta - t(\mu_1 + \mu_2 - N) \end{bmatrix}. \tag{6.157}$$

The time t_1 of interception on the horizontal axis $x_2 = 0$ is obtained, by extracting $x_2(t)$ from (6.157) and determining it from the equality $x_2(t_1) = 0$:

$$t_{x_2=0} = t_1 = \frac{\delta}{\mu_1 + \mu_2 - N}. \tag{6.158}$$

The point of interception on the horizontal axis $x_2 = 0$ is then computed, extracting $x_1(t)$ from (6.157):

$$x_1(t_1) = \frac{\delta^2}{2(\mu_1 + \mu_2 - N)} \tag{6.159}$$

Case 2: $\text{sign}(x_1) = 1$, $\text{sign}(x_2) = -1$: Repeating the same computation for the second quadrant, Eq. (6.153) takes the form

$$x(t) = e^{At} \begin{bmatrix} x_1(t_1) \\ 0 \end{bmatrix} + \left[\int_0^t e^{A(t-\tau)} d\tau\right] \begin{bmatrix} 0 \\ \mu_1 - \mu_2 - N \end{bmatrix}. \tag{6.160}$$

Further simplification leads to the following time estimate $t = t_2$ when the trajectory intercepts on the vertical axis $x_1 = 0$ and the intercept $x_2(t_2)$:

$$\begin{aligned} t_2 &= \frac{\delta}{\sqrt{(\mu_2 + \mu_1 - N)(\mu_2 - \mu_1 + N)}}, \\ x_2(t_2) &= -\frac{\delta\sqrt{(\mu_2 - \mu_1 + N)}}{\sqrt{(\mu_2 + \mu_1 - N)}}. \end{aligned} \tag{6.161}$$

Case 3: $\text{sign}(x_1) = -1$, $\text{sign}(x_2) = -1$: Repeating the same computation for the third quadrant, Eq. (6.153) takes the form

$$x(t) = e^{At}\begin{bmatrix} 0 \\ x_2(t_2) \end{bmatrix} + \left[\int_0^t e^{A(t-\tau)} d\tau\right]\begin{bmatrix} 0 \\ \mu_1 + \mu_2 - N \end{bmatrix}. \tag{6.162}$$

Further simplification leads to the following time estimate $t = t_3$ when the trajectory intercepts on the axis $x_2 = 0$ and the intercept $x_1(t_3)$:

$$\begin{aligned} t_3 &= \frac{\delta}{(\mu_2 + \mu_1 - N)} \frac{\sqrt{(\mu_2 - \mu_1 + N)}}{\sqrt{(\mu_2 + \mu_1 - N)}} \\ x_1(t_3) &= -\frac{\delta^2(\mu_2 - \mu_1 + N)}{2(\mu_2 + \mu_1 - N)^2}. \end{aligned} \tag{6.163}$$

Case 4: $\text{sign}(x_1) = -1$, $\text{sign}(x_2) = 1$: Repeating the same computation for the third quadrant, Eq. (6.153) takes the form

$$x(t) \le e^{At}\begin{bmatrix} x_1(t_3) \\ 0 \end{bmatrix} + \left[\int_0^t e^{A(t-\tau)} d\tau\right] \cdot \begin{bmatrix} 0 \\ \mu_2 - \mu_1 + N \end{bmatrix}. \tag{6.164}$$

Further simplification leads to the following time estimate $t = t_4$ when the trajectory intercepts on the axis $x_1 = 0$ and the intercept $x_2(t_4)$:

$$\begin{aligned} t_4 &= \frac{\delta}{(\mu_2 + \mu_1 - N)}, \\ x_2(t_4) &= \frac{\delta(\mu_2 - \mu_1 + N)}{(\mu_2 + \mu_1 - N)}. \end{aligned} \tag{6.165}$$

Hence, the time $\mathscr{T}_2$ taken by the trajectory to travel from the point O_4 on the ball B_δ to some point O_3 on the semi-axis e_1^+ is obtained by using (6.158), (6.161), (6.163) and (6.165) and it is as follows:

$$\begin{aligned} T_1 &= t_1 + t_2 + t_3 + t_4 \\ &= \frac{\delta}{\mu_2}\Delta, \end{aligned} \tag{6.166}$$

where

$$\eta = \frac{\mu_1 - N}{\mu_2}, \quad \Delta = \left[\frac{2}{1+\eta} + \frac{1}{\sqrt{(1+\eta)(1-\eta)}} + \frac{\sqrt{1-\eta}}{(1+\eta)\sqrt{1+\eta}}\right]. \tag{6.167}$$

Step 2: Time to travel from the point O_3 on the $e_1^+ = \{x \in \mathbf{R}^2 : x_1 = 0, x_2 > 0\}$ axis to the origin:

It can be seen that the time T_1 taken by one revolution depends on the initial condition δ, gain parameters (μ_1, μ_2), and the bound N on the uncertainty. Hence the time T_1 and time taken by the subsequent revolutions can be computed a priori. Furthermore, as shown by the last equality of (6.165), the closed-loop trajectory decays closer to the origin by a factor Ψ of the initial condition δ where

$$\Psi = \frac{1-\eta}{1+\eta}. \tag{6.168}$$

The computation of trajectories for four quadrants, repeated with the initial condition set at $x_2(t_4)$ to obtain the intersection of the trajectory with the axis $x_1 = 0$ at the end of the second revolution, is as follows:

$$x(T_2) = x_2(t_4)\,\Psi = \delta\Psi^2, \tag{6.169}$$

where T_2 is the time at which the second revolution is completed. Noting that one revolution takes $\frac{\Delta}{\mu_2}$ multiplied by the initial value on the vertical axis (see (6.166)), the total time taken for two revolutions is estimated as follows:

$$T_2 = T_1 + \frac{x_2(t_4)}{\mu_2}\Delta, \tag{6.170}$$

where $x_2(t_4)$ is treated as the new initial condition at the start of the second revolution. Equation (6.170), simplified by using (6.165) and (6.166), is as follows:

$$T_2 = \frac{\delta}{\mu_2}\Delta\,[1+\Psi]. \tag{6.171}$$

Similar arguments lead to the computation, corresponding to the third revolution:

$$\begin{aligned} x(T_3) &= x_2(T_2)\,\Psi = \delta\Psi^3 \\ T_3 &= T_2 + \frac{x_2(T_2)}{\mu_2}\Delta = \frac{\delta}{\mu_2}\Delta\left[1+\Psi+\Psi^2\right]. \end{aligned} \tag{6.172}$$

Proceeding further and generalizing the above results for the nth revolution, it is obtained that

$$\begin{aligned} x(T_n) &= x_2(T_{n-1})\,\Psi = \delta\Psi^n \\ T_n &= T_{n-1} + \frac{x_2(T_{n-1})}{\mu_2}\Delta = \frac{\delta}{\mu_2}\Delta\left[1+\Psi+\Psi^2+\cdots+\Psi^{n-1}\right]. \end{aligned} \tag{6.173}$$

The number of revolutions goes to ∞ as time t tends to ∞. Hence, the following holds true:

$$\lim_{t\to\infty} T_n = \lim_{n\to\infty} T_n = \lim_{n\to\infty} \quad \frac{\delta}{\mu_2}\Delta\left[1+\Psi+\Psi^2+\cdots+\Psi^{n-1}\right], \tag{6.174}$$

As can be noted from the definition (6.168) of Ψ that the inequality $0 < \Psi < 1$ always holds true because $\mu_2 > \mu_1 + N$. Hence, the infinite series in (6.174) is represented as follows:

$$1 + \Psi + \Psi^2 + \cdots + \Psi^{n-1} = \frac{1 - \Psi^n}{1 - \Psi}. \tag{6.175}$$

In turn, the time T_n, taken by an infinite number of revolutions, represents the second segment of the settling time $\mathscr{T}_s$ in question, and it is computed as

$$\begin{aligned}\mathscr{T}_2 = \lim_{n\to\infty} T_n &= \lim_{n\to\infty} \frac{\delta}{\mu_2}\Delta\left[\frac{1-\Psi^n}{1-\Psi}\right] \\ &= \frac{\delta\Delta}{\mu_2(1-\Psi)} \lim_{n\to\infty} (1-\Psi^n) = \frac{\delta\Delta}{\mu_2(1-\Psi)} < \infty.\end{aligned} \tag{6.176}$$

Relation (6.176) thus always gives a finite upper bound $\mathscr{T}_2$ on the settling time estimate of the twisting algorithm for finite values of gains μ_1, μ_2. The following can be obtained from (6.167) and (6.168):

$$\Delta = \frac{2\left[\sqrt{1-\eta^2}+1\right]}{(1+\eta)\left[\sqrt{1-\eta^2}\right]}, \quad 1 - \Psi = \frac{2\eta}{1+\eta}. \tag{6.177}$$

By substituting Δ and $(1 - \Psi)$ obtained from (6.177) and (6.168), respectively, the last equality of (6.176) is further simplified to

$$\mathscr{T}_2 = \frac{\delta\Delta}{\mu_2(1-\Psi)} = \frac{\delta\left[\sqrt{1-\eta^2}+1\right]}{\mu_2\eta\sqrt{1-\eta^2}} = \frac{\delta\left[\sqrt{1-\eta^2}+1\right]}{(\mu_1-N)\sqrt{1-\eta^2}}. \tag{6.178}$$

Then, letting

$$\mu_2 > \beta\mu_1 \tag{6.179}$$

for some $\beta > 1$, the following is obtained from the definition of η and Ψ in (6.167) and (6.168), respectively:

$$\lim_{\mu_1\to\infty} \eta \leq \frac{1}{\beta}, \quad \lim_{\mu_1\to\infty} \Psi \geq \frac{1-\beta^{-1}}{1+\beta^{-1}} = \frac{\beta-1}{\beta+1}. \tag{6.180}$$

Hence a finite $\delta > \frac{r_1}{\Psi}$ can always be chosen independent of μ_1 and it follows that

$$\lim_{\mu_1\to\infty} \mathscr{T}_2 = \lim_{\mu_1\to\infty} \frac{\delta\left[\sqrt{1-\eta^2}+1\right]}{(\mu_1-N)\sqrt{1-\eta^2}} \leq \lim_{\mu_1\to\infty} \frac{2\delta}{(\mu_1-N)\sqrt{1-\frac{1}{\beta^2}}} = 0. \tag{6.181}$$

Thus, the final objective, caused by the intuitive expectation, is achieved, namely, that increasing the gains of the twisting controller causes the settling time to decrease. Tuning rules to achieve an arbitrarily specified settling time are discussed next.

6.3.4 Settling Time Tuning

The closed-loop system (6.108), (6.3.1) is finite time stable in the presence of the persisting disturbances $\omega(t)$ with uniformly bounded magnitude $|\omega(t)| < N$. Using the reaching time estimate (6.117) and its settling time counterpart (6.176), the overall settling time

$$\mathcal{T}_s(\mathcal{T}_1, \mathcal{T}_2) = \begin{cases} \mathcal{T}_1 + \mathcal{T}_2, & 0 < \delta < r_0; \\ \mathcal{T}_2, & 0 < r_0 < \delta \end{cases} \tag{6.182}$$

is estimated with $r_0 = \sqrt{x_1^2(t_0) + x_2^2(t_0)}$, being the Euclidian norm of the system initial condition, and with the estimates $\mathcal{T}_1$ and $\mathcal{T}_2$, given by (6.117) and (6.176), respectively.

Constructive tuning rules are subsequently derived to specify the gain parameters l_1, l_2, μ_1, μ_2 such that an arbitrarily chosen desired settling time can be achieved. In addition, the parameter R, which defines the boundary of switching between the linear and the twisting control laws, is an independent parameter to be chosen by the user. Let the specified settling time be denoted as $T_s > 0$. Hence, in order to ensure that the closed-loop system (6.142) settles to the origin in specified time $\mathcal{T}_s$, the following must hold true:

$$\mathcal{T}_1 + \mathcal{T}_2 \leq \mathcal{T}_s, \tag{6.183}$$

where $\mathcal{T}_1$ and $\mathcal{T}_2$ are defined in (6.117) and (6.176), respectively.

The proposed tuning strategy is that of dividing the desired settling time $\mathcal{T}_s$ into two time segments which can be arbitrarily chosen by the user. Then each time segment is considered to be the permissible upper bound on $\mathcal{T}_1$ and $\mathcal{T}_2$, which then leads to explicit tuning rules for the gain parameters of both the linear controller and twisting one. Mathematically, the arbitrary allotment of the settling time can be obtained as follows:

$$\begin{aligned} \mathcal{T}_{s_1} &= \begin{cases} \epsilon_1 \mathcal{T}_s, & \text{if } r_0 > \delta; \\ 0, & \text{otherwise.} \end{cases} \\ \mathcal{T}_{s_2} &= \begin{cases} (1-\epsilon_1)\mathcal{T}_s, & \text{if } r_0 > \delta; \\ \epsilon_1\, \mathcal{T}_s, & \text{otherwise} \end{cases} \end{aligned} \tag{6.184}$$

where the scalar $\epsilon_1 \in (0, 1)$ can be chosen by the user arbitrarily. Observing (6.117) and (6.178), the objective is to find the tuning of the twisting and linear gains such that the following inequalities

$$\begin{aligned} &\frac{1}{\lambda} \ln \left[\frac{\mu_2(\alpha+1)|x_{1_0}| + \frac{2\mu_2}{\lambda}|x_{2_0}| + 2\alpha^2\lambda^2 x_{1_0}^2 + (\alpha^2+1)x_{2_0}^2}{R_1} \right] \\ &\leq \mathcal{T}_{s_1} \quad \frac{2\delta}{(\mu_1 - N)\left[\sqrt{1-\eta^2}\right]} \leq \mathcal{T}_{s_2} \end{aligned} \tag{6.185}$$

hold true. Then (6.183) is always ensured.

The application of twisting controller is invoked without linear feedback if the condition $r_0 < \delta$ holds true due to user's choice of R. As mentioned in Sect. 6.3.2, the gain μ_2 of twisting is seen as a parameter in the formula of $\mathscr{T}_1$. The requirements on the gains can be summarized from the previous sections as follows:

$$\begin{aligned} &r_1 = \max\left\{\frac{R}{\mu_2}, \sqrt{2R}\right\}, \quad r_2 = \frac{\rho R}{\mu_2}, \\ &\delta\Psi > r_1, \quad \mu_2 > \max\left\{\mu_1 + N, \quad \rho\sqrt{\frac{R}{2(1-\rho)}}, \quad \rho\right\}, \end{aligned} \tag{6.186}$$

where the scalar $\rho \in (0, 1)$ is at user's disposal.

6.3.4.1 Tuning Procedure

Provided that the initial conditions of the states x_1^0, x_2^0 and the uniform upper bound N are known a priori, the step-by-step tuning procedure is detailed.
Step 1: Choose an arbitrary $R > 0$. Let the condition

$$\mu_2 \geq \sqrt{\frac{R}{2}} \tag{6.187}$$

be imposed on the choice of μ_2. Then, r_1 is computed by combing (6.130) and (6.187):

$$r_1 = \sqrt{2}\,R. \tag{6.188}$$

Step 2: Next, select η to respect

$$0 < \eta \leq \frac{1}{\beta} \tag{6.189}$$

where $\beta > 1$ is a tuning variable. Such a selection is motivated by the fact that the variable η, thus defined, does not allow Ψ to equal either to unity or zero. It can thus be concluded that

$$\frac{1-\eta}{1+\eta} = \Psi \geq \frac{\beta-1}{\beta+1}. \tag{6.190}$$

Step 3: Select the scalar δ according to

$$\delta > \begin{cases} \frac{r_1(\beta+1)}{\beta-1}, & \text{if } r_0 > \frac{r_1(\beta+1)}{\beta-1}; \\ \frac{r_0(\beta+1)}{\beta-1}, & \text{if } r_0 < \frac{r_1(\beta+1)}{\beta-1}. \end{cases} \tag{6.191}$$

Then due to (6.191), the requirement $\delta\Psi > r_1$ is always satisfied.

Step 4: If $r_0 < \frac{r_1(\beta+1)}{\beta-1}$, the twisting controller tuning is invoked without linear feedback. Compute μ_1, μ_2 using the next step and apply the twisting controller. If $r_0 > \frac{r_1(\beta+1)}{\beta-1}$, go to *step 6*.

Step 5: Tuning for the twisting controller is carried out in this step using results of Sect. 6.3.3. The following requirement can be formulated from (6.178) and the second inequality of (6.185):

$$\frac{2\delta}{(\mu_1 - N)\sqrt{1-\eta^2}} \leq \mathscr{T}_{s_2}. \tag{6.192}$$

In turn, the following tuning rules on μ_1 and μ_2 can be obtained by combining (6.186) and (6.192):

$$\begin{aligned} \mu_1 &> \max\left\{ N, \quad \frac{2\delta}{\mathscr{T}_{s_2}\sqrt{1-\eta^2}} + N \right\}, \\ \mu_2 &> \max\left\{ \sqrt{\frac{R}{2}}, \quad \frac{\mu_1 - N}{\eta}, \quad \mu_1 + N, \quad \rho\sqrt{\frac{R}{2(1-\rho)}}, \quad \rho, \quad \beta\,\mu_1 \right\}. \end{aligned} \tag{6.193}$$

It is worth noticing that higher values of β (and in turn, lower values of η) may be required for very small settling time assignments. This is mainly due to the fact that the behavior of the settling time estimate (6.178) is fully known only in the limit $\mu_1 \to \infty$. Relationship of $\mathscr{T}_2$ with respect to μ_1 and β is highly nonlinear especially in the vicinity of $\mathscr{T}_2 = 0$ (see (6.178)). Hence (6.178) does not clarify as to the rate at which the settling time can be reduced. Nevertheless, due to (6.181), it can be deduced that there always exists a finite scalar β such that the second inequality of (6.185) is satisfied. In this sense, β is seen as a tuning variable.

Step 6: This step performs the tuning for the linear gains

$$l_1 = \alpha\,\lambda^2, \quad l_2 = (\alpha+1)\,\lambda,$$

where $\alpha > 1$ is an arbitrarily chosen scalar and the scalar λ can be obtained as follows. Perform the tuning procedure of *Step 5* to obtain the parameters μ_1, μ_2. Choose an arbitrary scalar $\rho \in (0, 1)$. Then choose an arbitrary scalar $r > 0$ such that $0 < r < r_2$ holds true where r_2 is computed from (6.137) using the above ρ. Next, compute $R_1 = R - r$ (it should be noted from Remark 6.2 that the relation $r < r_2 < R$ is guaranteed). To ensure that (6.123) holds true let us choose λ according to

$$\lambda \geq \sqrt{\frac{2\mu_2 N + (\alpha+1)^2 N^2}{r}}. \tag{6.194}$$

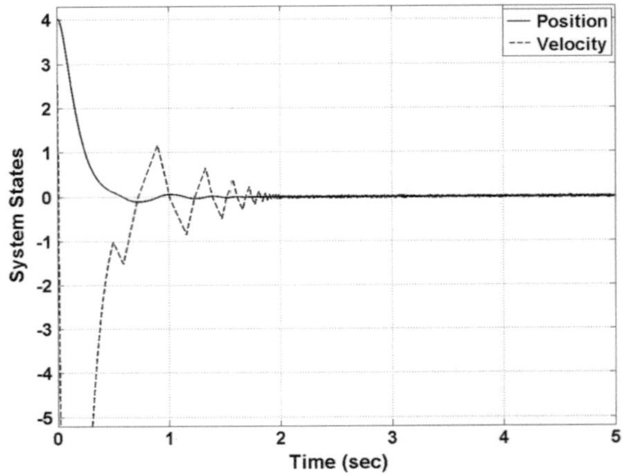

Fig. 6.3 Numerical simulation for the specification of $\mathscr{T}_s = 10\,\mathrm{s},\ N = 1$. © 2012 Oxford University Press. Reprinted from Oza et al. (2011) by permission of Oxford University Press

Next, utilizing the first inequality in (6.185), the following transcendental equation

$$\frac{1}{\gamma}\ln\left[\frac{\mu_2(\alpha+1)|x_{1_0}| + \frac{2\mu_2}{\gamma}|x_{2_0}| + 2\alpha^2\gamma^2 x_{1_0}^2 + (\alpha^2+1)x_{2_0}^2}{R_1}\right] - \mathscr{T}_{s_1} = 0 \tag{6.195}$$

is obtained. It can be noted from (6.117) and (6.122) that there always exists a solution γ such that (6.195) is satisfied. The basis for this is that $\lim_{\gamma\to\infty} \gamma^2 e^{-\gamma t} = 0$. A valid solution γ satisfying (6.195) can be obtained, for example, using numerical optimization routines. Hence the tuning rule for λ can be obtained as follows:

$$\lambda = \max\left\{\sqrt{\frac{2\mu_2 N + (\alpha+1)^2 N^2}{r}},\ \gamma\right\} \tag{6.196}$$

Then, (6.194) gives the tuning rules for the linear gains l_1, l_2.

Figure 6.3 illustrates the tuning procedure, applied numerically, for a specification of $\mathscr{T}_s = 10\,\mathrm{s}$ with the initial condition $x_0^1 = x_0^2 = 4$ and the upper bound $N = 1$. The aforementioned tuning steps have been carried out rendering the tuning variables as follows:

$$\rho = 0.99, r = 0.173, R = 1.5, \epsilon_1 = 0.3, \eta = 0.45, \beta = 2.11, \alpha = 1.01. \tag{6.197}$$

The tuning rules (6.193) and (6.196) then allow one to correctly compute the controller parameters $\mu_1 = 2.47, \mu_2 = 8.57, l_1 = 123.6, l_2 = 22.243$. A switch from the linear controller to the twisting controller is seen at $t = 0.5$ s.

6.4 ISS Point-Wise Feedback Synthesis of Parabolic Systems

The arsenal of Lyapunov function constructions is further illustrated in the infinite-dimensional setting. For this purpose, SM-based ISS synthesis of parabolic systems, developed in Pisano and Orlov (2017), is presented within the practical framework of in-domain embedded sensing and actuation. A reaction–diffusion–advection PDE with uncertain parameters and nonlinearities, representing external disturbances of both matched and mismatched nature, is chosen to exemplify a Lyapunov functional construction which is invoked to establish the global asymptotic stability in a Sobolev space, capturing spatial state derivatives of the same order as that of the plant equation. The proposed ISS stability synthesis is also accompanied with a set of simple tuning rules for the controller parameters.

The system of interest is governed by the parabolic boundary-value problem

$$\begin{aligned} z_t(x,t) &= \theta z_{xx}(x,t) + dz_x(x,t) + \lambda z(x,t) + f(x,t) \\ &\quad + \sum_{i=2}^{N-1} b_i(x)\,[u_i(t) + \psi_i(t)]\,, \end{aligned} \tag{6.198}$$

$$z_x(0,t) = -[u_1(t) + \psi_1(t)], \quad z_x(1,t) = u_N(t) + \psi_N(t) \tag{6.199}$$

of Neumann type. Hereinafter, $z(x,t)$ is the scalar temperature field with the spatial variable $x \in [0,1]$ and time variable $t \geq 0$, and it evolves in the space $L_2(0,1)$, the parameters θ, d, λ are the uncertain diffusion, advection, and reaction coefficients, and $f(x,t)$ is an uncertain distributed disturbance. The functions $u_i(t)$ and $\psi_i(t), i = 2, \ldots, N-1$ represent in-domain control inputs and disturbances, entering the in-domain control channels, respectively, whereas $b_i(x)$ denotes the spatial distribution of the ith in-domain control input. The system is equipped with the controlled and perturbed Neumann-type boundary conditions (6.199), where $u_1(t)$ and $u_N(t)$ denote manipulable boundary control inputs and $\psi_1(t)$, $\psi_N(t)$ are boundary disturbances. Finally, the associated initial condition is

$$z(x,0) = z^0(x) \in L_2(0,1). \tag{6.200}$$

Due to the simultaneous presence of the boundary control inputs and perturbations, it appears unacceptable to invoke the usual compatibility conditions $z^0(0) = u_1(0) + \psi_1(0)$ and $z^0(1) = u_N(1) + \psi_N(1)$ in the closed-loop setting, which is why the meaning of the boundary-value problem (6.198)–(6.199) is subsequently viewed in the mild sense (see Sect. 3.2 for this and other PDE solution concepts).

For later use, recall from Sect. 3.4.1 that the mild solutions of (6.198)–(6.199) coincide with the corresponding weak solutions of the so-called standardizing PDE in distributions (Butkovskii 1982)

$$
\begin{aligned}
z_t(x,t) &= \theta z_{xx}(x,t) + d z_x(x,t) + \lambda z(x,t) + f(x,t) \\
&+ \sum_{i=2}^{N-1} b_i(x)\,[u_i(t) + \psi_i(t)] + \theta[u_1(t) + \psi_1(t)]\delta(x) \\
&+ \theta[u_N(t) + \psi_N(t)]\delta(x-1),
\end{aligned} \tag{6.201}
$$

subject to the homogeneous boundary conditions

$$
z_x(0,t) = 0, \qquad z_x(1,t) = 0, \tag{6.202}
$$

and to the same initial condition (6.200). Since the right-hand side of (6.201) contains the Dirac distribution $\delta(\xi - 1)$, the meaning of (6.201) is defined indirectly according to the weak solution concept of Definition 3.2.

Let us now specify the form of the spatial distribution function of the in-domain actuators. Let the points $0 = x_1 < x_2 < \cdots < x_N = 1$ be taken equi-spaced in the spatial domain [0, 1], i.e.,

$$
x_i = (i-1)h, \quad i = 1, 2, \ldots, N, \quad h = \frac{1}{N-1}. \tag{6.203}
$$

Points x_i correspond to the fixed location of the collocated in-domain and boundary sensing and actuation devices. The present investigation focuses on point-wise actuation devices of Neumann type, similar to the considered boundary controllers. Hence, by following the same line of reasoning that has been used to derive (6.201)–(6.202) from (6.198)–(6.199), the spatial distribution function of the ith in-domain actuator takes the form

$$
b_i(x) = \theta\delta(x - x_i) \tag{6.204}
$$

of a Dirac distribution, located at the corresponding point x_i and pre-multiplied by the diffusivity parameter θ. Then, the corresponding standardizing PDE in distributions

$$
\begin{aligned}
z_t(x,t) &= \theta z_{xx}(x,t) + d z_x(x,t) + \lambda z(x,t) + f(x,t) \\
&+ \theta \sum_{i=1}^{N} \delta(x - x_i)\,(u_i(t) + \psi_i(t))\,,
\end{aligned} \tag{6.205}
$$

complemented by the homogenous boundary conditions (6.202) and by the initial condition (6.200), comes into play.

The control objective is to properly design control laws $u_i(t)$, rejecting the matched disturbances $\psi_i(t)$ while also attenuating the mismatched disturbance $f(x,t)$. Potential solutions of the closed-loop boundary-value problem (6.202), (6.205) should then satisfy the exponential ISS inequality (cf. that of Definition 3.15)

$$\|z(\cdot,t)\|^2_{L_2(0,1)} \le e^{-\beta t}\|z(\cdot,0)\|^2_{L_2(0,1)}$$
$$+\gamma_0\|f\|^2_{L_\infty(0,t;L_2(0,1))} + \sum_{i=1}^{N}\gamma_i\|\psi_i\|^2_{L_\infty(0,t)} \tag{6.206}$$

for any initial condition (6.200), for all $t \ge 0$, for all $f \in L^{loc}_\infty(L_2(0,1))$, for all $\psi_i \in L^{loc}_\infty$, and for some positive constants $\beta > 0$, γ_j, $j = 0, 1, \ldots, N$.

In addition, *a question is addressed* on how many collocated actuator–sensor pairs are needed to guarantee the exponential ISS property (6.206) with a prespecified arbitrarily large decay rate β.

The above ISS issues are subsequently treated under the following assumptions on the available information on the uncertain system parameters and on admissible magnitudes of the external disturbances.

Assumption 1 There exist a priori known constants $\theta_0 > 0$, $D \ge 0$, and Λ such that

$$0 < \theta_0 \le \theta, \quad \lambda \le \Lambda, \quad |d| \le D. \tag{6.207}$$

Assumption 2 There exist a nonnegative constant F and a priori known constants Ψ_i, $i = 1, 2, \ldots, N$ such that

$$\|f(\cdot,t)\|_{L_2(0,1)} \le F, \qquad |\psi_i(t)| \le \Psi_i \quad \forall\, t > 0. \tag{6.208}$$

Since under Assumption 2 the exponential ISS property (6.206) is enforced, the guaranteed closed-loop accuracy

$$\|z(\cdot,t)\|^2_{L_2(0,1)} \le \sigma_0\gamma_0 F^2 + \sigma\Sigma^n_{i=1}\gamma_i\Psi_i^2, \quad t \ge T \tag{6.209}$$

can be achieved in a finite transient time $T > 0$ with arbitrary $\sigma_0 > 1$ and $\sigma > 1$. Moreover, by employing sliding mode control components, it becomes possible to respect the closed-loop accuracy (6.209) with $\sigma = 0$.

6.4.1 *Control Synthesis*

The collocated feedback inputs

$$u_i(t) = -k_i z(x_i,t) - M_i \operatorname{sign}(z(x_i,t)), \; i = 1, \ldots, N \tag{6.210}$$

are involved to attain the stated control objective by properly tuning the proportional and switching gains $k_i \ge 0$ and $M_i \ge 0$.

The precise meaning of the solutions of the boundary-value problem (6.205), (6.202) with the discontinuous inputs (6.210) is adopted in the generalized sense of Definition 3.19. As established in Levaggi (2002), such a generalized solution is nothing else than a weak solution of the multivalued closed-loop system (6.202), (6.205) with the switched feedback (6.210) where the sign function is defined in the *Filippov* sense

$$\operatorname{sign}(z) \in \begin{cases} 1 & \text{if } z > 0, \\ [-1, 1] & \text{if } z = 0, \\ -1 & \text{if } z < 0. \end{cases} \tag{6.211}$$

The ISS analysis of the closed-loop system (6.202), (6.205) with the switched multivalued inputs (6.210), (6.211) is preceded by establishing specific solutions of the resulting boundary-value problem to globally exist. The uniqueness of such a solution is actually questionable for potential sliding modes even in the finite-dimensional setting with multivalued inputs (see Sect. 2.2) what is, however, irrelevant within the ISS analysis to be conducted where all (possibly, nonunique) plant trajectories are required to exponentially decay according to (6.206). The interested reader may refer to Guo and Jin (2015) for uniquely determined Filippov solutions in the PDE setting under a single-boundary control input.

6.4.2 *Existence of Closed-Loop Solutions*

Generalized solutions of the boundary-value problem (6.202), (6.205) are first shown to globally exist under the discontinuous input (6.210) for a particular case $N = 2$ which is addressed in depth. Passing back to the original plant representation, the closed-loop system (6.202), (6.205), (6.210) with $N = 2$ is given by

$$\begin{aligned} z_t(x,t) &= \theta z_{xx}(x,t) + dz_x(x,t) + \lambda z(x,t) + f(x,t) \\ &\quad -\theta\delta(x)\left[M_1 \operatorname{sign}(z(0,t)) - \psi_1(t)\right] \\ &\quad -\theta\delta(x-1)\left[M_2 \operatorname{sign}(z(1,t)) - \psi_2(t)\right], \end{aligned} \tag{6.212}$$

$$z_x(0,t) = k_1 z(0,t), \quad z_x(1,t) = -k_2 z(1,t). \tag{6.213}$$

The general treatment is then readily reconstructed by following the same line of arguing for the coupled plant PDEs, separately viewed over the plant localizations (x_i, x_{i+1}), $i = 1, \ldots, N-1$ and possessing two boundary inputs only.

In the sequel, potential dynamics (6.212), (6.213) are separately investigated out of the discontinuity surface and along it. In the former case, the functions $\operatorname{sign}(z(0,t))$ and $\operatorname{sign}(z(1,t))$ are single-valued whereas in the latter case of the multivalued sign function (6.211), sliding modes along $z(0,t) = 0$ and/or $z(1,t) = 0$, if any, are specified according to the equivalent control method of Sect. 2.2.5. Just in the latter case, the Neumann/Robin (mixed) boundary conditions (6.213) with $k_1 = 0$ and/or $k_2 = 0$ are, respectively, obtained with the equivalent input values $M_1 \operatorname{sign}(z(0,t)) = \psi_1(t)$ and/or $M_2 \operatorname{sign}(z(0,t)) = \psi_2(t)$, which are relevant to sliding modes at the left boundary and/or at the right boundary.

For demonstrating a generalized solution of the multivalued boundary-value problem (6.212), (6.213) to globally exist, the invertible state transformation

$$Q(x,t) = e^{\frac{d}{2\theta}x} z(x,t) \tag{6.214}$$

is applied to (6.202), (6.205) (properly specified outside and along the discontinuity surface) to simplify it to the one

$$\begin{aligned} Q_t(x,t) &= \theta Q_{xx}(x,t) + \left(\lambda - \frac{d^2}{4\theta}\right) Q(x,t) + e^{\frac{d}{2\theta}x} f(x,t) \\ &\quad -\theta\delta(x)\left[M_1 \mathrm{sign}(z(0,t)) - \psi_1(t)\right] \\ &\quad -\theta\delta(x-1)e^{\frac{d}{2\theta}}\left[M_2 \mathrm{sign}(z(1,t)) - \psi_2(t)\right] \end{aligned} \tag{6.215}$$

with no advection term, with the associated initial condition

$$Q(x,0) = e^{\frac{d}{2\theta}x} z^0(x) \in L_2(0,1), \tag{6.216}$$

and with Robin (mixed) boundary conditions

$$\begin{aligned} Q_x(0,t) &= \left(\frac{d}{2\theta} + k_1\right) Q(0,t), \\ Q_x(1,t) &= \left(\frac{d}{2\theta} - k_2\right) Q(1,t). \end{aligned} \tag{6.217}$$

It is worth recalling that the sliding modes along the discontinuity surface(s) $z(0,t) = 0$ and/or $z(1,t) = 0$ are governed by the PDE (6.215) with nullified terms $M_1 \mathrm{sign}(z(0,t)) - \psi_1(t) = 0$ and/or $M_2 \mathrm{sign}(z(1,t)) - \psi_2(t) = 0$, and with the boundary conditions (6.217), specified with $k_1 = 0$ and/or $k_2 = 0$. In any case, a generalized solution of such a boundary-value problem, is subsequently shown to admit the Fourier expansion

$$Q(x,t) = \sum_{m=1}^{\infty} q_m(t) r_m(x), \ x \in [0,1], t \geq 0 \tag{6.218}$$

in terms of the eigenfunctions $r_m(x) \in L_2(0,1), \ k = 1, \dots$ of the Sturm–Liouville problem

$$\begin{aligned} &\theta r_{xx}(x) = -\mu r(x), \\ &r_x(0) = \left(\frac{d}{2\theta} + k_1\right) r(0), \ r_x(1) = \left(\frac{d}{2\theta} - k_2\right) r(1). \end{aligned} \tag{6.219}$$

By Theorem 3.1, the Sturm–Liouville problem (6.219) possesses a $L_2(0,1)$-orthonormal basis of the uniformly bounded eigenfunctions

$$r_m(\cdot): \quad \max_{x\in[0,1]} |r_m(x)| \le R, \; m = 1, 2, \ldots \tag{6.220}$$

with some positive constant R and with corresponding real eigenvalues $0 < \mu_1 < \mu_2 < \cdots < \mu_m < \cdots$ such that

$$\sum_{m=1}^{\infty} \mu_m^{-1} < \infty. \tag{6.221}$$

Substituting (6.218) into (6.215) and taking into account the Fourier expansion of the shifted Dirac distribution

$$\delta(x-\xi) = \sum_{m=1}^{\infty} r_m(\xi) r_m(x) \tag{6.222}$$

yield the solution Fourier modes $q_m(t), \; m = 1, 2, \ldots$ to be governed by

$$\begin{aligned}\dot q_m(t) &= \left(-\mu_m + \lambda - \frac{d^2}{4\theta}\right) q_m(t) + f_m^e(t)\\ &-\theta r_m(0)\Big[M_1 \mathrm{sign}(< q(t), r(0) >) - \psi_1(t)\Big]\\ &-\theta r_m(1) e^{\frac{d}{2\theta}}\Big[M_2 \mathrm{sign}(< q(t), r(1) >) - \psi_2(t)\Big],\end{aligned} \tag{6.223}$$

where

$$f_m^e(t) = \int_0^1 e^{\frac{d}{2\theta}x} f(x,t) r_m(x) dx \tag{6.224}$$

is the corresponding Fourier coefficient of the external disturbance $e^{\frac{d}{2\theta}x} f(x,t)$ and

$$< q(\cdot), r(\cdot) >= \sum_{m=1}^{\infty} q_m(\cdot) r_m(\cdot). \tag{6.225}$$

In order to reproduce (6.222) it suffices to verify that the action

$$\int_0^1 \sum_{m=1}^{\infty} r_m(\xi) r_m(x) \varphi(x) dx = \sum_{k=1}^{\infty} \varphi_m r_m(\xi) = \varphi(\xi)$$

of the right-hand side of (6.222) on an arbitrary test function $\varphi(x) = \sum_{m=1}^{\infty} \varphi_m r_m(x)$, (as a matter of fact, admitting the standard Fourier representation) coincides with that of the shifted Dirac distribution $\delta(x-\xi)$.

It thus follows that a potential solution (6.218) of the boundary-value problem (6.215)–(6.217) can be represented in the integral form

$$Q(x,t) = \sum_{m=1}^{\infty} e^{\left(-\mu_m+\lambda-\frac{d^2}{4\theta}\right)t} q_m^0 r_m(x)$$
$$+\sum_{m=1}^{\infty} r_m(x) \int_0^t e^{\left(-\mu_m+\lambda-\frac{d^2}{4\theta}\right)(t-\tau)} \Big\{ f_m^e(\tau)$$
$$-\theta r_m(0)\Big[M_1 \text{sign}(Q(0,\tau)) - \psi_1(\tau)\Big]$$
$$-\theta r_m(1) e^{\frac{d}{2\theta}}\Big[M_2 \text{sign}(Q(1,\tau)) - \psi_2(\tau)\Big]\Big\} d\tau, \tag{6.226}$$

where

$$q_m^0 = \int_0^1 Q(x,0) r_m(x) dx, \ m = 1, 2, \ldots \tag{6.227}$$

are the Fourier coefficients of the initial distribution (6.216). In order to conclude that such a solution, formally satisfying (6.215)–(6.217), does globally exist it suffices to note that the right-hand side of (6.226) is straightforwardly verified to absolutely converge in $L_2(0,1)$ for any $t \ge 0$, for all $f \in L_\infty^{loc}(L_2(0,1))$, for all $\psi_1, \psi_2 \in L_\infty^{loc}$, and for any admissible value of the sign function (6.211). Indeed, taking into account that due to (6.220), (6.221), the series $\sum_{m=1}^{\infty} \left(\mu_m - \lambda + \frac{d^2}{4\theta}\right)^{-1} r_m(x)$ is absolutely convergent for any $x \in [0,1]$, one derives

$$\|Q(x,t)\|_{L_2(0,1)} \le \|Q(0,t)\|_{L_2(0,1)} e^{\left(-\mu_1+\lambda-\frac{d^2}{4\theta}\right)t}$$
$$+\|f\|_{L_\infty(0,t;L_2(0,1))} \sum_{m=1}^{\infty} \left(\mu_m - \lambda + \frac{d^2}{4\theta}\right)^{-1}$$
$$+\theta\Big(M_1 + \|\psi_1\|_{L_\infty(0,t)}\Big) \left[\sum_{m=1}^{\infty} \left(\mu_m - \lambda + \frac{d^2}{4\theta}\right)^{-1} r_m(0)\right]$$
$$+\theta e^{\frac{d}{2\theta}}\Big(M_2 + \|\psi_2\|_{L_\infty(0,t)}\Big) \left[\sum_{m=1}^{\infty} \left(\mu_m - \lambda + \frac{d^2}{4\theta}\right)^{-1} r_m(1)\right] < \infty \tag{6.228}$$

for all $t \ge 0$.

Extrapolating from $N = 2$ to an arbitrary N, the following result is summarized.

Theorem 6.7 *The closed-loop system* (6.202), (6.205) *with the discontinuous feedback* (6.210) *globally possesses a generalized solution for any initial condition* (6.200), *for any plant parameters* (6.207), *for any controller gains* k_i, $M_i \ge 0$, $i = 1, \ldots, N$, *and for any* $f \in L_\infty^{loc}(L_2(0,1))$, $\psi_1, \psi_2 \in L_\infty^{loc}$.

Proof In a particular case where only $N = 2$ boundary actuators were in play, it was explicitly shown that by means of the invertible transformation (6.214), a generalized solution of the closed-loop system (6.202), (6.205), (6.210) can be expressed in terms of a generalized solution of the auxiliary system (6.215), (6.217), which admits the integral representation (6.226) of the absolutely and uniformly convergent Fourier series (6.218). The extension of the proof to an arbitrary number N of the available actuators is straightforward. Such a technical extension is, however, rather lengthy and its details are left to the interested reader. ■

6.4.3 ISS Analysis and Tuning

Once the closed-loop system (6.202), (6.205), (6.210) is established to possess generalized solutions, its Lyapunov ISS analysis becomes eligible. Before applying such an analysis, suppose that the collocated proportional-discontinuous output feedback (6.210) is tuned with nonnegative switching gains M_i and with proportional gains such that

$$k_1 > 1 + \frac{1}{2}D, \ k_i > 1, \ i = 2, \dots, N-1, \ k_N \geq \frac{1}{2}D. \tag{6.229}$$

Provided that the number N of actuator–sensor pairs is large enough to meet the condition

$$N \geq \max\left\{1 + \frac{\beta + 2\Lambda}{\theta_0}, 0\right\} \tag{6.230}$$

with β being the decay rate from (6.206), the next result proves to be in force.

Theorem 6.8 *Under Assumption 1 on the plant parameters, consider the perturbed reaction–diffusion–advection process* (6.205) *with the boundary condition* (6.202) *and arbitrary initial condition* (6.200). *Let it be controlled by the output feedback* (6.210) *with proportional gains, tuned according to* (6.229), *with arbitrary switching magnitudes $M_i, i = 1, \dots, N$, and with sufficiently large actuators number N, satisfying* (6.230). *Then the closed-loop system is exponentially ISS so that* (6.206) *holds with*

$$\gamma_0 := \frac{\sqrt{2}}{\varepsilon\beta}, \quad \gamma_j := \frac{\theta_0}{\varepsilon\beta}, \quad j = 1, \dots, N, \tag{6.231}$$

and sufficiently small $\varepsilon > 0$ such that

$$\frac{\theta_0}{h} - 2\Lambda - \frac{\sqrt{2}\varepsilon}{2} > 0, \tag{6.232}$$

$$\varepsilon < 2\min\left\{k_1 - \frac{D}{2} - 1, k_i - 1, k_N - \frac{D}{2}\right\}, \tag{6.233}$$

for $i = 2, \dots, N-1$.

Proof Consider the Lyapunov functional candidate

$$V(z) = \frac{1}{2}\int_0^1 z^2(x,t)dx = \frac{1}{2}\|z(\cdot,t)\|^2_{L_2(0,1)}. \tag{6.234}$$

By applying the differentiation rule (3.102) (specified with $\varphi(\xi) = z(\xi, t)$, the spatial variable $\xi = x$, and the frozen time instant t) to the generalized (weak) solutions of the boundary-value problem (6.202), (6.205), the time derivative of $V(z) = V(z(\cdot,t))$ along these solutions, which is for simplicity referred to as $\dot{V}(t)$, is evaluated as follows:

$$\begin{aligned}\dot{V}(t) = &\int_0^1 z(x,t)z_t(x,t)dx = -\theta\int_0^1 z_x^2(x,t)dx\\ &+\frac{1}{2}d\left[z^2(1,t) - z^2(0,t)\right]\\ &+\lambda\int_0^1 z^2(x,t)dx + \int_0^1 z(x,t)f(x,t)dx\\ &+\theta\sum_{i=1}^N z(x_i,t)\left[u_i(t) + \psi_i(t)\right]dx.\end{aligned} \tag{6.235}$$

Estimating the right-hand side of the above relation (6.235) by applying the well-known Cauchy–Schwartz inequality (see Lemma 3.1) to its fourth term, then substituting the control law (6.210) into its fifth term, and finally employing the plant parameter estimates (6.207), one arrives at

$$\begin{aligned}\dot{V}(t) \le &-\theta_0\|z_x(\cdot,t)\|^2_{L_2(0,1)} + \Lambda\|z(\cdot,t)\|^2_{L_2(0,1)}\\ &+\|f(\cdot,t)\|_{L_2(0,1)}\|z(\cdot,t)\|_{L_2(0,1)} - \theta_0\sum_{i=1}^N \tilde{k}_i z^2(x_i,t)\\ &-\theta_0\sum_{i=1}^N\left[M_i - |\psi_i(t)|\right]|z(x_i,t)|,\end{aligned} \tag{6.236}$$

where

$$\begin{aligned}&\tilde{k}_1 = k_1 - \frac{1}{2}D, \quad \tilde{k}_N = k_N - \frac{1}{2}D,\\ &\tilde{k}_i = k_i, \quad i = 2,\ldots,N-1.\end{aligned} \tag{6.237}$$

By applying the straightforward norm decomposition

$$\|z(\cdot)\|^2_{L_2(0,1)} = \sum_{i=1}^{N-1}\|z(\cdot)\|^2_{L_2(x_i,x_{i+1})} \tag{6.238}$$

to the solution derivative $z_x(\cdot)$, it follows that

$$\dot V(t) \le -\theta_0 \sum_{i=1}^{N-1}\left[\|z_x(\cdot)\|^2_{L_2(x_i,x_{i+1})} + \tilde k_i z^2(x_i,t)\right]$$
$$+ \Lambda\|z(\cdot,t)\|^2_{L(0,1)} + \|f(\cdot,t)\|_{L_2(0,1)}\|z(\cdot,t)\|_{L_2(0,1)}$$
$$- \theta_0\tilde k_N z^2(x_N,t) - \theta_0\sum_{i=1}^{N}\left[M_i - |\psi_i(t)|\right]|z(x_i,t)|. \tag{6.239}$$

Now taking into account that the Poincaré inequality (3.7) under sampling (6.203) yields

$$\|z_x(\cdot)\|^2_{L_2(x_i,x_{i+1})} + \tilde k_i z^2(x_i,t) \ge h\|z_x(\cdot)\|^2_{L_2(x_i,x_{i+1})}$$
$$+ z^2(x_i,t) + \left(\tilde k_i - 1\right) z^2(x_i,t)$$
$$\ge \frac{1}{2h}\|z(\cdot)\|^2_{L_2(x_i,x_{i+1})} + \left(\tilde k_i - 1\right) z^2(x_i,t), \tag{6.240}$$

for $i = 1, 2, \ldots, N-1$, for $h = 1/(N-1) < 1$, and for the controller gains (6.229), inequality (6.239) is further manipulated to

$$\dot V(t) \le -\frac{\theta_0}{2h}\sum_{i=1}^{N-1}\|z(\cdot)\|^2_{L_2(x_i,x_{i+1})} + \Lambda\|z(\cdot,t)\|^2_{L_2(0,1)}$$
$$+ \|f(\cdot,t)\|_{L_2(0,1)}\|z(\cdot,t)\|_{L_2(0,1)} - \theta_0\sum_{i=1}^{N} k_i^* z^2(x_i,t)$$
$$- \theta_0\sum_{i=1}^{N}\left[M_i - |\psi_i(t)|\right]|z(x_i,t)|, \tag{6.241}$$

where

$$k_i^* = \tilde k_i - 1, \quad i = 1, \ldots, N-1, \text{ and } k_N^* = \tilde k_N. \tag{6.242}$$

Since (6.238) ensures that the Lyapunov functional candidate (6.234) satisfies the relation

$$\sum_{i=1}^{N-1}\|z(\cdot)\|^2_{L_2(x_i,x_{i+1})} = 2V(z), \tag{6.243}$$

inequality (6.241) is simplified to

$$\dot{V}(t) \leq -\left[\frac{\theta_0}{h} - 2\Lambda\right] V(z) + \|f(\cdot,t)\|_{L_2(0,1)}\sqrt{2V(z)} - \theta_0 \sum_{i=1}^{N} k_i^* z^2(x_i,t) - \theta_0 \sum_{i=1}^{N} [M_i - |\psi_i(t)|]\, |z(x_i,t)|. \tag{6.244}$$

The latter inequality verifies that relation (6.234) determines an exponential ISS Lyapunov functional (cf. Definition 4.3; also see Prieur and Mazenc 2012 for more details). Indeed, by taking into account the well-known inequality $2ab \leq \varepsilon a^2 + \varepsilon^{-1} b^2$, which is valid for arbitrary real numbers a and b and for an arbitrary $\varepsilon > 0$, (6.244) results in

$$\dot{V}(t) \leq -\left[\frac{\theta_0}{h} - 2\Lambda - \frac{\varepsilon}{\sqrt{2}}\right] V(z) - \theta_0 \sum_{i=1}^{N} \left[k_i^* - \frac{\varepsilon}{2}\right] z^2(x_i,t) + \frac{\sqrt{2}}{2\varepsilon}\|f(\cdot,t)\|^2_{L_2(0,1)} + \frac{\theta_0}{2\varepsilon} \sum_{i=1}^{N} \psi_i^2(t). \tag{6.245}$$

To complete the proof it remains to note that under conditions (6.229), (6.230), (6.233), coupled to notations (6.237), (6.242), the factors $\frac{\theta_0}{h} - 2\Lambda - \frac{\sqrt{2}\varepsilon}{2}$ and $k_i^* - \frac{\varepsilon}{2}$, $i = 1, \ldots, N$ are positive, and with this in mind, applying Comparison Lemma 2.1 to (6.245) yields the ISS property (6.206), specified with (6.231). ■

Theorem 6.8 presents the proportional gains tuning rules (6.229), (6.230) for the closed-loop system (6.202), (6.205), (6.210) to be ISS in the presence of mismatched disturbances, distributed over the plant domain. Properly tuning the magnitudes of the discontinuous components of the proposed feedback (6.210) allows one to additionally reject the matched disturbances, collocated with the available point-wise actuators. Such an extension of Theorem 6.8 is as follows.

Theorem 6.9 *Let along with the conditions of Theorem 6.8, Assumption 2 on the disturbance magnitudes be additionally in force and let the proportional gain tuning rules* (6.229), (6.230) *be accompanied with their counterpart*

$$M_i > \Psi_i, \qquad i = 1, 2, \ldots, N \tag{6.246}$$

on the discontinuous control components. Then starting from a finite time instant

$$T = \frac{\|z^0(\cdot)\|_{L_2(0,1)}}{(\sigma_0 - 1)F}, \tag{6.247}$$

dependent on the plant initial condition (6.200), *relation*

$$\|z(\cdot,t)\|_{L_2(0,1)} \le \gamma^2 F, \quad t \ge T \tag{6.248}$$

holds true with

$$\gamma^2 = \frac{2h}{\theta_0 - 2h\Lambda}\sigma_0, \tag{6.249}$$

and an arbitrary parameter $\sigma_0 > 1$ *for all generalized solutions* $z(\cdot,t)$ *of the closed-loop system* (6.202), (6.205), (6.210).

Proof Under Assumption 2, the differential inequality (6.244) on the Lyapunov functional (6.234) is simplified to

$$\dot{V}(t) \le -\left[\left(\frac{\theta_0}{h} - 2\Lambda\right)\sqrt{V(z)} - \sqrt{2}F\right]\sqrt{V(z)}. \tag{6.250}$$

The latter inequality ensures that the domain

$$\sqrt{V(z)} \le \sigma_0 \frac{\sqrt{2}hF}{\theta_0 - 2h\Lambda} \tag{6.251}$$

is invariant and it is reached in finite time with any $\sigma_0 > 1$. Indeed, (6.250) guarantees that the differential inequality

$$\dot{V}(t) \le -\sqrt{2}F(\sigma_0 - 1)\sqrt{V(z)} \tag{6.252}$$

holds true outside of domain (6.251) whereas (6.252) ensures the finite time attractiveness of this domain with a transient time T, estimated as follows:

$$T \le \frac{\sqrt{2V(z^0)}}{(\sigma_0 - 1)F}. \tag{6.253}$$

To complete the proof it suffices to note that by taking into account the explicit Lyapunov functional representation (6.234), the transient time estimate (6.253), and the attraction domain (6.251) itself are readily reproduced in terms of the state $z(\cdot,t)$ of the plant in the form of (6.247) and (6.248) respectively. ■

6.4.4 Supporting Simulation

In Pisano and Orlov (2017), capabilities of the ISS synthesis are numerically illustrated for the boundary-value problem (6.205), (6.202) with the parameters $\theta = 10$, $d = 1$, and $\lambda = 2$, distributed disturbance $f(x,t) = 20sin(2\pi x)sin(10\pi t))$ and collocated matched disturbances $\psi_i(t) = sin(2\pi t)$ $(i = 1, 2, \ldots, N)$. Due to the cho-

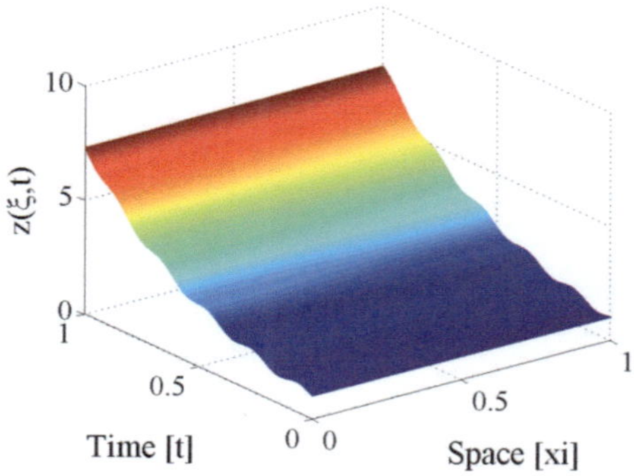

Fig. 6.4 Spatiotemporal solution profile in the open-loop test (Reprinted from Pisano and Orlov 2017, Copyright 2017, with permission from Elsevier)

sen disturbance profiles, the bounds in (6.208) take the value $F = 20/\sqrt{2} \approx 14.14$ and $\Psi_i = 1$. The initial condition is set as $z(\xi, 0) = 1$ and the upper bounds (6.207) for the uncertain parameters are specified as $\theta_0 = 5$, $D = 2$, $\Lambda = 3$. In accordance with (6.229) and (6.246), the controller parameters are set to $k_i = M_i = 2, \ i = 1, 2, \ldots, N$.

For solving the closed-loop PDE, a standard finite-difference approximation method is used by discretizing the spatial solution domain $x \in [0, 1]$ into a finite number of 200 uniformly spaced solution nodes. The resulting 200th-order discretized system is then solved by the fixed-step forward Euler method with step $T_s = 10^{-5}$.

To begin with, the performance of the open-loop system is numerically analyzed. Figure 6.4 depicts the unstable spatiotemporal profile of the solution $z(x, t)$ with $u_i(t) = 0, \ i = 1, 2, \ldots, N$.

The disturbance attenuation problem is then addressed under the prespecified attenuation level $\gamma^2 = 0.1$ for the attenuation factor (6.249) appearing in (6.248) and the decay rate $\beta = 10$. By applying relations (6.249), (6.203), (6.230), it turns out that at least $N = 32$ actuator–sensor pairs are required to guarantee the desired disturbance attenuation level and decay rate. The corresponding closed-loop behavior is shown in Fig. 6.5, which reports the long-term time evolution of the L_2 norm $\|z(\cdot, t)\|_{0,0,1}$ (left plot) and a zoom on the steady-state profile (right plot). The left plot shows clearly the finite time convergence toward the invariant domain (6.248). The zoom in the right plot shows that the steady-state accuracy remains within the guaranteed bound $\|z(\cdot, t)\|_{0,0,1} \leq \gamma^2 F \approx 1.4$, specified according to (6.248), (6.249). As typical in the variable structure control design, the actual accuracy appears to be much higher than that predicted by the theoretical computations, which is due to the worst-case nature of the underlying analysis.

To investigate how the number of actuator–sensor pairs varies with the desired level of attenuation γ^2, the computation of the minimal number N_{min} of required actuators has been made by considering relation (6.249) with different values of the desired attenuation coefficient γ^2. The resulting diagram is shown in Fig. 6.6 which highlights the inverse dependence of N_{min} on γ^2.

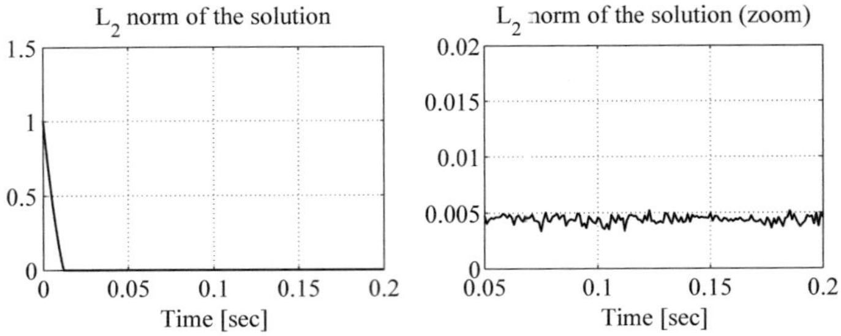

Fig. 6.5 Solution L_2 norm $\|z(\cdot, t)\|_{0,0,1}$ in the closed-loop test ($N = 32$) (Reprinted from Pisano and Orlov 2017, Copyright 2017, with permission from Elsevier)

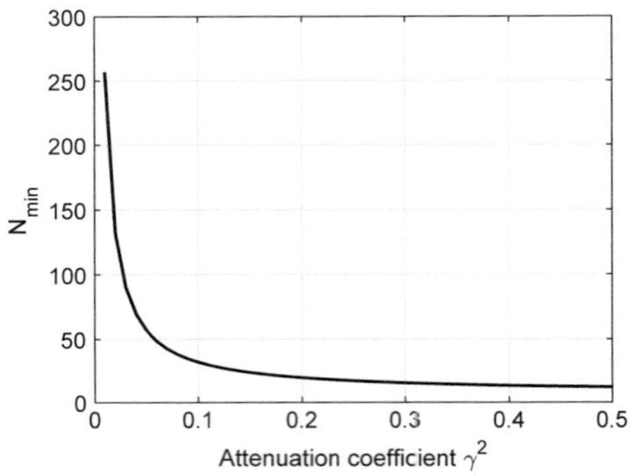

Fig. 6.6 Minimal number N_{min} of actuator–sensor pairs versus desired attenuation level (Reprinted from Pisano and Orlov 2017, Copyright 2017, with permission from Elsevier)

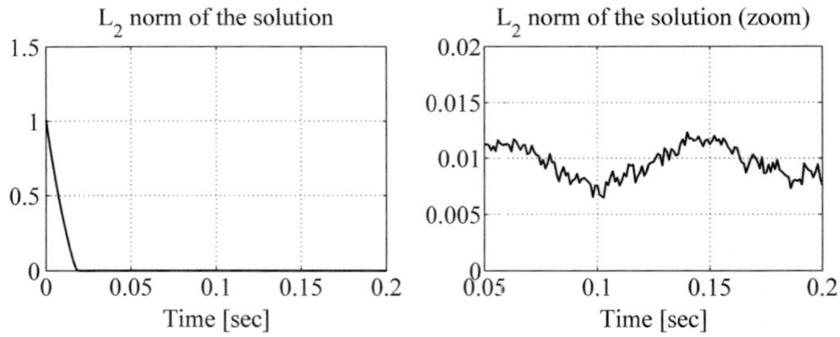

Fig. 6.7 Solution L_2 norm $\|z(\cdot, t)\|_{0,0,1}$ in the closed-loop test ($N = 19$) (Reprinted from Pisano and Orlov 2017, Copyright 2017, with permission from Elsevier)

Particularly, in order to attain the less restrictive disturbance attenuation level $\gamma^2 = 0.2$, compared to the value $\gamma^2 = 0.1$ used in the previous test, one needs to utilize at least $N = 19$ actuator–sensor pairs. While the left plot of Fig. 6.7 illustrates that using 19 actuator–sensor pairs, the transient time remains almost the same, the right plot of the figure shows that the actual accuracy decreases by nearly two times as compared with the previous test when 32 actuator–sensor pairs were used.

6.5 Concluding Remarks

Nonsmooth Lyapunov function constructions are utilized to tune FOSM and SOSM control algorithms for attenuating mismatched disturbances with a priori unknown bounds on their magnitudes. The proposed tuning paradigm is additionally illustrated with an uncertain reaction–diffusion–advection process, affected by matched and mismatched disturbances. A FOSM control system, composed of a finite number of actuation and sensing devices is tuned to achieve the ISS of the resulting closed-loop system. For the underlying DPS, it is shown that the level of the mismatched disturbance attenuation can be set arbitrarily, and the complete rejection of the distributed disturbance is theoretically achievable when the number of devices tends to infinity. Further PDE-flavored investigation on sampled-in-space SOSM controllers is among relevant problems to be tackled within the developed framework. Such SOSM controllers are expected to also attenuate the chattering phenomenon as it was the case of the first SOSM control applications (Orlov et al. 2010, 2011a, b) to DPS.

References

Boyd S, Ghaoui LE, Feron E, Balakrishnan V (1994) Linear matrix inequality in systems and control theory. SIAM Frontier Series, Philadelphia

Butkovskii AG (1982) Greens functions and transfer functions handbook. Horwood Ltd., Chichester

Byrnes C, Isidori A (1991) Asymptotic stabilization of minimum phase nonlinear systems. IEEE Trans Auto Control 36(10):1122–1137

Guo BZ, Jin FF (2015) Output feedback stabilization for one-dimensional wave equation subject to boundary disturbance. IEEE Trans Auto Control 60:824–830

Levant A (1993) Sliding order and sliding accuracy in sliding mode control. Int J Control 58:1247–1263

Levaggi L (2002) Sliding modes in banach spaces. Differ Integr Equ 15:167–189

Orlov Y, Aguilar L, Cadiou JC (2003) Switched chattering control vs. backlash/frcition phenomena in electrical servo motors. Int J Control 76:959–967

Orlov Y, Pisano A, Usai E (2010) Continuous state-feedback tracking of an uncertain heat diffusion process. Syst Control Lett 59:754–759

Orlov Y, Pisano A, Usai E (2011a) Tracking control of the uncertain heat and wave equation via power-fractional and sliding-mode techniques. SIAM J Control Optim 49:363–382

Orlov Y, Pisano A, Usai E (2011b) Exponential stabilization of the uncertain wave equation via distributed dynamic input extension. IEEE Trans Auto Control 56:212–216

Osuna T, Montano O, Orlov Y (2016) Nonlinear $\mathscr{L}_2$-gain analysis of hybrid systems in the presence of sliding modes and impacts. Math Probl Eng, Article ID 9074096, 10 p

Osuna T, Ponce IU, Orlov Y, Aguilar LT (2018a) $\mathscr{L}_2$-gain analysis of sliding mode dynamics. In: Fridman L, Barbot JP, Pleastan F (eds) Recent trends in sliding mode control. IET, London

Osuna T, Orlov Y, Aguilar LT (2018b) $\mathscr{L}_2$-gain tuning of variable structure SISO systems of relative degree n. Int J Control 91(11):2422–2444

Oza HB, Orlov Y, Spurgeon SK (2011) Lyapunov-based settling time estimate and tuning for twisting controller. IMA Journal of Mathematical Control and Information 29(4): 471–490

Perruquetti W, Floquet T, Orlov Y (2003) Finite time stabilization of interconnected second order nonlinear systems. In: Proceedings of the 42nd IEEE conference on decision and control, pp 4599–4604

Pisano A, Orlov Y (2017) On the ISS properties of a class of parabolic DPS with discontinuous control using sampled-in-space sensing and actuation. Automatica 81:447–454

Prieur C, Mazenc F (2012) ISS-Lyapunov functions for time-varying hyperbolic systems of balance laws. Math Control Signals Syst 24:111–134

Utkin VI (1992) Sliding modes in control and optimization. Springer, Berlin

Utkin V (2015) Mechanical energy-based lyapunov function design for twisting and supertwisting sliding mode control. IMA J Math Control Inf 32:675–688

Chapter 7
Lyapunov Approach to Adaptive Identification and Control in Infinite-Dimensional Setting

A standard approach to identifying a finite-dimensional dynamic system implies that the structure of the system is deduced by using physical laws and the problem is in finding the values of parameters in the state equation. The ability to ensure this objective is typically referred to as the parameter identifiability. For complex systems, however, it may not be possible to model all of the systems and a black-box approach should be brought into play. Here knowledge of the input–output map comes from controlled experiments. The questions then arise as to if the unknown parameters of the system can be reconstructed on-line and in which sense a state-space model, if any, is unique.

Once the parameter identifiability of the system is guaranteed, the task of the plant identification can be carried out by the Lyapunov redesign, which effectively utilizes the prior knowledge of the plant structure through its inclusion into the reference model, whose parameters evolve in time and converge to the unknown parameters of the plant. Many aspects of the resulting adaptive identifier design is well-understood by now, and due to the relative simplicity of the implementation and some degree of robustness with respect to small perturbations of the plant dynamics, these identifiers found practical applications both by itself and as a part of a model reference adaptive control (MRAC) system. There is already a large body of literature on this subject such as Åström and Wittenmark (1989), Goodwin and Sin (1984), Ioannou and Sun (1996), Landau (1979), Narendra and Annaswamy (1989), Sastry and Bodson (1989) to name a few.

In the present chapter, the effectiveness of the Lyapunov redesign is investigated side by side for the adaptive identification of linear time-delay and distributed parameter systems using Razumikhin and Krasovskii extensions.

Y. Orlov, *Nonsmooth Lyapunov Analysis in Finite and Infinite Dimensions*, Communications and Control Engineering,
https://doi.org/10.1007/978-3-030-37625-3_7

7.1 Lyapunov–Razumikhin Redesign and Identification of Linear Time-Delay Systems

The identifiability of linear delay-free systems is well-recognized (Miller and Michel 1990) to rely on their controllability. Unknown parameters of such a controllable system can be identified on-line, provided the system is persistently excited in a certain sense. To constructively address the persistent excitation notion a sufficiently rich input is prespecified independently of any particular underlying system.

The pioneering works (Nakagiri and Yamamoto 1995; Verduyn Lunel 2001) on the identification of linear TDS have demonstrated complexity of the problem, particularly, the parameter identifiability of a time-delay system was shown to place a restrictive condition on the structure of the system. The condition was defined through the characteristic matrix of the functional differential equation of the system in question and no indication was given on how to attain this condition using some accessible inputs.

Later on, constructively verifiable and enforceable identifiability conditions, and adaptive identification of system parameters and delays were proposed in Belkoura and Orlov (2002), Orlov et al. (2002, 2003) for general linear dynamic systems with delayed states, control inputs, and measured outputs. A delay-independent counterpart of the controllability condition, namely, the weak controllability condition, imposed in the above works on the delay system, guaranteed the identifiability of all unknown system parameters. If the latter condition was satisfied, sufficiently rich input signals were explicitly constructed in order to enforce the parameter identifiability of the system. It should be noted that the notion of a sufficiently rich system input did not relate to a system and it could, therefore, be verified independently of any particular underlying system, similar to that of Miller and Michel (1990) for the delay-free case.

The sufficiently rich input signals are subsequently utilized to construct both an adaptive parameter identifier and MRAC of the underlying delay system. However, for the combined plant/identifier and plant/controller systems these signals are introduced in conceptually different forms: for the former such a signal is straightforwardly applied to the plant whereas for the latter it is applied to the reference model.

The adaptive identifier and MRAC law are given in terms of ordinary differential equations, referred to as estimation systems, describing the evolution of the state estimate and the parameter estimate. The structure of the estimation systems is based on the Lyapunov–Razumikhin redesign extension, which is inspired from the finite-dimensional Lyapunov redesign (Ioannou and Sun 1996; Narendra and Annaswamy 1989), and it has what is called the series–parallel configuration. The justification of the convergence of the tunable parameters to their nominal values is based on the invariance principle extension to periodic TDS, given in Haddock and Terjeki (1983).

For the coupled plant/estimation (plant/identifier or plant/controller) system, a Lyapunov–Razumikhin function with a nonpositive time derivative along the solutions of the system is constructed. The time derivative of the Lyapunov–Razumikhin function takes zero values on a certain manifold in the state space. Applying a periodic sufficiently rich input signal (to the plant in the plant/identifier system and, respectively, to the model reference in the plant/controller system) guarantees the absence of the nontrivial trajectories on the manifold, thereby yielding the convergence of the tuned parameters to their nominal values.

For the sake of simplicity, the development of the MRAC design procedure is confined to SISO delay systems, unlike that of the adaptive identification where the general MIMO case is studied.

7.1.1 State-Space Representation and Weak Controllability

Linear dynamic systems with delayed states, control inputs, and measured outputs, all of finite dimension with a finite number of lumped delays, are governed by functional differential equations of the form

$$\dot{x}(t) = \sum_{i=0}^{r} [A_i x(t-\tau_i) + B_i u(t-\tau_i)], \tag{7.1}$$

$$y(t) = \sum_{i=0}^{r} C_i x(t-\tau_i),\ t \geq 0. \tag{7.2}$$

Hereinafter, $x(t) \in \mathbb{R}^n$ is a state vector, $u(t) \in \mathbb{R}^p$ is a piecewise continuous control input, $y(t) \in \mathbb{R}^q$ is a measured output, $0 = \tau_0 < \tau_1 < \cdots < \tau_r$ are time delays, $A_i \in \mathbb{R}^{n\times n}$, $B_i \in \mathbb{R}^{n\times p}$, $C_i \in \mathbb{R}^{n\times q}$, $i = 0, \ldots, r$ are matrix parameters. Since some matrix entries might be zero, without loss of generality system (7.1), (7.2) is assumed to possess the same state, input, and output delays.

For the autonomous homogeneous system

$$\dot{x}(t) = \sum_{i=0}^{r} A_i x(t-\tau_i) \tag{7.3}$$

with multiple delays and initial condition

$$x(-\theta) = \varphi(\theta) \in L_2(0, \tau_r; \mathbb{R}^n). \tag{7.4}$$

Lemma 3.6 is straightforwardly extended from the single-delay case. A unique solution

$$x_t(\theta) = x(t+\theta) = S(t)\varphi(\theta)\ t \geq 0 \tag{7.5}$$

of (7.3), (7.4) is thus concluded to exist where the shift operator $S(t),\ t \geq 0$ constitutes a C_0-semigroup of the autonomous homogeneous system (7.3) on the Hilbert space $H = \mathbb{R}^n \times L_2(0, \tau_r; \mathbb{R}^n)$ (see Sect. 3.6 for details).

Moreover (Hale 1971; Kolmanovskii and Nosov 1986), given $\varphi(\theta)$ and $w(t) = \sum_{i=0}^{r} B_i u(t - \tau_i)$, there exists a unique solution of (7.1), satisfying the initial condition (7.4). In terms of the aforementioned shift operator $S(t)$, this solution is given by

$$x_t = S(t)\varphi + \int_0^t S(t-s) w_s ds \tag{7.6}$$

where the input $w_s(\theta) = w(s - \theta)$ is viewed as a function of the argument $\theta \in [0, \tau_r]$. If the linear system (7.1) is asymptotically stable under $u \equiv 0$ then the semigroup $S(t)$ is exponentially stable. Just in the case, the Lyapunov–Krasovskii functional

$$V(x) =< Wx, x >$$

defined by means of the inner product $< \cdot, \cdot >$ in the Hilbert space $H = \mathbb{R}^n \times L_2(0, \tau_r; \mathbb{R}^n)$ and the operator

$$W = \int_0^\infty S^*(t) S(t) dt \tag{7.7}$$

has the negative definite time derivative

$$\dot{V}(t) = - < x_t, x_t >$$

along the trajectories of the homogeneous system (7.3) (cf. (7.7) with its finite-dimensional counterpart (5.5)).

In addition, if the function $u(t)$ is periodic, the steady-state behavior of the system is a periodic orbit of (7.1) given by

$$z_t = \int_{-\infty}^t S(t-s) w_s ds. \tag{7.8}$$

Along with the delay differential system (7.1), (7.2), consider its associated model

$$\dot{X}(t) = A(\lambda)X(t) + B(\lambda)U(t), \tag{7.9}$$
$$Y(t) = C(\lambda)X(t), \tag{7.10}$$

over the ring $\mathbb{R}[\lambda]$ of polynomials in a vector variable $\lambda = (\lambda_1, \dots, \lambda_k)^T$ with real coefficients. In the associated model (7.9), (7.10), the variables $X(\cdot)$, $U(\cdot)$, and $Y(\cdot)$ take values in the free finitely generated modules $\mathbb{R}^n[\lambda]$, $\mathbb{R}^p[\lambda]$, and $\mathbb{R}^q[\lambda]$ of $\mathbb{R}[\lambda]$-matrices (i.e., matrices whose entries are polynomials in λ) of the dimensions of $n \times 1$, $p \times 1$, and $q \times 1$, respectively. The above model was proposed in Sontag (1976) with k, standing for the maximal number of non-commensurable delay units $\eta_1, \dots, \eta_k,\ i = 1, \dots, k$ in (7.1), (7.2), and with $\lambda_1, \dots, \lambda_k$, being fixed delay operators $(\lambda_i x)(t) = x(t - \eta_i)$ of these non-commensurable units.

Recall that units $\eta_1, \dots, \eta_k$ are commensurable if there exist integers $\alpha_1, \dots, \alpha_k$ such that $\sum_{i=1}^k \alpha_i \eta_i = 0$ and $\sum_{i=1}^k \alpha_i^2 \neq 0$ and, respectively, $\eta_1, \dots, \eta_k$ are not commensurable if $\sum_{i=1}^k \alpha_i \eta_i = 0$ with integers $\alpha_1, \dots, \alpha_k$ implies that $\sum_{i=1}^k \alpha_i^2 = 0$.

In the case where all the system delays $(\lambda_i x)(t) = x(t - \tau_i),\ i = 1, \dots, r$ are non-commensurable one deal with the delay operators

$$A(\lambda) = \Sigma_{i=1}^r A_i \lambda_i, \quad B(\lambda) = \Sigma_{i=1}^r B_i \lambda_i \tag{7.11}$$

whereas for commensurable system delays $\tau_i = \alpha_i \tau_1$ with $\alpha_0 = 0, \alpha_1 = 1$, and some integers $\alpha_j,\ j = 2, \dots, r$, one has $(\lambda x)(t) = x(t - \tau_1)$ and

$$A(\lambda) = \Sigma_{i=0}^k A_i \lambda^{\alpha_i}, \quad B(\lambda) = \Sigma_{i=0}^k B_i \lambda^{\alpha_i}. \tag{7.12}$$

In general, setting of the non-commensurable delay units $\eta_1, \dots, \eta_k$ in (7.1), (7.2), one arrives at the fixed delay operators $(\lambda_i x)(t) = x(t - \eta_i),\ i = 1, \dots, k$ of these non-commensurable units.

If

$$T(s, \lambda) = C(\lambda)(sI - A(\lambda))^{-1} B(\lambda), \tag{7.13}$$

I is the identity operator, and $\lambda = e^{-s\eta} = (e^{-s\eta_1}, \dots, e^{-s\eta_k})^T$, then $T(s, e^{-s\eta})$ is the transfer function of the delay differential system (7.1), (7.2). Expanding (7.13) into Laurent series yields an equivalent representation of the transfer function:

$$T(s, e^{-s\eta}) = \sum_{j=1}^{\infty} s^{-j} C(e^{-s\eta}) A^{j-1}(e^{-s\eta}) B(e^{-s\eta}), \tag{7.14}$$

where

$$A(e^{-s\eta}) = \sum_{i=0}^r A_i e^{-s\tau_i},\ B(e^{-s\eta}) = \sum_{i=0}^r B_i e^{-s\tau_i},\ C(e^{-s\eta}) = \sum_{i=0}^r C_i e^{-s\tau_i}.$$

Thus, the transfer function of the delay differential system (7.1), (7.2) is completely defined by its Markov parameters $C(\lambda)A^{j-1}(\lambda)B(\lambda),\ j = 1, 2, \dots$.

The following definition of the weak controllability concept, initially introduced in Morse (1976) for commensurable delay systems, is given in Kalman matrix terms.

Definition 7.1 System (7.1) is said to be weakly controllable or controllable over the field $\mathbb{R}(\lambda)$ of rational functions in the vector variable λ with real coefficients iff

$$rank[B(\lambda) \mid A(\lambda)B(\lambda) \mid \cdots \mid A^{n-1}(\lambda)B(\lambda)] = n \tag{7.15}$$

over $\mathbb{R}[\lambda]$, or equivalently

$$rank\left[B(z) \mid A(z)B(z) \mid \cdots \mid A^{n-1}(z)B(z)\right] = n \tag{7.16}$$

for some $z \in \mathbf{C}^k$.

In the sequel, the weak controllability of linear TDS is shown to closely relate to their identifiability.

7.1.2 *Identifiability Analysis*

The identifiability concept is based on the comparison of system (7.1), (7.2), and its reference model

$$\dot{\hat{x}}(t) = \sum_{i=0}^{\hat{r}}[\hat{A}_i\hat{x}(t-\hat{\tau}_i) + \hat{B}_i u(t-\hat{\tau}_i)], \tag{7.17}$$

$$\hat{y}(t) = \sum_{i=0}^{\hat{r}} \hat{C}_i\hat{x}(t-\hat{\tau}_i),\ t \geq 0 \tag{7.18}$$

in which the system parameters A_i, B_i, C_i and delays τ_i, $i = 0, \ldots, r$ are replaced by $\hat{A}_i$, $\hat{B}_i$, $\hat{C}_i$ and $\hat{\tau}_i$, $i = 0, \ldots, \hat{r}$, respectively. The reference model (7.17), (7.18) is initialized with

$$\hat{x}(-\theta) = \hat{\varphi}(\theta) \in L_2(0, \tau_r; \mathbb{R}^n). \tag{7.19}$$

Definition 7.2 The Markov parameters of system (7.1), (7.2) are said to be identifiable, or equivalently, system (7.1), (7.2) admits the on-line identification of the transfer function iff there exists a control input $u(t)$ to be sufficiently rich for the system in the sense that the identity $y(t) \equiv \hat{y}(t)$ implies that

$$C(\lambda)A^{j-1}(\lambda)B(\lambda) = \hat{C}(\lambda)\hat{A}^{j-1}(\lambda)\hat{B}(\lambda),\ j = 1, 2, \ldots \tag{7.20}$$

(and consequently $T(s, \lambda) = \hat{T}(s, \lambda)$), regardless of a choice of the initial functions $\varphi(\theta)$, $\hat{\varphi}(\theta)$. In that case the identifiability is said to be enforced by the control input $u(t)$.

It is subsequently demonstrated how to explicitly construct a sufficiently rich control input, which enforces the identifiability of the Markov parameters of the system in question.

7.1.2.1 Sufficiently Nonsmooth Input Signals and On-Line Identifiability of the Transfer Function

Following Orlov et al. (2002), let us introduce notions of piecewise regular and sufficiently nonsmooth input signals to carry out that system (7.1), (7.2) is on-line transfer function identifiable under any sufficiently nonsmooth control input.

Definition 7.3 An input signal $u(t)$ is piecewise regular on $[-\tau_r, \infty)$ iff there exists an increasing sequence $\{t_j\}_{j=1}^{\infty}$ of time instants $-\tau_r = t_1 < t_2 < \cdots$ with $\lim_{j\to\infty} t_j = \infty$ such that $u(t)$ is of class $C^{\infty}(t_j, t_{j+1})$ on each subinterval (t_j, t_{j+1}), $j = 1, 2, \ldots$ and its lth-order derivative $u^{(l)}(t)$ has finite unilateral limits at the points t_j for all $l = 0, 1, \ldots$, i.e., $u(t+) = \lim_{t\to t_j+} u^{(l)}(t) < \infty$ and $u(t-) = \lim_{t\to t_j-} u^{(l)}(t) < \infty$.

Definition 7.4 A piecewise regular input signal $u(t)$ is said to be sufficiently discontinuous for system (7.1), (7.2) and its model (7.17), (7.18) if the following conditions are satisfied:

1. The set D of discontinuity points $t_j,\ j = 1, 2, \ldots$ of $u(t)$ contains a subset $D_0 \subset D$ which consists of at least $(p + k + \hat{k})$ non-commensurable points.
2. The jumps $\sigma_t = u(t+) - u(t-)$ at arbitrary p points $t_{\mu_i} \in D_0,\ i = 1, 2, \ldots, p$ form a matrix of the full rank, i.e.,

$$rank \left[\sigma_{t_{\mu_1}}, \ldots, \sigma_{t_{\mu_p}}\right] = p. \tag{7.21}$$

For ease of reference, the above conditions 1 and 2 are further referred to as the non-commensurability condition and the rank condition, respectively. Apparently, the rank condition does not relate to the plant equations whereas the non-commensurability condition explicitly utilizes the maximal number k of non-commensurable system delay units (as well as that $\hat{k}$ of the model) that can be unrealistic from a practical standpoint. To avoid this drawback one can modify the non-commensurability condition by requiring the set D_0 to consist of a countable number of discontinuity points of $u(t)$ so that both the non-commensurability condition and the rank condition can be verified independently of any underlying system. This feature allows one to subsequently ignore the underlying system and its model while constructing a sufficiently rich control input.

By definition, in order to construct a sufficiently discontinuous input signal one should, firstly, select a sufficient number of non-commensurable discontinuity points $t_j,\ j = 1, 2, \ldots$ (in particular, p points for non-delay systems when $k = 0$), secondly, construct corresponding input jumps σ_j which satisfy the rank condition, and thirdly, impose a smooth behavior on the input signal between the discontinuity points, while the prespecified input jumps are provided at these points.

The ability of construction of sufficiently many vectors $\sigma_j \in \mathbb{R}^p\ j = 1, 2, \ldots,$ which satisfy the rank condition, is demonstrated as follows. Construction of p such vectors $\sigma_j\ j = 1, \ldots, p,$ as a matter of fact, consists in specifying a basis

$\Omega = \{e_i\}_{i=1}^p$ in the Euclidean space $\mathbb{R}^p$. Since the union $\sum_p = \cup \mathbb{R}_{\sigma_{\mu_1},\ldots,\mu_{p-1}} \subset \mathbb{R}^p$ of all $(p-1)$-dimensional subspaces $\mathbb{R}_{\sigma_{\mu_1},\ldots,\mu_{p-1}} \subset \mathbb{R}^p$ spanned by arbitrary $p-1$ basis vectors $\sigma_{\mu_i} \in \Omega,\ i = 1,\ldots, p-1$ is of zero measure, it is always possible to find a vector $\sigma_{p+1} \in \mathbb{R}^p$ which does not belong to $\sum_p$. It follows that $p+1$ vectors $\sigma_j\ j = 1,\ldots, p+1$, thus constructed, satisfy the rank condition as well. An arbitrarily large number of vectors, satisfying the rank condition, is then obtained by iteration.

It is important from a practical standpoint to extend Definition 7.4 to the case when system (7.1), (7.2) is enforced by a continuous input signal, in particular, when the infinite slope of a sufficiently discontinuous input signal is approximated by a finite one.

Definition 7.5 A piecewise regular input signal $u(t)$ of class $C^l, l \geq 1$ is said to be sufficiently nonsmooth if its lth-order time derivative $u^{(l)}(t)$ is a sufficiently discontinuous input signal.

The following result on the Markov parameter identifiability is established in Orlov et al. (2002).

Theorem 7.1 *The identifiability of the system Markov parameters* $CA^{j-1}B$, $j = 1, 2, \ldots$ *of* (7.1), (7.2) *is enforced by any sufficiently nonsmooth control input* $u(t)$.

Proof The proof is based on the following observation. Since the solutions of TDS (7.1), (7.4) are well-known (Kolmanovskii and Nosov 1986) to become continuous on the period $[0, \tau_r]$, continuously differentiable on $[\tau_r, 2\tau_r]$, etc., the initial functions $\varphi(\theta)$, $\hat{\varphi}$ is subsequently assumed to be sufficiently smooth without loss of generality. With this in mind, the time derivatives $y^{(j)}(t)$ and $\hat{y}^{(j)}(t),\ j = 1, 2, \ldots$ of the state of the plant and that of the model are well-defined in the sense of Schwartz distributions. Taking into account that the system and model outputs coincide their time derivatives are concluded to coincide, too. While proving Theorem 7.1, these derivatives are computed through the well-known distribution formalism and convolutional calculus, especially developed for linear time-delay systems (Belkoura and Orlov 2002), and the equivalence of the output derivatives, thus computed, is then shown to result in the on-line identifiability of the transfer function of the system in question.

To begin with let us introduce n-vector functions $\varphi_l(t) = \varphi(\tau_l - t), l = 1, \ldots, r$ for $t \in [0, \tau_l]$ and $\varphi_l(t) = 0$, elsewhere. After that let us represent the state equation (7.1) in the form

$$\dot{X}(t) = \Sigma_{i=0}^r [A_i X(t-\tau_i) + B_i u(t-\tau_i)] + \delta\,\varphi(0) + \Sigma_{l=1}^r A_l \varphi_l(t), \quad (7.22)$$

$$Y(t) = \Sigma_{i=0}^r C_i X(t-\tau_i),\ t \geq 0 \quad (7.23)$$

with $\delta(t)$ being a Dirac function and

$$X(t) = \begin{cases} x(t) \text{ if } t \geq 0 \\ 0 \quad\ \text{if } t < 0 \end{cases}, \quad Y(t) = \begin{cases} y(t) \text{ if } t \geq 0 \\ 0 \quad\ \text{if } t < 0 \end{cases}. \quad (7.24)$$

In turn, system (7.22), (7.23) admits representation

$$\dot{X} = \mathscr{A} * X + \mathscr{B} * U + \delta\,\varphi(0) + \Phi, \tag{7.25}$$

$$Y = \mathscr{C} * X, \ t \geq 0, \tag{7.26}$$

where

$$\begin{aligned}
\mathscr{A}(t) &= \delta(t)A_0 + \delta_{\tau_1}(t)A_1 + \cdots + \delta_{\tau_r}(t)A_r,\\
\mathscr{B}(t) &= \delta(t)B_0 + \delta_{\tau_1}(t)B_1 + \cdots + \delta_{\tau_r}(t)B_r,\\
\mathscr{C}(t) &= \delta(t)C_0 + \delta_{\tau_1}(t)C_1 + \cdots + \delta_{\tau_r}(t)C_r,\\
\Phi(t) &= \sum\nolimits_{l=1}^{r} A_l\,\varphi_l(t),\\
U(t) &= \begin{cases} u(t) & \text{if } t \geq 0 \\ 0 & \text{if } t < 0 \end{cases},
\end{aligned} \tag{7.27}$$

$\delta_\tau(t) = \delta(t-\tau)$ is a Dirac function, atomized at $t = \tau$, symbol $*$ stands for the convolutional operator which is why

$$\begin{aligned}
(\mathscr{A} * X)(t) &= A_0\,X(t) + \cdots + A_r\,X(t-\tau_r),\\
(\mathscr{B} * U)(t) &= B_0\,u(t) + \cdots + B_r\,u(t-\tau_r),\\
(\mathscr{C} * X)(t) &= C_0\,X(t) + \cdots + C_r\,X(t-\tau_r).
\end{aligned} \tag{7.28}$$

Since the right-hand side of (7.25) involves impulsive inputs the meaning of this equation should be understood in the sense of distributions (see, e.g., Zemanian 1965), similar to PDE weak solutions of Sect. 3.2.4.

The nearest aim is to obtain analytical expressions for jth-order state derivatives $X^{(j)}(t),\ \ j = 1, 2, \ldots$ of (7.25) and decouple the resulting relations into a sum of regular and singular distributions.

For this purpose, the fundamental nxn-matrix $Q(t)$ of (7.25) is specified through the relations

$$Q^{(1)} = \mathscr{A} * Q + \delta,\ supp\ Q \subset [0, \infty) \tag{7.29}$$

where support $supp\ Q$ of $Q(t)$ is typically defined as closure of the set $\{t : Q(t) \neq 0\}$, and then analytical expressions are obtained for the time derivatives $Q^{(j)}(t)$, $j = 1, 2 \ldots$ of the fundamental matrix $Q(t)$.

Employing the well-known convolution property

$$(\mathscr{A} * Q)^{(1)} = \mathscr{A} * Q^{(1)} = \mathscr{A}^{(1)} * Q = \delta^{(1)} * \mathscr{A} * Q \tag{7.30}$$

and differentiating (7.29) yield

$$Q^{(2)} = \mathscr{A} * Q^{(1)} + \delta^{(1)} = \mathscr{A} * \mathscr{A} * Q + \mathscr{A} * \delta + \delta^{(1)} \tag{7.31}$$

where $\mathscr{A} * \delta = \Sigma_{l=0}^{r} A_l \delta_{\tau_l} * \delta = \Sigma_{l=0}^{r} A_l \delta_{\tau_l} = \mathscr{A}$ by definition

$$\delta_{\tau}^{i} * \delta_{\hat{\tau}}^{j} = \delta_{\tau+\hat{\tau}}^{i+j}, \ \tau, \hat{\tau} \in \mathbb{R}, \ i, j = 0, 1, \ldots \tag{7.32}$$

of distribution-dependent convolutions. Thus, by iterating on j, one derives

$$Q^{(j)} = \mathscr{A}^{*j} * Q + \sum_{i=0}^{j-1} \mathscr{A}^{*i} * \delta^{(j-1-i)}, \ j = 1, 2, \ldots \tag{7.33}$$

with $\mathscr{A}^{*i}$ recursively defined as $\mathscr{A}^{*0} = \delta$ and $\mathscr{A}^{*i} = \mathscr{A}^{*i-1} * \mathscr{A}, \ i = 1, 2, \ldots$.

Next let us note that for a sufficiently discontinuous input signal $u(t)$, the jth-order derivative $U^j(t)$ of $U(t)$ admits the following representation:

$$U^{(j)} = \{U\}^{(j)} + \sum_{l=1}^{j} \sum_{t_i} \delta_{t_i}^{(l-1)} \sigma_{t_i}^{(j-l)}, \tag{7.34}$$

where $t_i, \ i = 1, 2, \ldots$ are discontinuity points of $u(t)$, $\{U\}^{(j)}$ is a regular distribution, given by $\{U\}^{(j)}(t) = u^j(t)$ for $t \in (t_i, t_{i+1})$, and $\sigma^{(j)}(t) = u^{(j)}(t+) - u^{(j)}(t+), \ j = 0, 1, \ldots$. Since a sufficiently nonsmooth input $u(t)$ is obtained after differentiating an appropriate sufficiently discontinuous input, a similar representation is in force for such a signal as well, and without loss of generality, the proof focuses on the case as if a sufficiently discontinuous input applied to the system.

Finally, in order to collect all elements, necessary for an explicit representation of the jth-order derivatives of the solution $X(t)$ of (7.25), it remains to obtain time derivatives $\Phi^{(j)}, \ j = 1, 2 \ldots$ of Φ from (7.27). Due to the smoothness of the initial function $\varphi(\theta)$, that has additionally been imposed, also without loss of generality, the first-order derivative $(A_i\varphi_i)^{(1)}$ of $A_i\varphi_i, \imath = 1, \ldots, r$ is partitioned into a regular part, denoted by $\{A_i\varphi_i\}^{(1)}$, and a singular one

$$\begin{aligned}(A_i\varphi_i)^{(1)} &= \{A_i\varphi_i\}^{(1)} + \delta A_i\varphi_i(0+) - \delta_{\tau_i} A_i\varphi_i(\tau_i-) \\ &= \{A_i\varphi_i\}^{(1)} + \delta A_i\varphi(\tau_i-) - \delta_{\tau_i} A_i\varphi(0+).\end{aligned} \tag{7.35}$$

Similar to that, one derives

$$(A_i\varphi_i)^{(j)} = \{A_i\varphi_i\}^{(j)} + \sum_{\upsilon=0}^{j-1} \delta^{(j-1-\upsilon)} * (\delta A_i\varphi^{(\upsilon)}(\tau_i-) - \delta_{\tau_i} A_i\varphi^{(\upsilon)}(0+)) \tag{7.36}$$

that results in

$$\Phi^{(j)} = \{\Phi\}^{(j)} + \sum_{i=0}^{j-1} \delta^{(j-1-i)} * \Delta^{(i)}, \ j = 1, 2, \ldots \tag{7.37}$$

where $\{\Phi\}^{(j)}$ captures the corresponding regular distributions,

$$\Delta^{(i)} = \delta \sum_{j=1}^{r} A_j \varphi^{(i)}(\tau_j-) - \overline{\mathscr{A}}\, \varphi^{(i)}(0+), \tag{7.38}$$

and $\overline{\mathscr{A}} = \delta_{\tau_1} A_1 + \cdots + \delta_{\tau_r} A_r$. Clearly, the singular part of (7.38) is atomized at the time instants $0, \tau_1, \ldots, \tau_r$.

From now on, the solution $X(t)$ of (7.25) is also denoted as $X[\varphi, u](t)$ to emphasize the dependence of the solution on the exogenous inputs φ and u. It is clear that the solution X of the linear equation (7.25) is then given by

$$X = Q * (\delta\, \varphi(0) + \Phi) + Q * \mathscr{B} * U = X[\varphi, 0] + X[0, u]. \tag{7.39}$$

Taking into account (7.33), (7.34), (7.37), and (7.38), it follows that

$$\begin{aligned} X^{(j)}(\varphi, 0) &= Q^{(j)} * \delta\, \varphi(0) + Q * \Phi^{(j)} = \mathscr{A}^{*j} * Q * \delta\, \varphi(0) \\ &+ \sum_{i=0}^{j-1} \mathscr{A}^{*i} * \delta^{(j-1-i)} * \delta\, \varphi(0) + Q * \{\Phi\}^{(j)} + Q * \Delta^{(j-1)} \\ &+ \sum_{i=0}^{j-2} \mathscr{A}^{*j-1-i} * Q * \Delta^{(i)} + \sum_{i=0}^{j-2} \delta^{(j-2-i)} \sum_{\nu=0}^{i} \mathscr{A}^{*\nu} * \Delta^{(i-\nu)}. \end{aligned} \tag{7.40}$$

Since Q is a regular distribution, the second term and the last term are the only singular distributions in the right-hand side of the latter equality (7.40). These distributions are obviously atomized at the delay instants $\tau_i,\ i = 0, \ldots, r$.

In turn, differentiating $X(0, u)$ and using (7.30)–(7.34) yield

$$\begin{aligned} X^{(j)}(0, u) &= Q * \mathscr{B} * \{U\}^{(j)} + Q * \mathscr{B} * \sum_{l=1}^{j} \sum_{t_i} \delta_{t_i}^{(l-1)} \sigma_{t_i}^{(j-l)} \\ &= Q * \mathscr{B} * \{U\}^{(j)} + Q * \mathscr{B} * \sum_{t_i} \delta_{t_i}\, \sigma_{t_i}^{(j-1)} \\ &+ \sum_{l=2}^{j} \mathscr{A}^{*l-1} * Q * \mathscr{B} * \sum_{t_i} \delta_{t_i}\, \sigma_{t_i}^{(j-l)} \\ &+ \sum_{l=2}^{j} \sum_{\nu=0}^{l-2} \mathscr{A}^{*\nu} * \mathscr{B} * \sum_{t_i} \delta_{t_i}^{(l-2-\nu)}\, \sigma_{t_i}^{(j-l)}. \end{aligned} \tag{7.41}$$

It should be pointed out that in the right-hand side of the latter equality (7.41) just the last term is a singular distribution, atomized at the discrete set D_0 which by construction contains at least p points, say $t_{\mu_i},\ i = 1, 2, \ldots, p$, non-commensurable to the system delays $\tau_l,\ l = 0, \ldots, r$ and the model ones $\hat{\tau}_l,\ l = 0, \ldots, \hat{r}$.

Regrouping all regular distributions, which appear in the right-hand sides of (7.40) and (7.41), into $\{X(\varphi, 0)\}^{(j)}$ and $\{X(0, u)\}^{(j)}$, respectively, and taking into account (7.23), one arrives at

$$Y^{(j)}(\varphi, 0) = \mathscr{C} * \{X(\varphi, 0)\}^{(j)} + \mathscr{C} * \delta^{(j-1)}\varphi(0) + \sum_{l=0}^{j-2} \mathscr{C} * \delta^{(j-2-l)} * \mathscr{A}^{*l+1} * \delta\,\varphi(0) + \sum_{\nu=0}^{l} \mathscr{C} * \mathscr{A}^{*j} * \Delta^{(l-\nu)} \tag{7.42}$$

and

$$Y^{(j)}(0, u) = \mathscr{C} * \{X(0, u)\}^{(j)} + \sum_{l=2}^{j}\sum_{\nu=0}^{l-2} \mathscr{C} * \mathscr{A}^{*\nu} * \mathscr{B} * \sum_{t_i} \delta_{t_i}^{(l-2-\nu)}\, \sigma_{t_i}^{(j-l)}. \tag{7.43}$$

As a matter fact, while dealing with the reference model (7.17), (7.18), relations, similar to (7.42) and (7.43), are obtained.

Now the equality $y(t) = \hat{y}(t), t \geq 0$, or equivalently $Y(t) = \hat{Y}(t), t \geq 0$, ensures that the singular part of $Y^{(j)}$ is equivalent to that of $\hat{Y}^{(j)}$. Moreover, due to the non-commensurability condition, it follows that all multipliers, coming in (7.43) with the Dirac functions $\delta_{t_{\mu_i}}$, $i = 1, 2, \ldots, p$ and corresponding to the original system (7.1), (7.2), can separately be set to those corresponding to the model (7.17), (7.18):

$$\sum_{l=0}^{j-2} \mathscr{C} * \mathscr{A}^{*l} * \mathscr{B}\,\sigma_{t_{\mu_i}} = \sum_{l=0}^{j-2} \hat{\mathscr{C}} * \hat{\mathscr{A}}^{*l} * \hat{\mathscr{B}}\,\sigma_{t_{\mu_i}},\ j = 2, 3, \ldots. \tag{7.44}$$

Clearly, the latter relations, coupled to the rank condition (7.21), result in

$$\sum_{l=0}^{j-2} \mathscr{C} * \mathscr{A}^{*l} * \mathscr{B} = \sum_{l=0}^{j-2} \hat{\mathscr{C}} * \hat{\mathscr{A}}^{*l} * \hat{\mathscr{B}},\ j = 2, 3, \ldots. \tag{7.45}$$

From (7.45) subject to $j = 2$ it is concluded that

$$\mathscr{C} * \mathscr{B} = \hat{\mathscr{C}} * \hat{\mathscr{B}}. \tag{7.46}$$

In turn, from (7.45) subject to $j = 3$ and (7.46) it follows that

$$\mathscr{C} * \mathscr{A} * \mathscr{B} = \hat{\mathscr{C}} * \hat{\mathscr{A}} * \hat{\mathscr{B}}.$$

In general, the following relations

$$\mathscr{C} * \mathscr{A}^{*l} * \mathscr{B} = \hat{\mathscr{C}} * \hat{\mathscr{A}}^{*l} * \hat{\mathscr{B}},\ l = 0, 1, \ldots. \tag{7.47}$$

are obtained through iterations. Applying the Laplace transform $\mathscr{L}$ to these relations and employing the well-known Laplace transform property

$$\mathscr{L}[\mathscr{C} * \mathscr{A}^{*l} * \mathscr{B}](s) = \mathscr{L}[\mathscr{C}](s)\mathscr{L}^l[\mathscr{A}](s)\mathscr{L}[\mathscr{B}](s)$$

yield the equalities

$$C(z)A^l(z)B(z) = \hat{C}(z)\hat{A}^l(z)\hat{B}(z),\ l = 0, 1, \dots \tag{7.48}$$

which are satisfied for all vectors $z = e^{-s\eta} \in \Omega \subset \mathbf{C}^k$ with components $z_i = e^{-s\eta_i}$, $i = 1, \dots, k$ and $\Omega = \Im m\, e^{-s\eta}$. Identifiability of Markov parameters of system (7.1), (7.2) then straightforwardly results from (7.48). Thus, Theorem 7.1 is proved. ■

It should be noted that the on-line identifiability of the transfer function $T(s, \lambda)$ ($\lambda = e^{-s\eta}$) of system (7.1), (7.2) does not necessarily result in the identifiability of the system parameters $A(\lambda)$, $B(\lambda)$, $C(\lambda)$. If system (7.1), (7.2) is observable and weakly controllable then it represents a minimal realization of $T(s, \lambda)$ of least dimension n (Morse 1976). Such a realization $A(\lambda)$, $B(\lambda)$, $C(\lambda)$, however, is not unique because all similarity transforms

$$\tilde{A}(\lambda) = H^{-1}(\lambda)A(\lambda)H(\lambda),\ \tilde{B}(\lambda) = H^{-1}(\lambda)B(\lambda),\ \tilde{C}(\lambda) = C(\lambda)H(\lambda), \tag{7.49}$$

corresponding to unimodular matrices $H(\lambda)$, yield the same Markov sequence

$$C(\lambda)A^{j-1}(\lambda)B(\lambda) = \tilde{C}(\lambda)\tilde{A}^{j-1}(\lambda)\tilde{B}(\lambda),\ j = 1, 2, \dots .$$

For the convenience of the reader recall that $H(\lambda)$ is unimodular iff both $H(\lambda)$ and $H^{-1}(\lambda)$ are polynomial matrices in λ.

Thus, the parameter identifiability of system (7.1), (7.2) requires the number of unknown parameters to be reduced by specifying a form of (7.1), (7.2) such that if confined to this form the triple $(A(\lambda), B(\lambda), C(\lambda))$ is uniquely determined by the Markov parameters (7.49). A weakly controllable system is further established to be identifiable provided the entire state of the system is available for measurements, possibly, with delayed sensing.

7.1.2.2 Parameter and Delay Identifiability

To begin with, the parameter identifiability investigation is confined to system (7.1) with the full state measurement

$$y(t) = x(t - \tau),\ t \geq 0, \tag{7.50}$$

regardless of whether delayed sensing with $\tau > 0$ or delay-free sensing with $\tau = 0$ is available. Thus, (7.2) is specified with

$$C_l = I \; for \; some \; integer \; l \in [0, r] \qquad C_i = 0 \; for \; i \neq l. \tag{7.51}$$

A potential extension to incomplete measurements is far from being trivial and it is further exemplified for SISO TDS.

For TDS (7.1), (7.2) of the specific form (7.51) the parameter identifiability notion is addressed as follows.

Definition 7.6 System (7.1), (7.50) is said to be identifiable if there exists a control input $u(t)$ such that the identity $y(t) \equiv \hat{y}(t)$ results in

$$r = \hat{r}, \; \tau_i = \hat{\tau}_i, \; A_i = \hat{A}_i, \; B_i = \hat{B}_i \; for \; i = 0, \ldots, r,$$

regardless of a choice of the initial functions $\varphi(\theta)$, $\hat{\varphi}(\theta)$.

System (7.1), (7.50) turns out to be identifiable if and only if it is weakly controllable.

Theorem 7.2 *The time-delay system* (7.1), (7.50) *is identifiable if and only if it is weakly controllable. Moreover, if* (7.1), (7.50) *is weakly controllable then the parameter identifiability is enforced by any sufficiently discontinuous or nonsmooth control input* $u(t)$.

Proof *Sufficiency*: By Theorem 7.1 the identity $y(t) \equiv \hat{y}(t)$, enforced by a sufficiently discontinuous or nonsmooth control input $u(t)$, ensures that relations (7.20) are satisfied for all vectors $\lambda = e^{-s\eta} \in \Omega \subset \mathbf{C}^k$ with components $\lambda_i = e^{-s\eta_i}$, $i = 1, \ldots, k$ and $\Omega = \Im m \; e^{-s\eta}$. After removing the regular common factor $C(\lambda) = I\lambda_i^l$ with appropriate integers i and l these relations subject to (7.51) take the form

$$A^j(\lambda)\, B(\lambda) \;=\; \hat{A}^j(\lambda)\, \hat{B}(\lambda), \; j = 0, \ldots, n. \tag{7.52}$$

Relation (7.52), corresponding to $j = 0$, leads to

$$B(\lambda) = \hat{B}(\lambda) \; for \; all \; \lambda \in \Omega, \tag{7.53}$$

thereby yielding $B_0 + B_1\lambda^{\tau_1} + \cdots + B_r\lambda^{\tau_r} = \hat{B}_0 + \hat{B}_1\lambda^{\hat{\tau}_1} + \cdots + \hat{B}_{\hat{r}}\lambda^{\hat{\tau}_{\hat{r}}}$ for all $\lambda \in \Omega$. It follows

$$r = \hat{r}, \; \tau_i = \hat{\tau}_i, \; B_i = \hat{B}_i \; for \; i = 0, \ldots, r$$

and in order to complete the proof of the sufficiency part it remains to show that $A_i = \hat{A}_i, \; i = 0, \ldots, r$.

Coupled to (7.53), relation (7.52), corresponding to $i = 1$, is represented in the form

$$\Delta A(\lambda)\; B(\lambda) \;= 0 \tag{7.54}$$

where $\Delta A(\lambda) = A(\lambda) - \hat{A}(\lambda)$. By iterating on j, one derives

$$\Delta A(\lambda)\, A^j(\lambda)\, B(\lambda) = 0,\ j = 0, \ldots, n-1. \tag{7.55}$$

Due to the weak controllability of (7.1), there exists $\mu \in \Omega$ such that

$$rank\ \left[B(\mu) \mid \cdots \mid A^{n-1}(\mu)B(\mu)\right] = n.$$

By continuity this rank condition remains true in some domain $\Omega_0 \subset \Omega$. Thus, relation (7.55) ensures that $\Delta A(\lambda) = 0$ for all $\lambda \in \Omega_0 \subset \Omega$. Since $\Delta A(\lambda)$ is a polynomial function, one deduces that $A_i = \hat{A}_i$ for $i = 0, \ldots, r$. This completes the proof of the sufficiency part.

Necessity: If system (7.1), (7.50) is not weakly controllable then there exists a nonzero matrix $H(\lambda)$ with the property that

$$H(\lambda)(s - A(\lambda))^{-1} B(\lambda) = 0. \tag{7.56}$$

Now setting $z(t) = H(\lambda)x(t)$ one arrives at an uncontrollable submodule governed by $\dot{z}(t) = \tilde{A}(\lambda)z(t)$ with some matrix $\tilde{A}(\lambda)$ over the ring $\mathbb{R}[\lambda]$. Clearly, the entries of $\tilde{A}(\lambda)$ are unidentifiable under the zero initial conditions $z(t) = 0$. Thus, if system (7.1), (7.50) is not weakly controllable then it is not identifiable, and therefore an identifiable system (7.1), (7.50) is weakly controllable. Theorem 7.2 is completely proved. ■

Analyzing the proof of Theorem 7.2, one can see that the assertion of this theorem remains true if initial conditions (7.4), (7.19) and the control input $u(t)$, restricted to a regularity subinterval (t_j, t_{j+1}), $j = 1, 2, \ldots$, are only required to be of class C^{l+n+1} for some integer $l \geq 0$ where t_j is a discontinuity point of the lth-order time derivative of $u(t)$.

It is worth noticing that in a particular case of a delay-free system, the above result, Theorem 7.2, coincides with the classical result on the parameter identifiability. Indeed, the weak controllability of a delay-free linear dynamic system is nothing else than its controllability (indeed, (7.15) becomes the standard Kalman criterion of the controllability) whereas sufficiently discontinuous and nonsmooth input signals, expanded into appropriate Fourier series, are known (Miller and Michel 1990) to have enough frequencies and hence be sufficiently rich.

7.1.3 Razumikhin-Based Adaptive Identifier Design

Once the identifiability of system (7.1), (7.50) is guaranteed, the task of the identification of the unknown matrix parameters A_i, B_i, $i = 0, \ldots, r$ and delays τ_j, $j = 1, \ldots, r$ is achieved with the adaptive identifier proposed below. For a technical reason of dealing with bounded system trajectories, the unforced delay system (7.3) is assumed to be asymptotically stable.

The Lyapunov–Razumikhin approach to subsequently be used for the stability analysis of the overall plant-identifier system is an alternative to the Lyapunov–Krasovskii stability analysis of TDS

$$\dot{x}(t) = f(x_t) \tag{7.57}$$

with a continuous bounded operator $f : L_2(0, \tau; \mathbb{R}^n) \to \mathbb{R}^n$, and it relies on a positive definite function $V(x)$ rather than a functional $\mathscr{V}(x_t)$. The key idea behind the Razumikhin result (see, e.g., Gu et al. 2003; Haddock and Terjeki 1983) focuses on the Lyapunov–Krasovskii functional $\mathscr{V}(x_t) = \max_{\theta \in [-\tau,0]} V(x(t+\theta))$, which is constructed from a continuously differentiable positive definite function $V(x)$ and which does not grow provided that $\dot{V}(x(t))$ is nonpositive definite whenever $V(x(t)) = \mathscr{V}(x_t)$.

Theorem 7.3 (Razumikhin Invariance Principle) *Suppose there exists a continuously differentiable positive definite function $V(x)$ (referred to as a Razumikhin function) such that its time derivative $\dot{V}(x(t))$ along the solutions $x(t)$ of the system* (7.57) *is negative semidefinite whenever $V(x(t+\theta)) \le V(x(t))$ for all $\theta \in [-\tau, 0]$. Moreover, let the maximal invariant subset of the set $\dot{V}(x(t)) = 0$ consist of the origin only. Then* (7.57) *is asymptotically stable, and it is globally asymptotically stable if the function $V(x)$ is in addition radially unbounded.*

Proof Theorem 4.7 becomes applicable to (7.57) with the Lyapunov–Krasovskii functional $\mathscr{V}(x_t) = \max_{\theta \in [-\tau,0]} V(x(t+\theta))$, constructed from the Razumikhin function $V(x)$. By applying Theorem 4.7, the assertion of Theorem 7.3 is validated. ■

The above result is instrumental for the adaptive identifier design to be developed for linear TDS.

7.1.3.1 Synthesis Under A Priori Known Delays

For a moment, the time-delay values $\tau_j,\ j = 1, \dots, r$ are assumed to be known a priori. Let $G \in R^{n\times n}$ be a Hurwitz matrix and let positive definite symmetric matrices $P, Q \in R^{n\times n}$ solve the Lyapunov algebraic equation

$$G^T P + PG = -Q.$$

The following identification law was proposed in Orlov et al. (2003) to identify the matrix parameters $A_i,\ B_i,\ i = 0, \dots, r$:

$$\dot{\hat{x}}(t) = \sum_{i=0}^{r} [\hat{A}_i y(t-\tau_i) + \hat{B}_i u(t-\tau_i-\tau)] - G\Delta y(t),\ t \ge 0,$$

$$\hat{x}(-\theta) = \hat{\varphi}(\theta),\ 0 \le \theta \le \tau_r, \tag{7.58}$$

$$\dot{\hat{A}}_i = F_i P \Delta y(t) y^T(t-\tau_i),\ \hat{A}_i(0) = \hat{A}_i^0,$$
$$\dot{\hat{B}}_i = \Phi_i P \Delta y(t) u^T(t-\tau_i-\tau),\ \hat{B}_i(0) = \hat{B}_i^0, \tag{7.59}$$

where $\hat{A}_i^0 \in \mathbb{R}^{n\times n}$, $\hat{B}_i^0 \in \mathbb{R}^{n\times m}$, the output error $\Delta y(t) = y(t) - \hat{x}(t)$, the initial function $\hat{\varphi}(\theta) \in L_2(0, \tau_r; \mathbb{R}^n)$, the adaptation gain matrices F_i, Φ_i are positive definite. This law is inherited from Lyapunov redesign technique arguments (Ioannou and Sun 1996; Narendra and Annaswamy 1989), which are now extended to TDS.

The identification law is shown to ensure the convergence

$$\lim_{t\to\infty} \Delta y(t) = 0,\ \lim_{t\to\infty} \Delta A_i(t) = 0,\ \lim_{t\to\infty} \Delta B_i(t) = 0,\ i = 0, 1, \ldots, r \tag{7.60}$$

of the output error $\Delta y(t)$ and the parameter errors $\Delta A_i(t) = A_i - \hat{A}_i(t)$, $\Delta B_i(t) = B_i - \hat{B}_i(t)$ for arbitrary adaptation gains, initial conditions, and sufficiently nonsmooth periodic input signals. In order to construct a sufficiently nonsmooth periodic signal one can restrict any sufficiently nonsmooth signal to a finite time interval I such that the non-commensurability condition and the rank condition (7.21) hold for the nonsmoothness points $t_1, t_2, \ldots \in I$ of the signal and then continue this restricted signal periodically. A periodic signal, thus constructed, obeys Definition 7.5 and it is, therefore, sufficiently nonsmooth.

Theorem 7.4 *Let system* (7.1), (7.50) *to be identified satisfy the following conditions:*

1. *It is weakly controllable;*
2. *The delays τ, τ_i, $i = 1, \ldots r$ are known a priori;*
3. *The control input $u(t)$ is periodic and sufficiently nonsmooth.*

Then the limiting relations (7.99) *hold with the adaptive identification law* (7.58) *and parameters $\hat{A}_i(t)$ and $\hat{B}_i(t)$, $i = 0, \ldots, r$ tuned as* (7.59).

Proof For transparency, Theorem 7.4 is first proved in a particular, no-sensor delay case $\tau = 0$. The general result for $\tau \geq 0$ is then straightforwardly derived from this particular result.

(i) *The case* $\tau = 0$. Let us represent the overall system (7.1), (7.50), (7.58), (7.59) subject to $\tau = 0$ in terms of the output-parameter errors:

$$\dot{e}(t) = \sum_{i=0}^{r} A_i e(t-\tau_i),$$
$$\Delta\dot{x}(t) = \sum_{i=0}^{r} \{\Delta A_i [e(t-\tau_i) + z(t-\tau_i)] + \Delta B_i u(t-\tau_i)\} + G\Delta x(t),$$
$$\Delta\dot{A}_i(t) = -F_i P \Delta x(t) [e(t-\tau_i) + z(t-\tau_i)]^T,$$
$$\Delta\dot{B}_i = -\Phi_i P \Delta x(t) u^T(t-\tau_i),\ i = 0, \ldots, r \tag{7.61}$$

where $\Delta x(t) = x(t) - \hat{x}(t)$ is the state error, $e(t) = x(t) - z(t)$ is the state deviation with respect to the periodic orbit $z(t)$ of (7.1) which is given by (7.8). In the above equations, it has been used that $\Delta x(t) = \Delta y(t)$ when the delay-free state measurements are available.

In order to prove the global asymptotic stability (7.61) let us introduce the Lyapunov functional

$$V(e, \Delta x, \Delta A_0, \ldots, \Delta B_l) =< We, e > +\Delta x^T P \Delta x +$$
$$\sum_{i=0}^{r} [tr(\Delta A_i^T F_i^{-1} \Delta A_i) + tr(\Delta B_i^T \Phi_i^{-1} \Delta B_i)] \qquad (7.62)$$

composed of the quadratic Lyapunov–Krasovskii functional with W, determined by relation (7.7) for the exponentially stable semigroup $S(t)$ of the former equation of (7.61), and the Razumikhin function, specified by the remaining quadratic terms in the right-hand side of (7.62) where tr denotes the trace of a matrix. The computation of the time derivative of the Lyapunov functional V along the trajectories of (7.61) yields

$$\dot{V}(t) = - < e_t, e_t > -\Delta x^T(t) Q \Delta x(t).$$

Since this time derivative is only negative semidefinite, along with the overall system stability, it is concluded from Theorem 4.5 that only e_t and $\Delta x(t)$ escape to zero as $t \to \infty$.

The global asymptotic stability of (7.61), the right-hand side of which is clearly time-periodic, is additionally established by extending the Razumikhin invariance principle of Theorem 7.3 to periodic TDS. Such an extension to the periodic case is similar to the one obtained in Rouche et al. (1977) for non-delay systems.

According to the periodic Razumikhin invariance principle extension, there must be a convergence of the trajectories of (7.61) to the largest invariant subset of the set of solutions of (7.61) for which $\dot{V}(t) \equiv 0$, or equivalently

$$e(t) \equiv 0, \ \ \Delta x(t) \equiv 0. \qquad (7.63)$$

The manifold of (7.63), however, does not contain nontrivial trajectories of (7.61). Indeed, if (7.63) is in force, then taking into account (7.59), it follows that

$$\dot{\hat{A}}_i(t) = 0, \ \ \dot{\hat{B}}_i(t) = 0, \ \ i = 0, \ldots, r. \qquad (7.64)$$

In turn, coupled to (7.61), relations (7.63), (7.64) result in

$$\sum_{i=0}^{r} [\Delta A_i x(t - \tau_i) + \Delta B_i u(t - \tau_i)] \equiv 0$$

which by the identifiability of (7.1) (cf. Definition 7.6) ensures that

$$\Delta A_i = 0, \ \Delta B_i = 0, \ i = 0, \dots, r.$$

Due to Theorem 7.2, the identifiability of (7.1) is guaranteed under the conditions of the theorem, and it is actually enforced by the periodic input signal $u(t)$. Thus, by applying the invariance principle, system (7.61) is globally asymptotically stable and the required identifier convergence (7.99) is concluded. This proves Theorem 7.4 in the case of the delay-free sensing.

(ii) *The case* $\tau \geq 0$. To complete the proof let us note that in the general case of $\tau \geq 0$ system (7.1), rewritten in terms of the output vector $y(t) = x(t - \tau)$, is given by

$$\dot{y}(t) = \sum_{i=0}^{r} [A_i y(t - \tau_i) + B_i v(t - \tau_i)] \tag{7.65}$$

where the delayed control input $v(t) = u(t - \tau)$, and the transformed state vector $y(t)$ admits delay-free measurements. Apparently, the system, thus modified, still satisfies Conditions 1–3 of Theorem 7.4. Indeed, the verification of Conditions 1 and 2 is straightforward whereas Condition 3 for the delayed control input $v(t) = u(t - \tau)$ (where due to (7.51) $\tau = \tau_l$ for some integer $l \in [0, r]$) is verified by inspection of Definition 7.5. Thus, the earlier validated version of Theorem 7.4 with $\tau = 0$ is applicable to system (7.65). By applying this version to system (7.65) whose states are available to the delay-free measurements, the validity of Theorem 7.4 is established in the general case of $\tau \geq 0$. Theorem 7.4 is thus completely proved. ■

Since the Krasovskii–LaSalle invariance principle is not applicable to a general nonautonomous system (Khalil 2002) the periodicity condition imposed by Theorem 7.4 on the input signal can not be omitted. However, the assertion of Theorem 7.4 remains true if a periodic sufficiently nonsmooth input signal $u(t)$ is replaced by a nonperiodic one which saves the invariance principle (examples are almost-periodic signals and asymptotically autonomous signals Khalil 2002).

If some plant parameters are known a priori then the corresponding identifier equations can be omitted.

The above parameter identifier has numerically been tested (Orlov et al. 2003) to possess a favorable robustness property against small uncertainties on the time delays. This property and Theorem 7.4, coupled together, allow one to simultaneously identify, with a certain precision, all the matrix parameters and the delays in the system.

7.1.3.2 Synthesis Under A Priori Unknown Delays

The following identifier design procedure is proposed when the knowledge of the delay values is no longer available.

On the first step system (7.1) is modeled as an uncertain system whose a priori known delays $\hat{\tau}_i,\ i = 1, \ldots, \hat{r}$ are closely located to each other:

$$|\hat{\tau}_i - \hat{\tau}_{i-1}| < \varepsilon,\ i = 1, \ldots, \hat{r} \tag{7.66}$$

where $\varepsilon > 0$ is an identification precision. Since the earlier-obtained identifiability conditions do not depend on the commensurability of the delay values the model delays $\hat{\tau}_1, \ldots, \hat{\tau}_{\hat{r}}$ can uniformly be distributed over the time interval $[0, \mathscr{T}]$ with $\mathscr{T} \geq \tau_r$ being an upper estimate of the system delays. Certainly, in order to ensure the prescribed precision, the number $\hat{r}$ of the model delays is selected large enough so that some model matrix parameters (unknown, however, a priori) correspond to fictitious delays.

After that the parameter identifier (7.58), (7.59), being constructed for the aforementioned model, completes the design procedure. Indeed, by Theorem 7.4 the identifier parameters converge to the corresponding model parameters and due to the robustness of the identifier against small uncertainties on the delay values these parameters approximate those of the system whereas vanishing identifier parameters (i.e., those converging to zero as $t \to \infty$) carry out the fictitious delays.

In fact, the above procedure is nothing else than a discrete realization of the following adaptive identifier with a distributed delay:

$$\dot{\tilde{x}}(t) = \int_0^h [\tilde{\xi}_A(t,\theta)x(t-\theta) + \tilde{\xi}_B(t,\theta)u(t-\theta)]d\theta - G[x(t) - \tilde{x}(t)],\ t \geq 0,$$
$$\tilde{x}(-\theta) = \tilde{\varphi}(\theta),\ 0 \leq \theta \leq h, \tag{7.67}$$

$$\dot{\tilde{\xi}}_A(t,\theta) = F(\theta)P[x(t) - \tilde{x}(t)]x^T(t-\theta),\ \tilde{\xi}_A(0,\theta) = \tilde{\xi}_A^0(\theta),$$
$$\dot{\tilde{\xi}}_B(t,\theta) = \Phi(\theta)P[x(t) - \tilde{x}(t)]u^T(t-\theta),\ d\tilde{\xi}_B(0,\theta) = \tilde{\xi}_B^0(\theta), \tag{7.68}$$

where the initial conditions $\tilde{\varphi}(\theta), d\tilde{\xi}_A^0(\theta), d\tilde{\xi}_B^0(\theta)$ are square integrable functions of appropriate dimensions, the adaptation gains $F(\theta),\ \Phi(\theta)$ are continuous positive definite matrix functions of appropriate dimensions, the matrices P and G are the same as before.

In order to clarify the sense of the tuned functions $\tilde{\xi}_A(t,\theta), \tilde{\xi}_B(t,\theta)$ one should represent the state equation (7.1) in a similar form

$$\dot{x}(t) = \int_0^h [\xi_A(\theta)x(t-\theta) + \xi_B(\theta)u(t-\theta)]d\theta,\ t \geq 0, \tag{7.69}$$

where $\delta(t-\tau)$ is a Dirak function, atomized at $t = \tau$, and the distributions $\xi_A(\theta) = \sum_{i=0}^{r} A_i\delta(\theta - \tau_i)$, $\xi_B(\theta) = \sum_{i=0}^{r} B_i\delta(\theta - \tau_i)$ have a one-to-one relation to the system parameters and delays.

The identification law (7.67), (7.68) is shown to guarantee the state error convergence

$$\lim_{t\to\infty} \|\tilde{x}(t) - x(t)\| = 0 \tag{7.70}$$

as well as the weak* convergence

$$* - \lim_{t\to\infty} \tilde{\xi}_A(t,\theta) = \xi_A(\theta), \; * - \lim_{t\to\infty} \tilde{\xi}_B(t,\theta) = \xi_B(\theta) \tag{7.71}$$

of the tuned distributions to the nominal ones for arbitrary adaptation gains, initial conditions and periodic input signals, enforcing the identifiability of the system. Since the weak* convergence (7.71) ensures that

$$\lim_{t\to\infty} \tilde{\xi}_A(t,\theta) = 0, \; for\ all\ \theta \neq \tau_i, \; i = 0, \ldots, r \tag{7.72}$$

and

$$\lim_{t\to\infty} \int_{\tau_i-\varepsilon}^{\tau_i+\varepsilon} \tilde{\xi}_A(t,\theta)d\theta = A_i, \; \lim_{t\to\infty} \int_{\tau_i-\varepsilon}^{\tau_i+\varepsilon} \tilde{\xi}_B(t,\theta)d\theta = B_i, 1 = 0, \ldots, r \tag{7.73}$$

for all intervals $(\tau_i - \varepsilon, \tau_i + \varepsilon)$, $\varepsilon > 0$ with the only delay value τ_i inside, the identifier (7.67), (7.68) yields the nominal parameters and delay values as $t \to \infty$.

Summarizing, the following result is obtained.

Theorem 7.5 *Let the time-delay values* $\tau_j, \; j = 1, \ldots, r$ *be unknown a priori and let Conditions 2 and 3 of Theorem 7.4 be satisfied. Then the limiting relations* (7.70), (7.71) *hold with the adaptive identification law* (7.67) *and distributions* $\tilde{\xi}_A(t,\theta), \tilde{\xi}_B(t,\theta)$ *tuned as* (7.68).

Proof The proof is rather technical and it is based on the robustness of the approximation of the distributed-delay system (7.67), (7.68) by its counterpart with a finite number of discrete delays for which the result is established by Theorem 7.4. The detailed proof is left to the reader.

7.1.4 SISO Case Study

The Lyapunov–Razumikhin redesign approach to the adaptive identification of general linear TDS is subsequently generalized from the full state measurements to linear SISO TDS identification from output measurements. For this purpose, a sliding mode based state observer of the underlying SISO TDS with uncertain parameters is constructed and effectively utilized to arrive at the desired SISO TDS identifier that does not require any on-line differentiation of the measured data. As in Gomez et al. (2007), the development is confined to a class of nth-order linear TDS

$$\eta^{(n)}(t) = \sum_{i=0}^{r} \left[\sum_{j=0}^{n-1} \alpha_{ij}\eta^{(j)}(t-\tau_i) + \sum_{k=0}^{m} \beta_{ik}\nu^{(k)}(t-\tau_i) \right] \tag{7.74}$$

with the delays $0 = \tau_0 < \tau_1 < \cdots < \tau_r$, with the single input $v(t) \in \mathbb{R}^1$ of the order m, with at least one of the parameters $\beta_{im},\ i = 0, 1, \ldots, r$ to be nonzero, with the single output

$$z(t) = \eta(t) \in \mathbb{R}^1, \tag{7.75}$$

available for measurements, and with a reasonably smooth initial condition $\eta(\theta),\ \theta \in [-\tau_r, 0]$ of class C^{n+2}. Some of the parameters $\alpha_{ij}, \beta_{ik},\ i = 0, \ldots, r,\ j = 0, \ldots, n-1,\ k = 0, \ldots, m$ of the SISO TDS (7.74), (7.75) are permitted to be zero. Similarly to the general TDS (7.1), (7.2), the system (7.74), (7.75) is thus assumed to possess the same state, input, and output delay values $\tau_i,\ i = 0, 1, \ldots, r$.

Another assumption, made implicitly, is that the output measurement (7.75) is available without any delays. Even though the case of delayed measurements often occurs in practice, such a SISO TDS turns out to be unidentifiable if along with output delays it also exhibits input delays. For instance, a first-order system

$$\begin{aligned}\dot{\eta} &= \eta + v(t),\ \eta(0) = \eta_0,\\ z(t) &= \eta(t - \tau)\end{aligned}$$

is not identifiable because another system

$$\begin{aligned}\dot{\hat{\eta}} &= \hat{\eta} + v(t - \tau),\ \hat{\eta}(\tau) = \eta_0,\\ z(t) &= \hat{\eta}(t)\end{aligned}$$

generates the same output $z(t)$ for all $t \geq \tau$, regardless of whichever input $v(t)$ drives these systems. This example demonstrates that output delays are dual to input delays in the sense that any SISO TDS with no input delays can be represented, by rewriting the state equation in terms of the delayed output variable, as one with no output delays.

7.1.4.1 Identifiability of SISO TDS

Apparently, the SISO TDS (7.74), (7.75) is a particular case of the general TDS (7.1), (7.2) with the state vector

$$x(t) = [\eta(t), \ldots, \eta^{(n-1)}(t), v(t), \ldots, v^{(m-1)}(t)]^T, \tag{7.76}$$

the scalar input

$$u(t) = v^{(m)}(t), \tag{7.77}$$

and the vector output

$$y(t) = [\eta(t), v(t), \ldots, v^{(m-1)}(t)]^T, \tag{7.78}$$

whose state coordinates are governed by

$$\begin{aligned}
\dot{x}_1 &= x_2 \\
&\vdots \\
\dot{x}_{n-1} &= x_n \\
\dot{x}_n(t) &= \sum_{i=0}^{r}\left[\sum_{j=0}^{n-1}\alpha_{ij}x_{j+1}(t-\tau_i)\right. \\
&\quad \left. + \sum_{k=0}^{m-1}\beta_{ik}x_{n+k+1}(t-\tau_i) + \beta_{im}u(t-\tau_i)\right] \qquad (7.79) \\
\dot{x}_{n+1} &= x_{n+2}, \\
&\vdots \\
\dot{x}_{n+m-1} &= x_{n+m} \\
\dot{x}_{n+m} &= u \qquad (7.80) \\
y_1(t) &= x_1(t) \\
y_{s+1}(t) &= x_{n+s}(t),\ s = 1,\ldots,m. \qquad (7.81)
\end{aligned}$$

The aim of this section is to analyze the identifiability of the SISO TDS (7.74), (7.75). For this system the identifiability concept is introduced by means of specifying Definition 7.6 for the SISO TDS representation (7.79)–(7.81) and its reference model in which the system parameters and delays α_{ij}, β_{ik}, τ_i, $i = 0,\ldots,r$, $j = 0,\ldots,n-1$, $k = 0,\ldots,m$, are replaced by $\hat{\alpha}_{ij}$, $\hat{\beta}_{ik}$, $\hat{\tau}_i$, respectively.

Definition 7.7 System (7.74), (7.75) is said to be *identifiable* if there exists an input $v(t)$ such that the identity $\eta(t) \equiv \hat{\eta}(t)$ of the output of the system and that of the model, or, equivalently

$$\sum_{i=0}^{r}\left[\sum_{j=0}^{n-1}(\alpha_{ij}-\hat{\alpha}_{ij})\eta^{(j)}(t-\tau_i) + (\beta_{ik}-\hat{\beta}_{ik})v^{(k)}(t-\tau_i)\right] \equiv 0 \qquad (7.82)$$

results in

$$\alpha_{ij} = \hat{\alpha}_{ij},\ \beta_{ik} = \hat{\beta}_{ik},\ i = 0,\ldots,r,\ j = 0,\ldots,n-1,\ k = 0,\ldots,m \qquad (7.83)$$

regardless of a choice of the initial data. In that case the identifiability is said to be enforced by the input $v(t)$.

In turn, the notions of nonsmooth input signals are simplified as follows (cf. Definition 7.5).

Definition 7.8 A scalar piecewise regular input signal $v(t)$ is said to be *sufficiently nonsmooth* if it is of class $C^{\nu-1}$ for some $\nu \geq \max\{1, m\}$ and the set D of discontinuity points t_j, $j = 0, 1, \ldots$ of $v^{(\nu)}(t)$ contains an infinite countable number of incommensurable points.

Clearly, the above definition is verifiable independently of any underlying SISO TDS. To construct a sufficiently nonsmooth signal $v(t)$ of a class $C^{\nu-1}$, $\nu \geq \max\{1, m\}$, one should, firstly, select a sequence of incommensurable points t_j, $j = 0, 1, \ldots$, where the νth-order input derivative $v^{(\nu)}(t)$ undergoes discontinuities, and secondly, impose a smooth behavior on the signal between these points. Examples are periodic functions $v(t)$ whose νth-order derivative is piecewise constant:

$$v^{(\nu)}(t) = \begin{cases} u_0 \text{ if } t \in [jT, t_0 + jT), \\ u_1 \text{ if } t \in [t_0 + jT, (j+1)T), \end{cases} \quad j = 0, 1, \ldots \tag{7.84}$$

where u_0, u_1 are some constants and periodic discontinuity points $t_j = t_0 + jT$, $j = 0, 1, \ldots$ are of a period $T > 0$, incommensurable to t_0 (i.e., the ratio t_0/T is not a rational number).

The following result extends Theorem 7.2 to the present case.

Theorem 7.6 *The SISO TDS* (7.74), (7.75) *is identifiable and its identifiability is enforced by any sufficiently nonsmooth input signal $v(t)$ of class $C^{\nu-1}$ with $\nu \geq \max\{1, m\}$.*

Proof Since only SISO TDS are considered and their state-space representation (7.80)–(7.81) is, therefore, fixed, successive differentiations of the output vector (7.78) yield the full information on the state (7.76) of the system. Apart from this, the system in question is straightforwardly verified to be weakly controllable due to the presence of at least one of the parameters β_{im}, $i = 0, 1, \ldots, r$, different from zero. Thus, Theorem 7.2 becomes applicable to the normal form (7.80)–(7.81) of the SISO TDS. By applying Theorem 7.2, the validity of Theorem 7.6 is then established. ∎

7.1.4.2 SM State Observer-Based Adaptive Identification

An adaptive identifier design over output measurements becomes available for linear internally asymptotically stable SISO TDS. The underlying system (7.74) is, therefore, assumed to be asymptotically stable under $v \equiv 0$. Such an assumption complies with a typical requirement for any automatic control system, and if it is not the case, an adaptive stabilization should additionally be addressed. A particular adaptive synthesis is separately illustrated in the next section.

Once a sufficiently nonsmooth input signal $v(t)$ is chosen to enforce the system identifiability, the task of the identification of the unknown parameters, including the delay values, can be carried out by an adaptive identifier which utilizes the prior knowledge of the system structure through its inclusion into the reference model, whose parameters evolve in time and converge to the unknown parameters of the SISO TDS. For a moment, the time-delay values $\tau_i,\ i = 1, \ldots, r$, are assumed to be known a priori.

In order to identify the system parameters $\alpha_{ij}, \beta_{ik},\ i = 0, \ldots, r,\ j = 0, \ldots, n-1,\ k = 0, \ldots, m$, the following identification law

$$\dot{\hat{x}}_1(t) = \hat{x}_2(t) + w_1(t),\ \hat{x}_1(-\theta) = \hat{\varphi}_1(\theta)$$

$$\vdots$$

$$\dot{\hat{x}}_{n-1}(t) = \hat{x}_n(t) + w_{n-1}(t),\ \hat{x}_{n-1}(-\theta) = \hat{\varphi}_{n-1}(\theta), \tag{7.85}$$

$$\begin{aligned}\dot{\hat{x}}_n(t) = &\sum_{i=0}^{r}\left\{\sum_{s=0}^{n-2}\hat{\alpha}_{is}\hat{x}_{s+1}(t-\tau_i)\right.\\ &+ \hat{\alpha}_{i(n-1)}[\hat{x}_n(t-\tau_i) + w_{n-1}(t-\tau_i)]\\ &+ \left.\sum_{k=0}^{m}\hat{\beta}_{ik}v^{(k)}(t-\tau_i)\right\} + \gamma w_{n-1}(t),\end{aligned} \tag{7.86}$$

$$\hat{x}_n(-\theta) = \hat{\varphi}_n(\theta),\ 0 \le \theta \le \tau_r$$

$$\dot{\hat{\alpha}}_{is}(t) = \lambda_{is}\hat{x}_{s+1}(t-\tau_i)w_{n-1}(t),\ \hat{\alpha}_{is}(0) = \hat{\alpha}^0_{is},\quad s = 0, 1, \ldots, n-2$$

$$\dot{\hat{\alpha}}_{i(n-1)}(t) = \lambda_{i(n-1)}[\hat{x}_n(t-\tau_i) + w_{n-1}(t-\tau_i)]w_{n-1}(t), \tag{7.87}$$

$$\hat{\alpha}_{i(n-1)}(0) = \hat{\alpha}^0_{i(n-1)} \tag{7.88}$$

$$\dot{\hat{\beta}}_{ik}(t) = \mu_{ik}v^{(k)}(t-\tau_i)w_{n-1}(t),\ \hat{\beta}_{ik}(0) = \hat{\beta}^0_{ik} \tag{7.89}$$

is proposed. Hereinafter, $\hat{\alpha}^0_{ij},\ \hat{\beta}^0_{ik}$ are the initial values of the parameter estimates, the initial model functions $\hat{\varphi}_{j+1}(\theta)$ are of class C^{n+2}, the adaptation gains $\gamma, \lambda_{ij}, \mu_{ik}$ are positive, the identifier inputs are given by

$$w_1 = M_1(t)\text{sign}\ (x_1 - \hat{x}_1),\ M_1(t) = L + |\hat{x}_2(t)| \tag{7.90}$$

$$w_s = M_s(t)\text{sign}\ w_{s-1},\ M_s(t) = L + |\hat{x}_{s+1}(t)|,\quad s = 2, \ldots, n-1 \tag{7.91}$$

with L being an upper bound of the norm $\|\eta(t)\|_{C^{n-1}} = \max_{j=0,\ldots,n-1}\max_{t\ge-\tau_r}|\eta^{(j)}(t)|$ of the state trajectory $\eta(t)$. Since only asymptotically stable TDS (7.74) are studied, their solutions $\eta(t)$ are always bounded. A norm bound L, such that

$$L \ge \|\eta(\cdot)\|_{C^{n-1}} + \delta \tag{7.92}$$

for some positive δ, can readily be established in practical applications.

Thus, the identifier dynamics are governed by the functional differential equations (7.85)–(7.91) with discontinuous right-hand side. The meaning of these equations remains conventional beyond the discontinuity manifold

$$(x_1 - \hat{x}_1) \times w_1(x, \hat{x}) \times \cdots \times w_{n-2}(x, \hat{x}) = 0 \tag{7.93}$$

of the input functions (7.90), (7.91), whereas along this manifold it is defined in the Filippov sense.

Recall the concept of sliding mode solutions of delay differential equations, inherited from Sect. 2.2.5, where the regularization of discontinuous systems is developed in a Hilbert space.

In a vicinity of the discontinuity manifold the original system is regularized by substituting a related system, whose solutions exist in the conventional sense. A sliding mode solution, if any, is then obtained by making characteristics of the new system approach those of the original one. A particular regularization of (7.90), (7.91) is obtained when low-pass filters

$$\begin{aligned} w_s &= M_s \operatorname{sign} \xi_s, \ \varepsilon_s \dot{\xi}_s + \xi_s = w_{s-1}, \\ &s = 2, \ldots, n-1 \end{aligned} \tag{7.94}$$

with sufficiently small $\varepsilon_s > 0$ are involved into (7.91) to replace the input variables w_{s-1} by their averaged values ξ_s. After that, the signum functions can be approximated by hyperbolic tangents to get a smooth realization of the input signals (7.90), (7.94).

It should be noted that from the implementation standpoint, the presence of the fast switching (7.90), (7.91) in the update law (7.85)–(7.89) is not as serious as the controller switching because the identifier inputs (7.90), (7.91) do not drive an actuator.

The idea behind the identifier synthesis (7.85)–(7.91) is based on the use of the sliding mode observer (7.85) that generates sliding motions on the surfaces

$$x_s - \hat{x}_s = 0, \ s = 1, \ldots, n-1. \tag{7.95}$$

Once the sliding mode occurs on (7.95), the equivalent control value $w_s^{eq} = x_{s+1} - \hat{x}_{x+1}, \ s = 1, \ldots, n-1$ is substituted for the input w_s into the update law (7.85)–(7.89) to simplify it to

$$\begin{aligned} \dot{\hat{x}}_n(t) = \sum_{i=0}^{r} \Bigg[&\sum_{j=0}^{n-1} \hat{\alpha}_{ij} x_{j+1}(t - \tau_i) \\ &+ \sum_{k=0}^{m} \hat{\beta}_{ik} v^{(k)}(t - \tau_i) \Bigg] + \gamma(x_n(t) - \hat{x}_n(t)) \end{aligned} \tag{7.96}$$

$$\begin{aligned} \dot{\hat{\alpha}}_{ij} &= \lambda_{ij} x_{j+1}(t - \tau_i)[x_n(t) - \hat{x}_n(t)], \ i = 0, \ldots, r, \\ &\quad j = 0, \ldots, n-1 \end{aligned} \tag{7.97}$$

$$\dot{\hat{\beta}}_{ik} = \mu_{ik} v^{(k)}(t - \tau_i)[x_n(t) - \hat{x}_n(t)], \ k = 0, \ldots, m. \tag{7.98}$$

The resulting sliding mode update law (7.96)–(7.98) represents the parameter identification algorithm over the full information on the state vector, that has been developed in Sect. 7.1.3.

As shown below, the identification law (7.85)–(7.91) ensures the convergence

$$\lim_{t\to\infty} \Delta x_s(t) = 0, \ \lim_{t\to\infty} \Delta\alpha_{ij}(t) = 0, \ \lim_{t\to\infty} \Delta\beta_{ik}(t) = 0 \tag{7.99}$$

of the state estimate and parameter errors

$$\Delta x_s(t) = x_s(t) - \hat{x}_s(t), \ s = 1, \ldots, n, \tag{7.100}$$

$$\begin{aligned} &\Delta\alpha_{ij}(t) = \alpha_{ij} - \hat{\alpha_{ij}}(t), \\ &\Delta\beta_{ik}(t) = \beta_{ik} - \hat{\beta_{ik}}(t), \ i = 0, \ldots, r, \ j = 0, \ldots, n-1, \ k = 0, \ldots, m, \end{aligned} \tag{7.101}$$

whenever a sufficiently nonsmooth periodic input signal drives the system.

Theorem 7.7 *Consider a linear SISO TDS* (7.74), (7.75) *with the assumptions above. Let the input signal $v(t)$ be periodic and sufficiently nonsmooth. Then the limiting relation* (7.99) *holds with the adaptive identification law* (7.85)–(7.91), *with arbitrary initial conditions and with arbitrary positive adaptation gains.*

Proof Due to (7.79), (7.85), (7.90), the dynamics of the state estimate error Δx_1 are governed by

$$\Delta\dot{x}_1(t) = \Delta x_2(t) - M_1(t)\text{sign } \Delta x_1(t). \tag{7.102}$$

Differentiating the quadratic function $V_1(\Delta x_1) = (\Delta x_1)^2$ along the trajectories $\Delta x_1(t)$ of (7.102) subject to (7.90), (7.92) results in

$$\begin{aligned} \dot{V}_1(t) &= 2\Delta x_1(t)[\Delta x_2(t) - (L + |\hat{x}_2(t)|)\text{sign } \Delta x_1(t)] \\ &\le -2\delta|\Delta x_1(t)| = -2\delta\sqrt{V_1(t)}. \end{aligned} \tag{7.103}$$

By Lemma 4.1, any solution of the differential inequality (7.103) converges to zero in finite time $T = \delta^{-1}\sqrt{V(0)}$.

Thus, starting from a finite time instant $T_1 \in [0, T]$, the model system (7.85)–(7.89) evolves in a sliding mode along the surface $\Delta x_1 = 0$. The so-called sliding mode dynamics on the surface $\Delta x_1 = 0$ are then described by applying the equivalent control method, presented in Sect. 2.2.5.

According to the equivalent control method, the sliding mode equation

$$\dot{\hat{x}}_2(t) = \hat{x}_3(t) + M_2(t)\text{sign } \Delta x_2(t), \ t \ge T_1 \tag{7.104}$$

on the surface $\Delta x_1 = 0$ is obtained by substituting the solution $w_1^{eq} = \Delta x_2$ of the equation $\Delta\dot{x}_1 = 0$ into (7.85) for w_1.

By analogy to (7.102), the dynamics of the state estimation error Δx_2 are governed by

$$\Delta\dot{x}_2(t) = \Delta x_3(t) - M_2(t)\text{sign } \Delta x_2(t), \ t \ge T_1, \tag{7.105}$$

and starting from a finite time moment $T_2 > T_1$ they evolve in a sliding mode on the surface $\Delta x_2 = 0$. While evolving on the intersection of the surfaces $\Delta x_s = 0$, $s = 1, 2$, the sliding mode dynamics

$$\dot{\hat{x}}_3(t) = \hat{x}_4(t) + M_3(t)\text{sign } \Delta x_3(t), \ t \geq T_2 \tag{7.106}$$

similar to (7.104), are described according to the equivalent control method.

By iterating on s, one concludes that starting from a finite time moment $T_{n-1} > 0$ there appears a sliding mode on the intersection of the surfaces (7.95) and the sliding mode dynamics (7.96)–(7.98) are derived according to the equivalent control method by substituting $x_s, \ s = 1, \ldots, n-1$ and the solution $w_{n-1}^{eq} = \Delta x_n$ of the equation $\Delta \dot{x}_{n-1} = 0$ into (7.86)–(7.89) for $\hat{x}_s$ and w_{n-1}, respectively.

To complete the proof it remains to note that the sliding mode dynamics (7.96)–(7.98) represent the adaptive identifier synthesis of Theorem 7.4, specified in terms of the SISO TDS (7.74) with the available state measurements. Thus, after a finite time instant, the state feedback identifier synthesis is enforced in the sliding mode, and by applying Theorem 7.4, the assertion of Theorem 7.7 is validated. ■

The knowledge of some plant parameters allows one to skip the corresponding update laws similar to that of the general TDS identifier design of Sect. 7.1.3. Moreover, for the case where the delay values are no longer known a priori, the distributed-delay-flavored procedure of Sect. 7.1.3.2 applies here as well. Instead of specifying Theorem 7.5 to the present case, the effectiveness of the proposed adaptive identification of unknown plant parameters and delay values is illustrated in a numerical study of Sect. 7.1.5.

7.1.4.3 Model Reference Adaptive Control

On-line identification methods proposed can also be used in many direct and direct–indirect adaptive control schemes such as those developed in Miller and Michel (1990) for non-delay dynamic systems. To facilitate the exposition, the model reference adaptive synthesis to be developed is confined to a SISO TDS (7.74), specified as follows:

$$\begin{aligned} \dot{x}(t) &= = \sum_{i=0}^{r} [a_i x(t-\tau_i) + b_i u(t-\tau_i)], \\ y(t) &= = x(t), \ t \geq 0 \end{aligned} \tag{7.107}$$

when the state and input delays $0 = \tau_0 < \tau_1 < \cdots < \tau_r$ are known a priori, and delay-free state measurements are available. An extension to higher-order SISO TDS with unknown delays is actually possible and repeats the details of the adaptive identifier design, developed before.

In the sequel, the scalar parameters $a_0, \ldots, a_r, b_0, \ldots, b_r$, unknown a priori, are to be identified while a control signal u is synthesized to guarantee the state deviation

$$\Delta x(t) = x(t) - x^*(t)$$

to converge to zero as $t \to \infty$. The output $x^*(t)$ to be tracked is produced by a reference model

$$\dot{x}^*(t) = \sum_{i=0}^{r} [a_i^* x^*(t - \tau_i) + b_i^* f(t - \tau_i)],\ t \geq 0 \tag{7.108}$$

of the same structure as that of the plant. Simplifying the reference model by making $a_i^* = 0,\ b_i^* = 0,\ i = 1, \ldots, r$ it becomes possible to design the system to behave as a non-delay system.

The following assumptions are imposed on the reference model:

1. The nominal model (7.108) with $f \equiv 0$ is asymptotically stable.
2. The reference input $f(t)$ is bounded.
3. The reference model (7.108) is weakly controllable.

The former two assumptions comply with typical requirements for any automatic control system whereas by Theorem 7.2, the latter assumption ensures the identifiability of the reference model, thereby admitting the on-line parameter identification in the MRAC algorithm presented below.

In order to guarantee all MRAC inputs to be bounded lower and upper parameter bounds

$$\underline{a}_i \leq a_i \leq \overline{a}_i,\ \underline{b}_i \leq b_i \leq \overline{b}_i,\ i = 0, \ldots, r,$$

are assumed to be available for their subsequent use. It is also assumed that b_0 differs from zero and

$$\min\{|\underline{b}_0|, |\overline{b}_0|\} > \sum_{i=1}^{r} max\{|\underline{b}_i|, |\overline{b}_i|\}. \tag{7.109}$$

Clearly, (7.109) implies

$$\text{either } \underline{b}_0 > 0 \text{ or } \overline{b}_0 < 0, \tag{7.110}$$

that by Theorem 7.2, ensures Assumption 3 to be in force.

In the proposed MRAC algorithm, the parameter estimates are updated as follows:

$$\dot{\hat{a}}_i(t) = \begin{cases} 0 & \begin{array}{l} \text{if } \hat{a}_i(t) = \underline{a}_i \text{ and } e(t)x(t - \tau_i) < 0 \\ \text{or} \\ \text{if } \hat{a}_i(t) = \overline{a}_i \text{ and } e(t)x(t - \tau_i) > 0 \end{array} \\ \mu_i e(t) x(t - \tau_i) & \text{otherwise} \end{cases},$$

$$\hat{a}_i(0) = \hat{a}_i^0 \in [\underline{a}_i, \overline{a}_i],$$

$$\dot{\hat{b}}_j(t) = \begin{cases} 0 & \begin{array}{l}\text{if } \hat{b}_i(t) = \underline{b}_i \text{ and } e(t)u(t-\tau_i) < 0 \\ \text{or} \\ \text{if } \hat{b}_i(t) = \overline{b}_i \text{ and } e(t)u(t-\tau_i) > 0\end{array} \\ \nu_i e(t)u(t-\tau_i) & \text{otherwise} \end{cases},$$

$$\hat{b}_i(0) = \hat{b}_i^0 \in [\underline{b}_i, \overline{b}_i],\ i = 0, \dots, r \tag{7.111}$$

where $\mu_i > 0,\ \nu_i > 0,\ i = 0, \dots, r$ are adaptation gains. As in the finite-dimensional case (Ioannou and Sun 1996; Narendra and Annaswamy 1989), in order to restrict the parameter estimates to remain between the upper and lower bounds for the actual parameters, the ordinary differential equations (7.111) with discontinuous right-hand sides are involved. Given initial conditions, these equations prove to have a unique solution to the right (Filippov 1988, Theorem 1, p. 106), the precise meaning of which is defined in the Filippov sense, addressed in Sect. 2.1. Due to (7.110) the parameter estimate $\hat{b}_0(t)$ remains invertible while it is marching to the nominal value.

Now, to guarantee the desired state and parameter convergence

$$\lim_{t\to\infty} [\Delta x(t)]^2 = 0, \tag{7.112}$$

$$\lim_{t\to\infty} \sum_{i=0}^{r} \{[a_i - \hat{a}_i(t)]^2 + [b_i - \hat{b}_i(t)]^2\} = 0 \tag{7.113}$$

the adaptive control law is chosen in the form

$$u(t) = \hat{b}_0^{-1}(t) \left\{ \sum_{i=0}^{r} [a_i^* x^*(t-\tau_i) - \hat{a}_i(t)x(t-\tau_i)] \right. \\ \left. -\nu \Delta x(t) + \sum_{i=0}^{r} b_i^* f(t-\tau_i) - \sum_{i=1}^{r} \hat{b}_i(t)u(t-\tau_i) \right\},\ t \ge 0 \tag{7.114}$$

where the adaptation parameter $\nu > 0$ and

$$u(t) = 0\ for\ t \le 0.$$

Theorem 7.8 *Consider a linear delay system* (7.107) *and a reference model* (7.108) *with the assumptions above. Then the adaptive algorithm* (7.111), (7.114) *ensures the state convergence* (7.112) *for arbitrary admissible adaptation parameter and gains, initial conditions and reference input. Along with this, all signals in the closed-loop system are bounded. If, in addition, the reference input $f(t)$ is periodic and sufficiently discontinuous/nonsmooth then the parameter convergence* (7.113) *holds as well.*

Proof Substituting the control law (7.114) into the state equation (7.107), let us represent the resulting system in terms of the state error $\Delta x(t)$ and parameter errors $\Delta a_i(t) = a_i - \hat{a}_i(t),\ \Delta b_i(t) = b_i - \hat{b}_i(t),\ i = 0, \dots, r:$

$$\Delta\dot{x} = -\nu\Delta x(t) + \sum_{i=0}^{r}[\Delta a_i(t)x(t-\tau_i) + \Delta b_i(t)u(t-\tau_i)]. \tag{7.115}$$

First, let us prove the state convergence (7.112). For this purpose, consider the Razumikhin function

$$V(\Delta x, \Delta a_0, \dots, \Delta b_r) = [\Delta x]^2 + \sum_{i=0}^{r}\{\mu_i^{-1}[\Delta a_i]^2 + \nu_i^{-1}[\Delta b_i]^2\} \tag{7.116}$$

and compute its time derivative along the trajectories of the closed-loop system (7.111), (7.115). The straightforward differentiation yields

$$\dot{V}(t) \le -2\nu[\Delta x(t)]^2 \tag{7.117}$$

that results in (7.112) by applying Barbalat's Lemma 2.2.

Next let us verify the boundedness of the control signal (7.114). Since according to the integral representation (7.6), the output $x^*(t)$ of the reference model (7.108), driven by a bounded input $\sum_{i=0}^{r} b_i^* f(t-\tau_i)$, is uniformly bounded for all $t \ge 0$ the state $x(t)$ of the plant (7.107) is also uniformly bounded by virtue of (7.117). Hence, the right-hand side of (7.114) is bounded on a finite time segment $[0, T]$, $T > 0$ by some constant $M > 0$, possibly dependent on T. Under assumption (7.109) this constant can definitely be selected independent on T. Indeed, let us fix $T > 0$ and let $|u(t)| \le M$ for all $t \in [0, T]$. Then employing (7.114) yields $|u(t)| \le M + cM = M_1$ for all $t \in [0, T + \tau_r]$ where

$$c = \frac{\sum_{i=1}^{r} max\{|\underline{b}_i|, |\overline{b}_i|\}}{min\{|\underline{b}_0|, |\overline{b}_0|\}} < 1$$

by virtue of (7.109). Following the same procedure, one derives that $|u(t)| \le M + cM_1 = M + cM + c^2M = M_2$ for all $t \in [0, T + 2\tau_r]$. Then by iterating on i, the upper estimate $|u(t)| \le M\sum_{i=0}^{k} c^i$ is obtained for all $t \in [0, T + \tau_r]$ and $k = 1, 2, \dots$. Taking into account that $M\sum_{i=0}^{\infty} c^i = \frac{M}{1-c}$ it is concluded that $|u(t)| \le \frac{M}{1-c}$ for all $t \ge 0$. Thus, the control signal (7.114) is uniformly bounded by the absolute constant $\frac{M}{1-c}$ for all $t \ge 0$.

To complete the proof it remains to show that the desired parameter convergence (7.113) holds under an appropriate choice of the reference input $f(t)$. Proof of this part invokes the extension of the Invariance Principle, Theorem 4.7 to time-periodic TDS which is made in analogy to that of Rouche et al. (1977).

To formally apply this principle, the closed-loop system (7.111), (7.115) is augmented with the reference model (7.108), thereby viewing the term $x(t) = \Delta x(t) + x^*(t)$ as a state variable rather than a time-varying input. Following the same trick that has been used in the proof of Theorem 7.5, the above Razumikhin function (7.116) is then modified as follows:

$$V_0(e, \Delta x, \Delta a_0, \ldots, \Delta b_r) = V(\Delta x, \Delta a_0, \ldots, \Delta b_r) + < We, e > \qquad (7.118)$$

where W is determined by (7.7) with the exponentially stable C_0-semigroup T of the reference model (7.108), $e(t) = x^*(t) - y(t)$ is the reference state deviation from the steady state periodic orbit $y(t)$ of (7.108), given by (7.8) with the reference input f, being substituted for w.

Differentiating V_0 along the trajectories of (7.108), (7.111), (7.115) yields

$$\dot{V}_0(t) \leq -2\nu[\Delta x(t)]^2 - < e_t, e_t > . \qquad (7.119)$$

Due to the aforementioned invariance principle extension, applied to the periodic system (7.108), (7.111), (7.115) with the periodic reference input $f(t)$, there must be a convergence of the trajectories of this system to the largest invariant subset of the set of solutions of (7.108), (7.111), (7.115) for which $\dot{V}_0 \equiv 0$. Taking into account (7.111), (7.115) this leads to the expressions

$$\Delta a_i(t) = \Delta a_i = const, \ \Delta b_i(t) = \Delta b_i = const, \ i = 0, \ldots, r,$$
$$\sum_{i=0}^{r} [\Delta a_i x^*(t - \tau_i) + \Delta b_i u(t - \tau_i)] = 0 \qquad (7.120)$$

on the set where the relation $\Delta x \equiv 0$, resulting from $\dot{V}_0 \equiv 0$, holds. Using (7.114) restricted to the manifold $\Delta x \equiv 0$, the latter equality can be rewritten as

$$\sum_{i=0}^{r} \Delta a_i x^*(t - \tau_i) + \Delta b_0 \hat{b}_0^{-1} \left\{ \sum_{i=0}^{r} [a_i^* - \hat{a}_i] x^*(t - \tau_i) \right.$$
$$\left. + \sum_{i=0}^{r} b_i^* f(t - \tau_i) - \sum_{i=1}^{r} \hat{b}_i u(t - \tau_i) \right\} + \sum_{i=1}^{r} \Delta b_i u(t - \tau_i) = 0. \quad (7.121)$$

If the reference input $f(t)$ is sufficiently discontinuous/nonsmooth it follows that $\Delta b_0 = 0$ because, otherwise, relation (7.121) does not hold at a discontinuity/nonsmoothness point t of the reference input. Indeed, in this case all terms in the left-hand side of (7.121) are continuous or, respectively, smooth enough with the only exception for $\Delta b_0 \hat{b}_0^{-1} b_i^* f(t)$. Recall in this regard that any solution $x^*(t)$ of the reference model (7.121) is well-known (Hale 1971) to be continuous for all $t \geq 0$. Thus by Definition 7.4, the sufficiently discontinuous reference input $f(t)$ is continuous at the time instants $t - t_i$, $i = 1, \ldots, r$, whereas the continuity of the control input $u(t)$ at the same time moments $t - t_i$ results from its recursive construction (7.114) to be proved by induction. A relevant conclusion in terms of smooth solutions for t large enough is reproduced if the reference input is sufficiently nonsmooth rather than discontinuous.

Hence, if $\Delta b_0 \neq 0$ the left-hand side of (7.121) is discontinuous/nonsmooth at the time moment t that contradicts (7.121). Thus, $\Delta b_0 = 0$ and (7.121) takes the form

$$\sum_{i=0}^{r}[\Delta a_i x^*(t-\tau_i)+\Delta b_i u(t-\tau_i)]=0. \tag{7.122}$$

Now substituting the control law (7.114), restricted to the manifold $\Delta x \equiv 0$ and computed at the time moment $t-\tau_1$ into (7.122), yields

$$\sum_{i=0}^{r}\Delta a_i x^*(t-\tau_i)+\Delta b_1\hat{b}_0^{-1}\left\{\sum_{i=0}^{r}[a_i^*-\hat{a}_i]x^*(t-\tau_1-\tau_i)\right.$$
$$\left.+\sum_{i=0}^{r}b_r^* f(t-\tau_1-\tau_i)-\sum_{i=1}^{r}\hat{b}_i u(t-\tau_1-\tau_i)\right\}+\sum_{i=2}^{l}\Delta b_i u(t-\tau_i)=0 \tag{7.123}$$

that by the same reasoning ensures $\Delta b_1=0$.

In general, the following relations

$$\Delta b_i=0,\ i=0,\ldots,r \tag{7.124}$$

are deduced from (7.121). Due to (7.124), relations (7.121) result in

$$\sum_{i=0}^{r}\Delta a_i x^*(t-\tau_i)=0 \tag{7.125}$$

that by virtue of the parameter identifiability of the reference model yields

$$\Delta a_i=0,\ i=0,1,\ldots,r. \tag{7.126}$$

Thus, by applying the invariance principle extension, the overall system (7.108), (7.111), (7.115) is globally asymptotically stable and hence, the required parameter convergence (7.113) holds under a periodic sufficiently discontinuous/nonsmooth reference input $f(t)$. Theorem 7.8 is proven. ■

7.1.5 Application to Engine Transient Fuel Identification

The air-to-fuel ratio dynamics in a port-fuel injected internal combustion engine (see Fig. 7.1) are represented by the following differential equations (Orlov et al. 2009):

$$\begin{aligned}\frac{dm_p}{dt}&=-\frac{m_p}{\rho}+XW_{f,i},\\ W_{f,c}&=(1-X)W_{f,i}+\frac{m_p}{\rho},\\ \lambda&=\frac{W_{a,c}}{W_{f,c}}.\end{aligned} \tag{7.127}$$

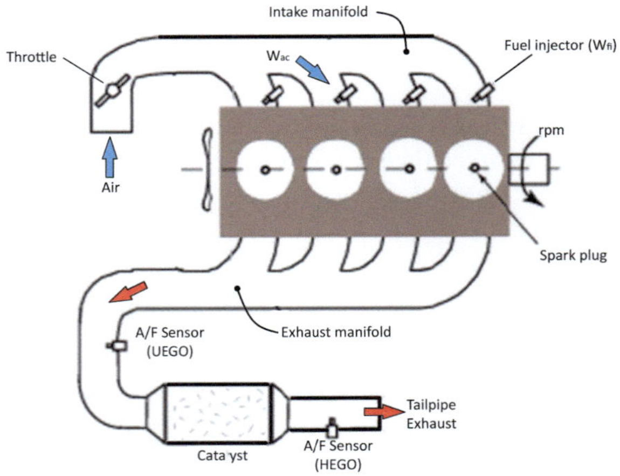

Fig. 7.1 Schematics of a gasoline spark ignition engine. © 2008 *WILEY*. Reprinted, with permission, from Orlov et al. (2009)

Here $W_{f,i}$ is the injected fuel flow rate, $W_{f,c}$ is the fuel flow rate into the engine cylinders, $0 < X < 1$ is a fraction of the injected fuel that replenishes the liquid fuel puddle in the intake ports of the engine, and $\rho > 0$ is the time constant of fuel evaporation from this liquid puddle. The mass of fuel in the puddle (summed over all engine cylinders) is m_p, the in-cylinder air-to-fuel is λ, and $W_{a,c}$ denotes the airflow into the cylinder.

In order to accurately control the engine air-to-fuel ratio, the parameters X and ρ need to accurately be known. Indeed, if estimates of X and ρ, $\hat{X}$ and $\hat{\rho}$, respectively, are known, and $W_{f,c}^d$ is the desired in-cylinder fuel flow rate computed to match the airflow, $W_{a,c}$, then injecting the fuel flow rate is determined according to

$$W_{f,i} = \frac{W_{f,c}^d - \frac{\hat{m}_p}{\hat{\rho}}}{1 - \hat{X}},$$

where $\hat{m}_p$ is an estimate of fuel puddle mass, results in $W_{f,c}$ matching $W_{f,c}^d$. The accurate air-to-fuel ratio control is critically important for engine emissions, drivability, and fuel economy. Following Orlov et al. (2009), capabilities of the developed adaptive TDS identification are experimentally tested to account for system delays as well as for the dynamics of the fuel–air ratio sensor, "lost" fuel, and for the exhaust mixing.

Note that

$$\begin{aligned}\dot{W}_{f,c} &= (1 - X)\dot{W}_{f,i} + \frac{1}{\rho}\left(XW_{f,i} - \frac{m_p}{\rho}\right) - \dot{X}W_{fi} - \frac{m_p}{\rho^2}\dot{\rho} \\ &= (1 - X)\dot{W}_{f,i} + \frac{1}{\rho}XW_{f,i}\end{aligned}$$

$$
\begin{aligned}
&-\frac{1}{\rho}(W_{f,c}-(1-X)W_{f,i})-\dot{X}W_{fi}-\frac{m_p}{\rho^2}\dot{\rho} \\
&=(1-X)\dot{W}_{f,i}+\frac{1}{\rho}(W_{f,i}-W_{f,c})-\dot{X}W_{fi}-\frac{m_p}{\rho^2}\dot{\rho}.
\end{aligned}
$$

The parameters X and ρ change with the engine speed and airflow, $W_{a,c}$, and hence their time derivatives are included in the above expression.

Let μ_{ug} denote the measured *fuel–air* ratio with a linear universal exhaust gas oxygen sensor. Accounting for the first-order sensor and exhaust mixing dynamics, and for the delay, due to time elapse between fuel injection and exhaust stroke of the engine as well as due to time it takes the exhaust gas to reach the sensor location, one obtains

$$
\dot{\mu}_{ug}+a_{ug}\mu_{ug}=a_{ug}\frac{W_{f,c}(t-\tau)}{W_{a,c}(t-\tau)}(1-l), \tag{7.128}
$$

where $1/a_{ug}$ is the time constant, τ is the delay, and l (typically, $0\le l<1$) accounts for "lost fuel" in the engine cylinders at cold temperatures or for miscalibration in the airflow estimation. By differentiating (7.128) with respect to time, and assuming constant τ and l, it follows that

$$
\begin{aligned}
&\ddot{\mu}_{ug}+\dot{\mu}_{ug}\left(a_{ug}+\frac{1}{\rho(t-\tau)}+\frac{\dot{\rho}(t-\tau)}{\rho(t-\tau)}+\frac{\dot{W}_{a,c}(t-\tau)}{W_{a,c}(t-\tau)}\right) \\
&+\mu_{ug}a_{ug}\left(\frac{1}{\rho(t-\tau)}+\frac{\dot{\rho}(t-\tau)}{\rho(t-\tau)}+\frac{\dot{W}_{a,c}(t-\tau)}{W_{a,c}(t-\tau)}\right) \\
&=a_{ug}(1-l)\frac{\dot{W}_{f,i}(t-\tau)}{W_{a,c}(t-\tau)}(1-X(t-\tau)) \\
&+a_{ug}(1-l)\frac{W_{f,i}(t-\tau)}{W_{a,c}(t-\tau)}\left(\frac{1}{\rho(t-\tau)}-\dot{X}(t-\tau)+(1-X(t-\tau))\frac{\dot{\rho}(t-\tau)}{\rho(t-\tau)}\right).
\end{aligned} \tag{7.129}
$$

The values of X and ρ depend on the engine operating variables such as airflow (or manifold pressure), engine speed, and engine coolant temperature. To illustrate the applicability of the developed adaptive identification approach in a simplest possible form the parameterizations of X and ρ as functions of the engine operating variables is avoided. The approach is, therefore, restricted to the application of the proposed adaptive identification algorithm to only time intervals where the air flow, manifold pressure, engine speed and engine coolant temperature are assumable to be constant. This is typically the case in the engine dynamometer calibration phase. In such a case, $\dot{W}_{a,c}=\dot{X}=\dot{\rho}=0$ and (7.129) simplifies to

$$
\begin{aligned}
\ddot{\mu}_{ug}+\dot{\mu}_{ug}\left(a_{ug}+\frac{1}{\rho}\right)+\frac{a_{ug}}{\rho}\mu_{ug}&=a_{ug}(1-X)\frac{\dot{W}_{f,i}(t-\tau)}{W_{a,c}(t-\tau)}(1-l) \\
&+\frac{a_{ug}}{\rho}\frac{W_{f,i}(t-\tau)}{W_{a,c}(t-\tau)}(1-l),
\end{aligned} \tag{7.130}
$$

where X and ρ can now be treated as constant parameters corresponding to a particular operating point of the engine, while $W_{f,i}$ may be time varying. Generating the dependence of X and ρ on engine operating variables results in the adaptive identification to separately be applicable for different operating points.

Letting $x_1 = \mu_{ug}$, $x_2 = \dot{\mu}_{ug}$ $v = \frac{W_{f,i}}{W_{a,c}}$, $\alpha = a_{ug} + \frac{1}{\rho}$, $\alpha^0 = \frac{a_{ug}}{\rho}$, $\beta^0 = \alpha^0(1-l)$, $\beta = a_{ug}(1-X)(1-l)$, the second-order SISO TDS

$$\begin{aligned} \dot{x}_1 &= x_2 \\ \dot{x}_2(t) &= \beta\dot{v}(t-\tau) + \beta^0 v(t-\tau) - \alpha x_2(t) - \alpha^0 x_1(t), \\ z(t) &= x_1(t) \end{aligned} \tag{7.131}$$

is obtained with the positive parameters $\beta, \beta^0, \alpha, \alpha^0, \tau$ to be identified, with the asymptotically stable internal dynamics, and with the measured output z.

It is first assumed that the delay value τ is known a priori. This assumption is reasonable for the application where the delay is well correlated to engine speed and airflow (both are constant) and can be identified beforehand by looking at a delay between a step change in fueling rate and a change in fuel-to-air measured signal. Thus one can apply Theorem 7.7 for the adaptive identification of the parameters $\beta, \beta^0, \alpha, \alpha^0$. For the resulting system (7.131), the adaptive identification law (7.85)–(7.91) is specified to

$$\begin{aligned} \dot{\hat{x}}_1(t) &= \hat{x}_2(t) + w(t) \\ \dot{\hat{x}}_2(t) &= \gamma w(t) + \hat{\beta}(t)\dot{v}(t-\tau) + \hat{\beta}^0 v(t-\tau) \\ &\quad -\hat{\alpha}(t)[\hat{x}_2(t) + w(t)] - \hat{\alpha}^0(t)\hat{x}_1(t) \\ \dot{\hat{\beta}}(t) &= \lambda_1 \dot{v}(t-\tau)w(t) \\ \dot{\hat{\beta}}^0(t) &= \lambda_2 v(t-\tau)w(t) \\ \dot{\hat{\alpha}}(t) &= -\lambda_3[\hat{x}_2(t) + w(t)]w(t) \\ \dot{\hat{\alpha}}^0(t) &= -\lambda_4 \hat{x}_1(t)w(t) \\ w(t) &= (L + |\hat{x}_2(t)|)\text{sign}\,[x_1(t) - \hat{x}_1(t)], \quad L \geq \delta + \|z(\cdot)\|_{C^1} \end{aligned} \tag{7.132}$$

where $\gamma, \lambda_1, \lambda_2, \lambda_3, \lambda_4 > 0$ are adaptation parameters, L is an upper bound of the C^1-norm of the output variable z, and $\delta > 0$ is a constant.

In order to find the delay value τ when it is unknown a priori one can apply the distributed-delay-flavored identification procedure of Sect. 7.1.3.2 and consider (7.131) as a system with a large number r of fictitious delays ih, $i = 1, \ldots, r$, uniformly distributed over a time interval $[0, \mathcal{T}]$ that covers an upper bound $\mathcal{T} = rh$ of the real system delay τ:

$$\ddot{z}(t) = \sum_{i=1}^{N}\{\beta_i \dot{v}(t-ih) + \beta_i^0 v(t-ih)\} - \alpha\dot{z}(t) - \alpha^0 z(t). \tag{7.133}$$

According to the procedure, the identification law (7.132) is modified to

$$\begin{aligned}
\dot{\hat{x}}_1(t) &= \hat{x}_2(t) + w(t)\\
\dot{\hat{x}}_2(t) &= \gamma w(t) + \sum_{i=1}^{r}\{\hat{\beta}_i(t)\dot{v}(t-ih) + \hat{\beta}_i^0(t)v(t-ih)\}\\
&\quad -\hat{\alpha}(t)[\hat{x}_2(t) + w(t)] - \hat{\alpha}^0(t)\hat{x}_1(t)\\
\dot{\hat{\beta}}_i(t) &= \lambda_{1i}\dot{v}(t-ih)w(t)\\
\dot{\hat{\beta}}_i^0(t) &= \lambda_{2i}v(t-ih)w(t), \quad i = 1,\ldots,r\\
\dot{\hat{\alpha}}(t) &= -\lambda_3[\hat{x}_2(t) + w(t)]w(t)\\
\dot{\hat{\alpha}}^0(t) &= -\lambda_4\hat{x}_1(t)w(t)\\
w(t) &= (L + |\hat{x}_2(t)|)\operatorname{sign}[x_1(t) - \hat{x}_1(t)], \quad L \geq \delta + \|z(\cdot)\|_{C^1}
\end{aligned} \tag{7.134}$$

where $\gamma, \lambda_{1i}, \lambda_{2i}, \lambda_3, \lambda_4$ are positive parameters. Then the limiting relations

$$\begin{aligned}
&\lim_{t\to\infty}\hat{x}_1(t) = z(t),\ \lim_{t\to\infty}\hat{x}_2(t) = \dot{z}(t),\ \lim_{t\to\infty}\hat{\alpha}(t) = \alpha,\ \lim_{t\to\infty}\hat{\alpha}^0(t) = \alpha^0,\\
&\lim_{t\to\infty}\hat{\beta}_{i_0}(t) = \beta,\ \lim_{t\to\infty}\hat{\beta}_{i_0}^0(t) = \beta^0,\ \lim_{t\to\infty}\hat{\beta}_i(t) = 0,\ \lim_{t\to\infty}\hat{\beta}_i^0(t) = 0
\end{aligned} \tag{7.135}$$

hold for some $i_0 \in \overline{1,\ldots,N}$ and $i = 1,\ldots,i_0-1, i_0+1,\ldots,N$. A delay estimate $\tau = i_0 h$ is thus obtained.

7.1.5.1 Simulation Results

In the simulations, performed in Gomez et al. (2007) with SIMNON, the system parameters were specified as follows:

$$\beta = 4,\ \beta^0 = \alpha^0 = 12.5,\ \alpha = 7.5,\ \tau = 0.3. \tag{7.136}$$

These parameter values corresponded to those encountered during typical engine operation. In the simulations, the adaptive identification law (7.132) was evaluated for the a priori known delay value whereas its counterpart (7.134) was evaluated when the delay value is unknown yet, and it can assume only one of a finite number of known values.

In order to better present simulation results the measured state of the plant and that of the identifier were initialized with the same conditions, $x_1(0) = \hat{x}_1(0) = 0.068$. Note that since x_1 is the measured fuel-to-air ratio this assumption is reasonable. Due to this assumption, the sliding mode was enforced on the surface $x_1 - \hat{x}_1 = 0$ by the switching observer input (7.132) of the small amplitude $L = 0.2$ and it appeared from the very beginning of the identification process. Other state variables were arbitrarily initialized as $x_2(0) = 0,\ \hat{x}_2(0) = 0.1$.

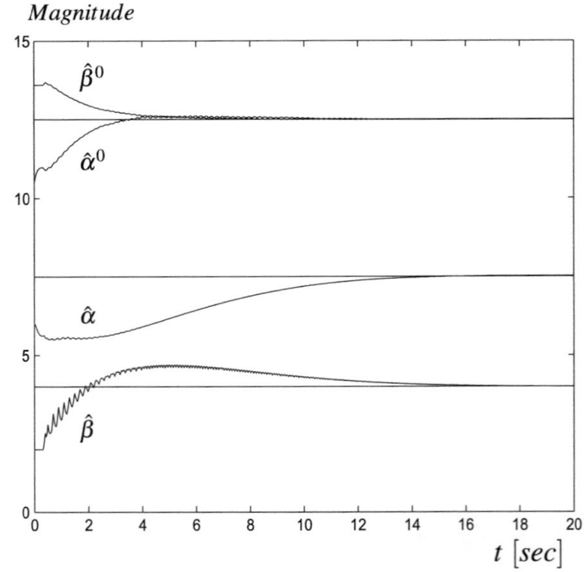

Fig. 7.2 Time histories of the parameter estimates in the single-delay case

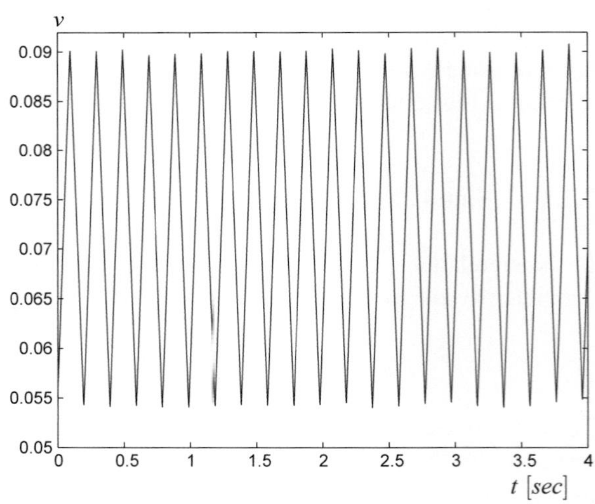

Fig. 7.3 Time history of the input signal (7.137)

The results in Figs. 7.2 and 7.3 show that fast convergence of the parameter estimates is attained with the identification law (7.132) if the adaptation parameters and fuel excitation are sufficiently aggressive. In the simulations, parameters estimates were initialized as $\hat{\beta}(0) = 2$, $\hat{\beta}^0(0) = 13.6$, $\hat{\alpha}^0(0) = 10.5$, $\hat{\alpha}(0) = 6$, and adaptation algorithm settings were $\gamma = 10, \lambda_1 = 750, \lambda_2 = 430, \lambda_3 = 900, \lambda_4 = 760$ whereas the sufficiently nonsmooth input $v(t)$ was generated by the dynamic system

$$\dot{v}(t) = 0.37 \cdot \text{sign } \sin(0.1 + 31.71t), \; v(0) = 0.06. \tag{7.137}$$

Even faster convergence can be attained if $v(t)$ is allowed to vary within a larger range, yet in order to maintain combustion in an engine the range of $v(t)$ must be limited to [1/18, 1/11].

In turn, the adaptive identification law (7.134) was evaluated in simulations for the delay generator $h = 0.1$, the upper bound $N = 5$, and delay values

$$\tau_i = ih, \; i = 1, \ldots, 5. \tag{7.138}$$

The nominal delay value corresponded to $\tau_3 = 0.3$, and the fictitious delay values were $\tau_1 = 0.1, \; \tau_2 = 0.2, \; \tau_4 = 0.4, \; \tau_5 = 0.5$. Figure 7.4 illustrates that the parameter estimates converge reasonably fast to the true parameter values in accordance with (7.134), if the input signal, depicted in Fig. 7.5, is deliberately chosen to be sufficiently nonsmooth, and the adaptation algorithm settings are appropriately assigned. In the simulation run, made for the multi-delay case, the sufficiently nonsmooth input $v(t)$ used, was generated by the dynamic system

$$\dot{v}(t) = 0.03 \cdot \text{sign} \; \sin(0.1 + 6.28t) + 0.16 \cdot \text{sign} \; \sin(0.1 + 28.26t), \; v(0) = 0.06. \tag{7.139}$$

The parameter estimates were initialized with

$$\hat{\beta}_i(0) = 2, \; \hat{\alpha}(0) = 6, \; \hat{\beta}_i^0(0) = 13.6, \; \hat{\alpha}^0(0) = 10.5, \; i = 1, \ldots, 5,$$

whereas the adaptation algorithm settings were

$$\begin{aligned} &\lambda_{11} = 565, \; \lambda_{12} = 235, \; \lambda_{13} = 350, \; \lambda_{14} = 470, \; \lambda_{15} = 452, \\ &\lambda_{21} = 145, \; \lambda_{22} = 145.5, \; \lambda_{23} = 12.2, \; \lambda_{24} = 146.5, \; \lambda_{25} = 146.5, \\ &\lambda_3 = 470, \; \lambda_4 = 22.2, \; \gamma = 95. \end{aligned} \tag{7.140}$$

It is concluded from Fig. 7.4 that the parameters $\hat{\beta}_j(t), \; \hat{\beta}_j^0(t), \; j = 1, 2, 4, 5$ become negligible, thereby establishing both the fictitious delay values $\tau_1 = 0.1, \tau_2 = 0.2, \tau_4 = 0.4, \tau_5 = 0.5$, and the nominal delay value $\tau_3 = 0.3$. In turn, other parameters $\hat{\beta}_3(t), \; \hat{\beta}_3^0(t), \; \hat{\alpha}(t), \hat{\alpha}^0(t)$ yield an appropriate approximation of the system parameters $\beta, \; \beta^0, \; \alpha, \; \alpha^0$.

Finally, in order to test the identifier robustness against uncertainties on the system delay the simulation was repeated in a mismatched case with an approximate knowledge of the delay value τ. For this purpose, the system delay was reset to $\tau = 0.305$ while the delay values (7.138) that appeared in the adaptive identifier (7.134) remained the same as in the matched case where in contrast to the mismatched case one of the identifier delays, namely τ_3, was perfectly matched to the system delay. The mean square error

$$J(t) = \frac{1}{12}\left\{[\alpha - \hat{\alpha}(t)]^2 + [\alpha^0 - \hat{\alpha}^0(t)]^2 + \sum_{i=1}^{5}[\beta_i - \hat{\beta}_i(t)]^2 + \sum_{i=1}^{5}[\beta_i^0 - \hat{\beta}_i^0(t)]^2\right\}, \tag{7.141}$$

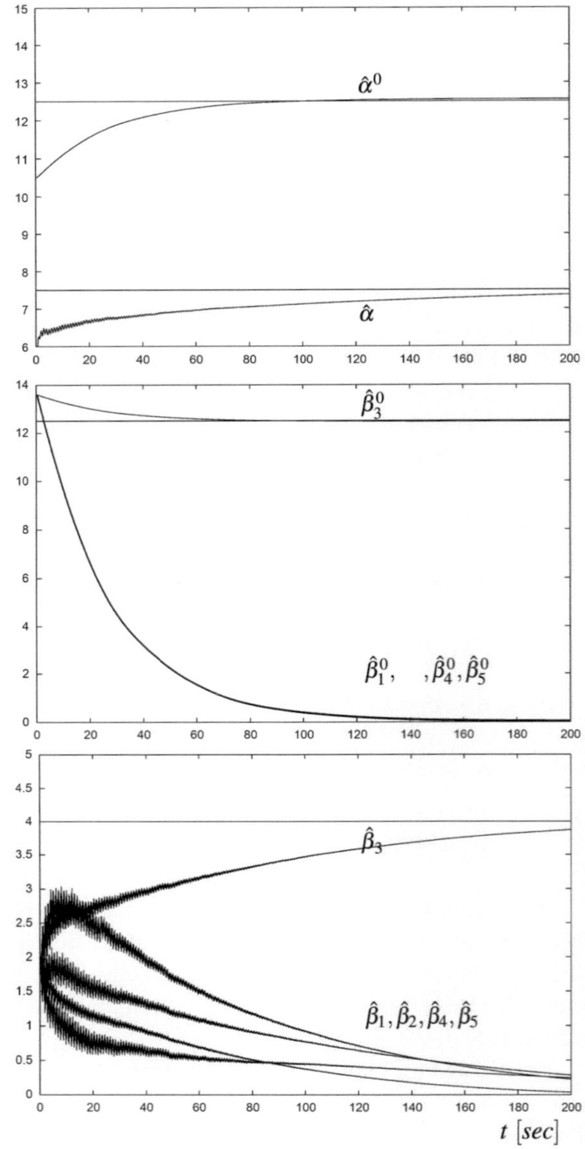

Fig. 7.4 Time histories of the parameter estimates in the multi-delay case (Reprinted from Gomez et al. 2007, Copyright 2007, with permission from Elsevier)

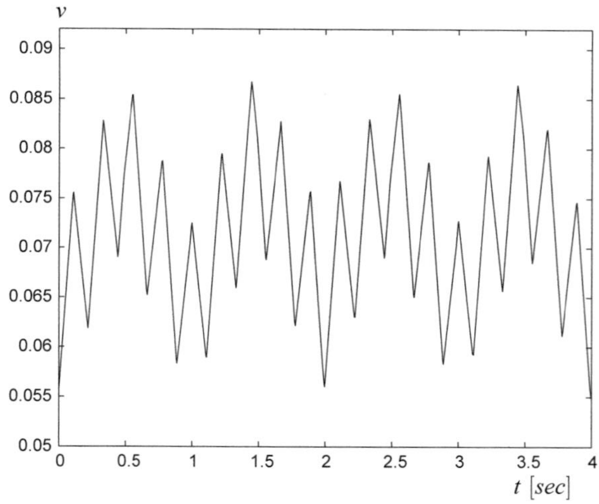

Fig. 7.5 Time history of the input signal (7.139) (Reprinted from Gomez et al. 2007, Copyright 2007, with permission from Elsevier)

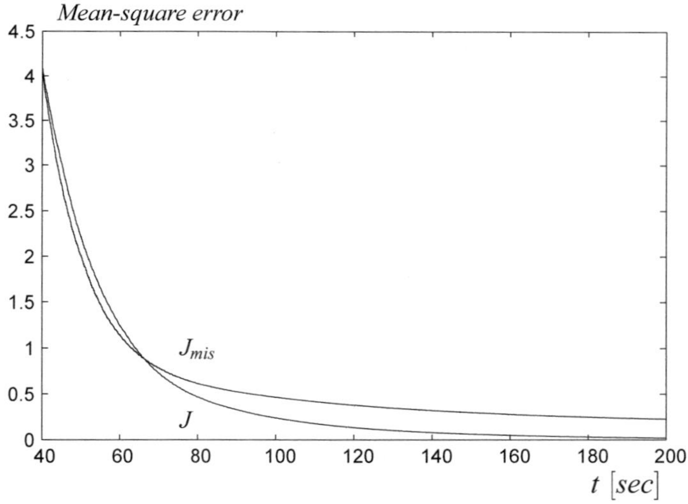

Fig. 7.6 Identifier performance (7.141) in the matched case (J) and mismatched case (J_{mis}) (Reprinted from Gomez et al. 2007, Copyright 2007, with permission from Elsevier)

evaluated with respect to the deviations of the parameter estimates

$$\hat{\alpha}(t), \hat{\alpha}^0(t), \hat{\beta}(t), \hat{\beta}^0(t)\hat{\beta}_i(t), \hat{\beta}_i^0(t), \ i = 1, \ldots, 5,$$

from the true values of the parameters

$$\alpha = 7.5, \ \alpha^0 = \beta^0 = \beta_3^0 = 12.5, \ \beta = \beta_3 = 4,$$
$$\beta_1 = \beta_2 = \beta_4 = \beta_5 = \beta_1^0 = \beta_2^0 = \beta_4^0 = \beta_5^0 = 0,$$

and computed in the successive simulation runs is presented in Fig. 7.6 both for the matched case and the mismatched case (because of the scale, chosen for presentation, the initial stages of the monotonous decays of the mean square errors remained beyond the figure). This figure clearly demonstrates a favorable robustness property of the developed identifier against potential mismatching between the model delay and its nominal value.

An experimental assessment, adding practical value to the algorithm performance, may be found in Orlov et al. (2009).

7.2 Lyapunov–Krasovskii Redesign and Identification of Linear Parabolic PDE

The parameter identification of DPS fundamentally differs from that of the finite-dimensional and time-delay settings because of the spatial parameter variability and the presence of the spatial derivatives in the plant equation.

The key problem in the identification of the spatially varying plant parameters is the development of the constructive spatial identifiability conditions. The general constructive criteria for spatial identifiability such as (Courdesses et al. 1981; Kitamura and Nakagiri 1977; Nakagiri 1983; Pierce 1979) encompass spatially varying parameters in their formulation, they are, however, verifiable only in the constant parameter case. As a result, current nonadaptive methods of identification of space-varying parameters in DPS, such as output least squares, maximum likelihood estimators, and method of characteristics (cf. Banks and Kunish 1989; Omatu and Seinfeld 1989, and references therein), are not accompanied by the easily enforceable identifiability conditions that could guarantee their convergence. In comparison to these algorithms, model reference adaptive identifiers of the spatially varying parameters in DPS, which by construction effectively utilize the prior knowledge of the plant structure through its inclusion into the reference model, could potentially offer physically natural enforceable identifiability conditions for a broad class of systems. Into adaptive identification, the first richness-like identifiability condition for plants with spatially varying coefficients was introduced in Baumeister (1987). There were, however, two problems with this condition: (i) it was defined through the solution of the equation of the plant whose coefficients were unknown, and therefore it was unverifiable, and (ii) it was passive in a sense that no indication was given on how to attain it using some manipulatable quantity. The identifiability conditions in Baumeister et al. (1997), Demetriou and Rosen (1994a, b, 1995), Utkin and Orlov (1990) suffer from the same two problems.

Another important aspect of the adaptive identification of the spatially varying plant parameters is the reduction of the order of the spatial differentiation of the measurement data in the identification law from that of the highest spatial derivative of the plant to decrease the identifier sensitivity to measurement errors. The pioneering works in the area (Baumeister 1987; Baumeister et al. 1997; Demetriou and Rosen 1994a, b, 1995), with the exception of Alt et al. (1984), Hoffmann and Sprekels (1984), offer no spatial derivative order reduction because of the repetition of the spatial derivative structure of the plant in the adaptive identification law. This, clearly, admits large amplification of even the smallest measurement errors present in most situations of practical interest and, therefore, makes these algorithms of limited utility. Although the works (Alt et al. 1984; Hoffmann and Sprekels 1984) reduce the order of spatial derivatives of the data, the proposed algorithms give only L_2-norm convergence of the spatially varying parameter estimates and do not demonstrate how to identify plant parameters point-wise. Furthermore, the results of Alt et al. (1984), Hoffmann and Sprekels (1984) are formulated without the persistency of excitation requirement, limiting thereby the identification capability of the algorithms given there to only single parameter.

For DPS with spatially invariant parameters, the first constructive identifiability concept of persistency of excitation was attained in Hong and Bentsman (1994a, b) via explicit conditions on the manipulatable in-domain and boundary inputs. Subsequently, preliminary results were presented in Solo and Bentsman (1993) for spatially varying parameter identification with identifiability conditions, formulated in terms of the spatial spectral input characteristics using averaging theory, although a persistently exciting input was not explicitly constructed. The spatially varying parameter identification was first constructively addressed in Orlov and Bentsman (1995, 2000) where persistently exciting inputs were presented explicitly. Construction of persistent excitation given in the latter works is further presented for a linear parabolic PDE, chosen to exemplify the Krasovskii-based identifier design in the PDE setting. The resulting identifier synthesis gives rise to the algorithm structure that falls under the so-called series–parallel/parallel configuration (Ioannou and Sun 1996; Landau 1979; Narendra and Annaswamy 1989). Both distributed in-domain sensing and boundary sensing are successively addressed.

7.2.1 Identification over In-Domain Sensing

The following mathematical model of DPS of parabolic type is used throughout this section:

$$\begin{aligned}
\rho(x)\frac{\partial Q}{\partial t} &= \frac{\partial[k(x)\frac{\partial Q}{\partial x}]}{\partial x} - q(x)Q, \quad 0 < x < 1, \ t > 0, \\
\frac{\partial Q(0,t)}{\partial x} &= \beta_0(t), \quad \frac{\partial Q(1,t)}{\partial x} = \beta_1(t), \quad t > 0, \\
Q(x,0) &= Q_0(x), \quad 0 < x < 1.
\end{aligned} \tag{7.142}$$

Equation (7.142) describes the propagation of heat in a one-dimensional rod where $Q(x,t)$ is the value of the temperature field of the plant at point $x \in [0,1]$ at time moment $t \geq 0$, $\rho(x) \geq \rho_0 > 0$ is a heat capacity coefficient, $k(x) \geq k_0 > 0$ is a heat conduction coefficient, $q(x) \geq 0$ is a heat exchange coefficient, $Q_0(x)$ is an initial state, $\beta_0(t)$, $\beta_1(t)$, and $u(x,t)$ are boundary and external inputs, respectively. All functions $\rho(x), q(x), k(x), Q_0(x), \beta_0(t), \beta_1(t), u(x,t)$ are assumed to be sufficiently smooth.

As discussed in Sect. 3.4, the above assumptions guarantee existence, uniqueness, and differentiability of the solutions of the boundary-value problems the meaning of which is adopted in the strong or weak sense dependent on whether homogeneous ($\beta_0, \beta_1 \equiv 0$) or nonhomogeneous boundary conditions are in force. Recall, that for the nonhomogeneous boundary conditions with nonzero β_0 and/or β_1, the weak solutions of the above parabolic boundary-value problem are defined (see Definition 3.3) as a result of the convolutions of the input and Green functions of the corresponding boundary-value problems.

Following Orlov and Bentsman (2000), it is shown that it is possible to simultaneously identify all spatially distributed plant parameters provided that the system in question is persistently excited by sufficiently rich external and/or boundary inputs. An adaptive identifier is then represented as error systems describing the evolution of the state error and the parameter error. The state and the parameter error subsystems take the form of PDE and ODE, respectively. Lyapunov–Krasovskii functional with nonpositive time derivative along the solutions of the coupled plant-identifier system is constructed. It takes zero values on a certain manifold in the state space. An explicitly constructed input signal, generating persistent plant excitation, is shown to guarantee the absence of the nontrivial trajectories on this manifold, and thereby to ensure the existence of a unique zero steady state of the error system. The justification of the asymptotic stability of the overall error system is finalized by applying the infinite-dimensional version of the Krasovskii–LaSalle invariance principle, given by Theorem 4.7.

For certainty, the exposition is confined to Neumann boundary conditions and to one spatial variable, however, appropriate extensions to Dirichlet and Robin (mixed) boundary conditions and to several spatial variables are rather straightforward.

7.2.1.1 Parameter Identifiability

To ensure the parameter convergence, the adaptive identification setup should guarantee that the trivial steady-state solution of the state-parameter error equation for the coupled plant-identifier system is isolated, i.e., it is unique in a sufficiently large vicinity around it. The ability to ensure this objective is usually referred to as the parameter identifiability. This ability becomes enforceable if it is straightforwardly related to the explicitly given manipulatable external and/or boundary inputs. The parameter convergence is then attained through the asymptotic stabilization of the isolated trivial solution of the error equation. Thus, the identifiability and stabilization together provide a sufficient condition for the parameter convergence.

To introduce the identifiability concept for the system in question let us along with the boundary-value problem (7.142), consider its reference model of the same form in which the unknown coefficients $\rho(x), k(x), q(x)$ are replaced by the known quantities $\tilde{\rho}(x), \tilde{k}(x), \tilde{q}(x)$, respectively. Let the output error between the system state and the model state and the parameter mismatches be given by

$$\begin{aligned} \Delta Q(x,t) &= Q(x,t) - \tilde{Q}(x,t),\ \Delta\rho(x) = \rho(x) - \tilde{\rho}(x), \\ \Delta k(x) &= k(x) - \tilde{k}(x),\ \Delta q(x) = q(x) - \tilde{q}(x), \quad x \in [0,1], t \geq 0. \end{aligned} \tag{7.143}$$

Definition 7.9 The spatially varying parameters $\rho(x), k(x), q(x)$ of the DPS (7.142) are said to be identifiable under the external and boundary inputs $u(x,t)$ and $\beta_0(t), \beta_1(t)$ iff given an arbitrary reference model and the corresponding output and parameter mismatch (7.143), the relation

$$\Delta Q(x,t) \equiv 0 \tag{7.144}$$

or, equivalently

$$\Delta\rho(x)\dot{Q} = [\Delta k(x) Q']' - \Delta q(x) Q, \quad 0 < x < 1,\ t > 0 \tag{7.145}$$

ensures that

$$\Delta\rho(x) = \Delta k(x) = \Delta q(x) = 0 \quad \forall x \in [0,1]. \tag{7.146}$$

The identifiability of a subset of the parameters $\rho(x), k(x), q(x)$ are defined similarly.

7.2.1.2 Persistent Excitation

Identification of the spatially varying parameters $\rho(x), k(x), q(x)$ places additional requirements on the the external and boundary inputs $u(x,t)$ and $\beta_0(t), \beta_1(t)$. The following definition is due.

Definition 7.10 External and boundary inputs $u(x,t)$ and $\beta_0(t), \beta_1(t)$ are said to generate persistent excitation of the boundary-value problem (7.142) iff the Fourier coefficients $l_n(t),\ n = 0, 1, \ldots$ of the solution

$$Q(x,t) = \Sigma_{n=0}^{\infty} l_n(t) g_n(x) \tag{7.147}$$

of the boundary-value problem (7.142) and their time derivatives $\dot{l}_n(t)$ are linearly independent functions.

The definition given above does not depend of a particular orthonormal basis of the functions

$$g_n(x) \in \overset{\circ}{H}_2\ (0,1) = \left\{ g(x) : \frac{\partial g(0)}{\partial x} = \frac{\partial g(1)}{\partial x} = 0,\ \frac{\partial^2 g}{\partial x^2} \in L_2(0,1) \right\}, \quad n = 0, 1, \ldots$$

used in the Fourier series. As shown below, the choice of the basis can be arbitrary.

Lemma 7.1 *If $u(x,t)$ and $\beta_0(t)$, $\beta_1(t)$ generate persistent excitation of the underlying system* (7.142)) *with respect to the orthonormal basis of functions $g_n(x) \in \overset{\circ}{H}_2$ $(0,1)$, $n = 0, 1, \ldots$, then it generates persistent excitation with respect to an arbitrary orthonormal basis of functions $\tilde{g}_m(x) \in \overset{\circ}{H}_2$ $(0,1)$, $m = 0, 1, \ldots$.*

Proof Along with (7.147) let there be a representation

$$Q(x,t) = \Sigma_{m=0}^{\infty} \tilde{l}_m(t) \tilde{g}_m(x).$$

Then

$$\tilde{l}_m(t) = \int_0^1 Q(x,t)\tilde{g}_m(x)dx = \int_0^1 \Sigma_{n=0}^{\infty} l_n(t) g_n(x) \tilde{g}_m(x)dx = \Sigma_{n=0}^{\infty} \alpha_{mn} l_n(t)$$

and

$$\tilde{l}(t) = A l(t), \tag{7.148}$$

where

$$A = \{\alpha_{mn}\}_{m,n=0}^{\infty},\ \tilde{l} = (\tilde{l}_1, \tilde{l}_2, \ldots)^T,\ l = (l_1, l_2, \ldots)^T,\ \alpha_{mn} = \int_0^1 g_n(x)\tilde{g}_m(x)dx.$$

Analogously to (7.148), one obtains

$$l(t) = A^T \tilde{l}(t)$$

which obviously implies that $A^{-1} = A^T$.

Assume now that the functions $\tilde{l}_m(t)$, $\dot{\tilde{l}}_m(t)$, $m = 0, 1, \ldots$ are linearly dependent, i.e.,

$$\Sigma_{m=0}^{\infty}[\tilde{\omega}_m' \tilde{l}_m(t) + \tilde{\omega}_m'' \dot{\tilde{l}}_m(t)] = 0,$$

where constants $\tilde{\omega}_m'$, $\tilde{\omega}_m''$ are not equal to zero simultaneously. Then, taking into account (7.148) yields

$$\Sigma_{m=0}^{\infty}[\tilde{\omega}_m^{'}\tilde{l}_m(t) + \tilde{\omega}_m^{''}\dot{\tilde{l}}_m(t)] = \Sigma_{m,n=0}^{\infty}\alpha_{mn}[\tilde{\omega}_m^{'}l_n(t) + \tilde{\omega}_m^{''}\dot{l}_n(t)]$$
$$= \Sigma_{n=0}^{\infty}[\omega_n^{'}l_n(t) + \omega_n^{''}\dot{l}_n(t)] = 0,$$

where not all constants $\omega_n^{'} = \Sigma_{m=0}^{\infty}\alpha_{mn}\tilde{\omega}_m^{'}$, $\omega_n^{''} = \Sigma_{m=0}^{\infty}\alpha_{mn}\tilde{\omega}_m^{''}$, $n = 0, 1, \ldots$ are equal to zero by the invertibility of the infinite-dimensional matrix A. This, however, contradicts the linear independence of functions $l_n(t), \dot{l}_n(t)$, $n = 0, 1, \ldots$. Consequently, functions $\tilde{l}_m(t), \dot{\tilde{l}}_m(t)$, $m = 0, 1, \ldots$ are also linearly independent. Lemma 7.1 is thus proved. ∎

The persistent excitation of the heat process (7.142) is recognized to be sufficient for the parameter identifiability.

Theorem 7.9 *If the external and boundary inputs $u(x, t)$ and $\beta_0(t), \beta_1(t)$ generate persistent excitation of the boundary-value problem* (7.142) *then the parameters $k(x), \rho(x), q(x)$ are identifiable under $u(x, t), \beta_0(t), \beta_1(t)$.*

Proof Representing solution of the plant equation (7.142) as in Sect. 3.4 in the form of the Fourier series

$$Q(x, t) = \Sigma_{n=0}^{\infty}l_n(t)\cos \pi nx \tag{7.149}$$

and substituting it into (7.145) yield

$$\Sigma_{n=0}^{\infty}\Delta\rho(x)\cos\pi nx\ \dot{l}_n(t) = -\Sigma_{n=0}^{\infty}\Bigg\{[\Delta q(x) +$$
$$(\pi n)^2\Delta k(x)]cos\ \pi nx + \pi n\frac{\partial\Delta k(x)}{\partial x}\sin\ \pi nx\Bigg\}l_n(t),$$

where Fourier coefficients $l_n(t)$, $n = 0, 1, \ldots$ and their derivatives $\dot{l}_n(t)$ are linearly independent by virtue of Lemma 7.1 and by the persistent excitation condition of the theorem, imposed on the plant equation. Hence,

$$\pi n\frac{\partial\Delta k(x)}{\partial x}\sin\ \pi nx = -[\Delta q(x) + (\pi n)^2\Delta k(x)]\cos\ \pi nx$$
$$\Delta\rho(x)\cos\ \pi nx = 0, \quad n = 0, 1, \ldots$$

and the validity of (7.146) follows due to the fact that the function sets $\{\cos\ \pi nx\}$, $\{\sin\pi nx\}$ have nonintersecting everywhere dense zero sets. This proves Theorem 7.9. ∎

In order to test whether the external input $u(x, t)$ and boundary inputs $\beta_0(t), \beta_1(t)$ generate persistent excitation of the boundary-value problem (7.142), one can use the explicit solution representation (7.147) which is in accordance with (3.96) and (3.106) specified by

$$
\begin{aligned}
l_n(t) &= \int_0^1 \rho(x)Q_0(x)g_n(x)dxe^{-\lambda_n^2 t} + \int_0^t e^{-\lambda_n^2(t-\tau)} \int_0^1 \rho(x)u(x,\tau)g_n(x)dxd\tau \\
&+ \Big[\rho(z)g_n(z)\Big]'\Big|_{z=0} \int_0^t e^{-\lambda_n^2(t-\tau)}\beta_0(\tau)d\tau \\
&- \Big[\rho(z)g_n(z)\Big]'\Big|_{z=1} \int_0^t e^{-\lambda_n^2(t-\tau)}\beta_1(\tau)d\tau, \quad n = 0, 1, \ldots
\end{aligned} \tag{7.150}
$$

where $\{\lambda_n^2\}_{n=0}^{\infty}$ and $\{g_n(x)\}_{n=0}^{\infty}$ form the set of the eigenvalues and orthonormal basis of the eigenfunctions of the corresponding Sturm–Liouville problem

$$
\frac{\partial[k(x)\frac{\partial Q}{\partial x}]}{\partial x} - q(x)g_n(x) = -\lambda_n^2\rho(x)g_n(x),
$$

$$
\frac{\partial g_n(0)}{\partial x} = \frac{\partial g_n(1)}{\partial x} = 0.
$$

In a particular case of zero initial condition $Q_0(x) = 0$, the following simple inputs are straightforwardly verified to generate persistent excitation of (7.142):

1. An arbitrary time-invariant input

$$
u(x,t) = u(x) = \Sigma_{n=0}^{\infty} u_n g_n(x)
$$

with nonzero Fourier coefficients $u_n \neq 0$, $n = 0, 1, \ldots$ and zero boundary inputs $\beta_0(t) \equiv 0,\ \beta_1(t) \equiv 0$. Indeed, the functions

$$
l_n(t) = u_n\lambda_n^{-2}(1 - e^{-\lambda_n^2 t}),\ \dot{l}_n(t) = u_n e^{-\lambda_n^2 t},
$$

$n = 0, 1, \ldots$ are linearly independent.

2. Time-invariant inputs

$$
\beta_0(t) = \nu^0, \quad \beta_1 = \nu^1
$$

and zero external input $u(x,t) \equiv 0$ such that

$$
\omega_n = \nu^0\Big[\rho(z)g_n(z)\Big]'\Big|_{z=0} - \nu^1\Big[\rho(z)g_n(z)\Big]'\Big|_{z=1} \neq 0, \quad n = 0, 1, \ldots \tag{7.151}
$$

Just in the case,

$$
l_n(t) = \omega_n\lambda_n^{-2}\Big(1 - e^{-\lambda_n^2 t}\Big), \quad \dot{l}_n(t) = \omega_n e^{-\lambda_n^2 t}, \quad n = 0, 1, \ldots
$$

are linearly independent.

For arbitrary initial conditions $Q_0(x)$ and zero boundary inputs, a time-invariant generator $u(x,t) = u(x)$ of persistent excitation of (7.142) is constructed in a similar manner to the above item 1 to ensure that all Fourier coefficients of the function $Q_0(x) + u(x)$ are nonzero.

In general, persistent excitation generators $r(x,t)$ and $\beta_0(t), \beta_1(t)$ to be independent of any specific initial function $Q_0(x)$ are constructed to ensure the linear independence of the Fourier coefficients (7.150) of the solution of (7.142) and their time derivatives. For zero boundary inputs $\beta_0(t) \equiv 0$, $\beta_1(t) \equiv 0$, it is particularly the case of a persistently exciting external input $u(x,t)$ to be independent of initial conditions of the boundary-value problem (7.142) provided that $u(x,t)$ is time-periodic and it is such that all Fourier coefficients of the corresponding steady-state periodic solution are nonzero.

7.2.1.3 Krasovskii-Based Adaptive Identifier Design

Once a persistently excited input has been constructed, and hence, the identifiability of the unknown distributed plant parameters is guaranteed, the plant parameter identification is achieved with the adaptive identifier design proposed in Orlov and Bentsman (2000). In order to identify the spatially varying plant parameters $\rho(x), k(x), q(x)$ of the heat process (7.142), the proposed design is as follows:

$$\begin{aligned}
\tilde{\rho}(x,t)\frac{\partial \tilde{Q}}{\partial t} &= \frac{\partial[\tilde{k}(x,t)\frac{\partial \tilde{Q}}{\partial x}]}{\partial x} - \tilde{q}(x,t)\tilde{Q} + u(x,t) + \nu_0(Q - \tilde{Q}),\\
\frac{\partial \tilde{Q}(0,t)}{\partial x} &= \beta_0(t), \quad \frac{\partial \tilde{Q}(1,t)}{\partial x} = \beta_1(t), \quad t > 0,\\
\tilde{Q}(x,0) &= \tilde{Q}_0(x), \quad 0 < x < 1
\end{aligned} \tag{7.152}$$

$$\begin{aligned}
\dot{\tilde{k}} &= -\nu_1 \frac{\partial (Q - \tilde{Q})}{\partial x}\frac{\partial \tilde{Q}}{\partial x}, \quad \tilde{k}(x,0) = \tilde{k}_0(x),\\
\dot{\tilde{q}} &= -\nu_2(Q - \tilde{Q})\tilde{Q}, \qquad \tilde{q}(x,0) = \tilde{q}_0(x),\\
\dot{\tilde{\rho}} &= -\nu_3(Q - \tilde{Q})\frac{\partial \tilde{Q}}{\partial x}, \quad \tilde{\rho}(x,0) = \tilde{\rho}_0(x)
\end{aligned} \tag{7.153}$$

where $\nu_i,\ i = 0,1,2,3$ are adaptation parameters, and $\tilde{Q}_0(x)$, $\tilde{\rho}_0(x) \geq \rho_0 > 0$, $\tilde{k}_0(x) \geq k_0 > 0$, $\tilde{q}_0(x) \geq 0$ are smooth functions.

As shown below, the adaptive identification (7.152) ensures the L_2 convergence

$$\lim_{t\to\infty} \int_0^1 \{[\Delta Q(x,t)]^2 + [\Delta\rho(x,t)]^2 + [\Delta k(x,t)]^2 + [\Delta q(x,t)]^2\}dx = 0 \tag{7.154}$$

of the deviations

$$\Delta Q = Q - \tilde{Q},\ \Delta\rho = \rho - \tilde{\rho},\ \Delta k = k - \tilde{k},\ \Delta q = q - \tilde{q}. \tag{7.155}$$

It should be noted that the L_2 convergence, generally speaking, does not result in the point-wise convergence. Therefore, the tunable parameters $\tilde{k}(x,t)$, $\tilde{\rho}(x,t)$, $\tilde{q}(x,t)$ are not guaranteed by (7.154) to converge point-wise to the plant parameters $k(x)$, $\rho(x)$, $q(x)$. However, assumptions on the plant parameters, coupled with the convergence given by (7.154), are sufficient to identify these parameters point-wise. Indeed, due to the continuous differentiability, each of the plant parameters can be represented point-wise, except possibly at the boundary points (Kwakernaak and Sivan 1991), in terms of the Fourier series

$$\begin{aligned}\rho(x) &= \lim_{N\to\infty} \Sigma_{n=-N}^{N} \rho_n exp(j2\pi nx),\\ k(x) &= \lim_{N\to\infty} \Sigma_{n=-N}^{N} k_n exp(j2\pi nx),\\ q(x) &= \lim_{N\to\infty} \Sigma_{n=-N}^{N} q_n exp(j2\pi nx),\end{aligned} \tag{7.156}$$

where $j = \sqrt{-1}$, and ρ_n, k_n, q_n are the corresponding Fourier coefficients. Moreover, due to the isomorphism between $L_2(0, 1)$ and the space l_2 of the Fourier coefficients, convergence (7.154) leads to the uniform in n componentwise convergence

$$\lim_{t\to\infty} \tilde{\rho}_n(t) = \rho_n, \ \lim_{t\to\infty} \tilde{k}_n(t) = k_n, \ \lim_{t\to\infty} \tilde{q}_n(t) = q_n, \quad n = 0, 1, -1, 2, -2\ldots \tag{7.157}$$

of the coefficients

$$\begin{aligned}\tilde{\rho}_n(t) &= \int_0^1 \tilde{\rho}(x,t) exp(j2\pi nx)dx,\\ \tilde{k}_n(t) &= \int_0^1 \tilde{k}(x,t) exp(j2\pi nx)dx,\\ \tilde{q}_n(t) &= \int_0^1 \tilde{q}(x,t) exp(j2\pi nx)dx\end{aligned} \tag{7.158}$$

of the orthogonal expansions of the tunable controller parameters $\tilde{k}(x,t)$, $\tilde{\rho}(x,t)$, $\tilde{q}(x,t)$ to the corresponding Fourier coefficients ρ_n, k_n, q_n of the plant parameters. Hence, the unknown plant parameters becomes identifiable point-wise as follows:

$$\begin{aligned}\rho(x) &= \lim_{N\to\infty} \lim_{t\to\infty} \Sigma_{n=-N}^{N} \tilde{\rho}_n exp(j2\pi nx),\\ k(x) &= \lim_{N\to\infty} \lim_{t\to\infty} \Sigma_{n=-N}^{N} \tilde{k}_n exp(j2\pi nx),\\ q(x) &= \lim_{N\to\infty} \lim_{t\to\infty} \Sigma_{n=-N}^{N} \tilde{q}_n exp(j2\pi nx), \quad x \in (0, 1).\end{aligned} \tag{7.159}$$

The summations in the right-hand sides of (7.159) represent parameter estimates, converging to the plant parameters point-wise. Whenever the parameter values at the end points $z = 0$ and $z = 1$ are distinct the parameter convergence fails to hold at the ends because of the Gibbs phenomenon (Kwakernaak and Sivan 1991). It is worth noticing that the order of limits in (7.159) is not in general interchangeable and the series may diverge as $N \to \infty$ for all finite t. Summarizing, the following result is obtained.

Theorem 7.10 *Let the external and boundary inputs* $r(x,t)$ *and* $\beta_0(t)$, $\beta_1(t)$ *be time-periodic and generate persistent excitation of the heat process* (7.142). *Then, the limiting relation* (7.154) *holds with the adaptive identification law* (7.152), *and with the identifier parameters tuned as* (7.153). *Apart from this, the unknown plant parameters are computed point-wise according to* (7.158), (7.159).

Proof First, let us prove the local existence of a unique solution of the overall system (7.142), (7.152), (7.153). For this purpose, we integrate (7.153) and substitute the outputs of the integrators into the plant equation (7.142). Since the outputs $\tilde{k}$, $\tilde{\rho}$, $\tilde{q}$ are positive at initial time moment $t = 0$ and Lipschitz continuous in $(x, t, \tilde{Q}, \tilde{Q}')$, the resulting equation falls under Friedman (1969, Theorem 16.2) with $\delta = 0.5$, $\sigma = 1$, and according to this theorem there exists a unique local solution of (7.142), (7.152), (7.153).

Now let us introduce the function $\delta Q(x,t) = Q(x,t) - Q_{ss}(x,t)$ where $Q_{ss}(x,t)$ is a steady-state periodic solution of (7.142). Since the boundary-value problem (7.142) generates an analytic exponentially stable semigroup in the Hilbert space $L_2(0,1)$, such a solution $Q_{ss}(x)$ has been shown to exist and be given by (7.8). Using the Lyapunov functional

$$V(t) = \frac{1}{2}\int_0^1 \Big\{\rho(x)[\delta Q(x,t)]^2 + \rho(x)[\Delta Q(x,t)]^2 + \frac{1}{\nu_1}[\Delta k(x,t)]^2 + \frac{1}{\nu_2}[\Delta q(x,t)]^2 + \frac{1}{\nu_3}[\Delta \rho(x,t)]^2\Big\}dx$$

where, according to (7.142), (7.152), (7.153), (7.155), variables ΔQ, δQ, Δk, Δq, $\Delta\rho$ satisfy equations

$$\begin{aligned}\rho(x)\frac{\partial \Delta Q}{\partial t} + \Delta\rho(x,t)\frac{\partial \tilde{Q}}{\partial t} &= \frac{\partial[k(x)\frac{\partial \Delta Q}{\partial x}]}{\partial x} + \frac{\partial[\Delta k(x,t)\frac{\partial \tilde{Q}}{\partial x}]}{\partial x} - \Delta q(x,t)\tilde{Q}\\ &\quad -[q(x)+\nu_0]\Delta Q, \quad 0<x<1, \quad t>0,\\ \frac{\partial \Delta Q(0,t)}{\partial x} &= \frac{\partial \Delta Q(1,t)}{\partial x} = 0, \quad t>0,\end{aligned} \tag{7.160}$$

$$\begin{aligned}\rho(x)\frac{\partial \delta Q}{\partial t} &= \frac{\partial[k(x)\frac{\partial \delta Q}{\partial x}]}{\partial x} - q(x)\delta Q, \quad 0<x<1, \quad t>0,\\ \frac{\partial \delta Q(0,t)}{\partial x} &= \frac{\partial \delta Q(1,t)}{\partial x} = 0, \quad t>0,\end{aligned} \tag{7.161}$$

$$\begin{aligned}\Delta\dot{k} &= \nu_1\frac{\partial \Delta Q}{\partial x}\frac{\partial \tilde{Q}}{\partial x}, && 0<x<1, \quad t>0,\\ \Delta\dot{q} &= \nu_2\Delta Q\tilde{Q}, && 0<x<1, \quad t>0,\\ \Delta\dot{\rho} &= \nu_3\Delta Q\dot{\tilde{Q}}, && 0<x<1, \quad t>0,\end{aligned} \tag{7.162}$$

the local solution of (7.142), (7.152), (7.153), (7.155) are readily established to be well-posed for all $t \geq 0$.

Indeed, differentiating V with respect to t along the trajectories of (7.160)–(7.162), employing integration by parts and applying boundary conditions yield

$$\dot{V}(t) = \int_0^1 \Delta Q \left\{ -\Delta\rho \frac{\partial \tilde{Q}}{\partial t} + \frac{\partial [k \frac{\partial \delta Q}{\partial x}]}{\partial x} - q\delta Q \right.$$

$$\left. + \frac{\partial [k \frac{\partial \Delta Q}{\partial x}]}{\partial x} + \frac{\partial [\Delta k \frac{\partial \tilde{Q}(x)}{\partial x}]}{\partial x} - q\delta Q - [q + \nu_0]\Delta Q - \Delta q \tilde{Q} \right\} dx$$

$$+ \int_0^1 \left\{ \Delta k \frac{\partial \Delta Q}{\partial x} \frac{\partial \tilde{Q}}{\partial x} + \Delta q \Delta Q \tilde{Q} + \Delta\rho \Delta Q \dot{\tilde{Q}} \right\} dx$$

$$= - \int_0^1 k \left(\frac{\partial \delta Q}{\partial x} \right)^2 dx - \int_0^1 q(\delta Q)^2 dx$$

$$- \int_0^1 k \left(\frac{\partial \Delta Q}{\partial x} \right)^2 dx - \int_0^1 (q + \nu_0)(\Delta Q)^2 dx \leq 0, \quad (7.163)$$

which ensures the uniform boundedness $V(t) \leq V(0) < \infty$ of the Lyapunov functional for all $t \geq 0$, and consequently, the uniform L_2-boundedness of the solutions of (7.160)–(7.162) and their stability.

Since the principal term $(\partial/\partial x)[k(x)\partial/\partial x]$ in the parabolic PDE (7.160)–(7.162) has a compact resolvent in $L_2(0, 1)$ and the output of dynamical model (7.152) is a smooth function, every trajectory of the error system (7.160)–(7.162) is precompact due to its boundedness (see Remark to Theorem 4.3.4, Henry 1981). In addition, system (7.160)–(7.162) turns out to be time-periodic because $\tilde{Q}(x, t) = \delta Q(x, t) - \Delta Q(x, t) + Q_{ss}(x, t)$ can be viewed as a linear combination of the state variable δQ and the time-periodic function $Q_{ss}(x, t)$. Therefore, due to the extension of the Invariance Principle, Theorem 4.7 to time-periodic DPS (which is made in analogy to that of Rouche et al. 1977), there must be a convergence of the trajectories of the periodic system (7.160)–(7.162) to the maximal invariant subset of a set of solutions of (7.160)–(7.162) for which

$$\dot{V}(t) = - \int_0^1 \left\{ k(x) \left[\frac{\partial \delta Q(x, t)}{\partial x} \right]^2 + q(x)[\delta Q(x, t)]^2 \right.$$

$$\left. + k(x) \left[\frac{\partial \Delta Q(x, t)}{\partial x} \right]^2 + [q(x) + \nu_0][\Delta Q(x, t)]^2 \right\} dx \equiv 0. \quad (7.164)$$

Taking into account (7.160), (7.162), this leads to the expressions

$$\delta Q(x, t) = 0, \quad \Delta Q(x, t) = 0, \quad (7.165)$$

$$\Delta k(x, t) = \Delta k(x), \quad \Delta q(x, t) = \Delta q(x), \quad \Delta\rho(x, t) = \Delta\rho(x), \quad (7.166)$$

$$\Delta\rho(x)\frac{\partial Q}{\partial t} = \frac{\partial[\Delta k(x)\frac{\partial Q}{\partial x}]}{\partial x} - \Delta q(x)Q, \quad 0 < x < 1, \ t > 0 \tag{7.167}$$

on the set where $\dot{V}(t) = 0$. Since the external and boundary inputs $u(x,t)$, $\beta_0(t)$, $\beta_1(t)$ generate persistent excitation of (7.142), the parameters $k(x)$, $\rho(x)$, $q(x)$ prove to be identifiable under $u(x,t)$, $\beta_0(t)$, $\beta_1(t)$ due to Theorem 7.9. Hence by Definition 7.9, it follows from (7.167) that

$$\Delta k(x) \equiv \Delta q(x) \equiv \Delta\rho(x) \equiv 0. \tag{7.168}$$

Thus, the maximal invariant subset of (7.164) is composed of (7.165) and (7.168), and by applying the aforementioned extension of the invariance principle, the desired L_2-convergence (7.154) is established to hold true.

To complete the proof it remains to note that (7.154) validates the point-wise parameter estimation (7.159) through the Fourier representation (7.158) of the identifier parameters (7.153). ■

If some of the plant parameters are known a priori or are known to be spatially invariant, then the corresponding Eq. (7.153) can be omitted or, respectively, replaced by the equations

$$\begin{aligned}
\dot{\bar{k}}(t) &= -\nu_1 \int_0^1 \frac{\partial(Q-\tilde{Q})}{\partial x}\frac{\partial \tilde{Q}}{\partial x}dx, \ \bar{k}(0) = \bar{k}_0 > 0\\
\dot{\bar{q}}(t) &= -\nu_2 \int_0^1 (Q-\tilde{Q})\tilde{Q}dx, \qquad \bar{q}(0) = \bar{q}_0 > 0,\\
\dot{\bar{\rho}}(t) &= -\nu_3 \int_0^1 (Q-\tilde{Q})\frac{\partial \tilde{Q}}{\partial x}dx, \quad \bar{\rho}(0) = \bar{\rho}_0 > 0
\end{aligned} \tag{7.169}$$

with respect to the corresponding lumped parameter estimates

$$\bar{k}(t) = \int_0^1 \tilde{k}(x,t)dx, \quad \bar{q}(t) = \int_0^1 \tilde{q}(x,t)dx, \quad \bar{\rho}(t) = \int_0^1 \tilde{\rho}(x,t)dx$$

of the unknown plant parameters.

Since the identifier, governed by (7.152), (7.153), (7.158), (7.159), requires neither temporal derivatives nor spatial derivatives of the plant output of the same order as that of the highest spatial plant derivative, it has reduced sensitivity to measurement errors. The heat conduction coefficient identification problem is, however, ill-posed (Banks and Kunish 1989; Omatu and Seinfeld 1989) and an appropriate regularization is, therefore, required. In the identification law (7.153), the ill-posedness manifests itself by the presence of the first-order spatial plant derivative $Q'(x,t)$. For ensuring the well-posedness of the algorithm proposed the regularization should be carried out at the implementation stage in a manner standard for the ill-posed problems such as the regularizing replacement of the spatial differentiation by the finite difference with a sufficiently small step (Kravaris and Seinfeld 1985) or by the SM first-order spatial derivative observer output (Orlov 2000).

7.2.1.4 MRAC Synthesis

MRAC of the heat process (7.142) to be developed aims to track a reference signal $Q^*(x,t)$ to guarantee the L_2 convergence

$$\lim_{t\to\infty}\int_0^1 [\Delta Q^*(x,t)]^2 dx = 0 \tag{7.170}$$

of the state deviation

$$\Delta Q^*(x,t) = Q(x,t) - Q^*(x,t)$$

and to simultaneously identify the parameters $\rho(x)$, $k(x)$, and $q(x)$. Hereinafter, the reference $Q^*(x,t)$ is the desired system output and it is governed by

$$\begin{aligned}
\rho^*(x)\frac{\partial Q^*}{\partial t} &= \frac{\partial[k^*(x)\frac{\partial Q^*}{\partial x}]}{\partial x} - q^*(x)Q^* + r(x,t), \quad 0 < x < 1,\ t > 0,\\
\frac{\partial Q^*(0,t)}{\partial x} &= \beta_0^*(t), \quad \frac{\partial Q^*(1,t)}{\partial x} = \beta_1^*(t), \quad t > 0,\\
Q^*(x,0) &= Q_0^*(x), \quad 0 < x < 1.
\end{aligned} \tag{7.171}$$

where $\rho^*(x) \ge \rho_0^* > 0,\ q^*(x) \ge 0,\ k^*(x) \ge k_0^* > 0,\ Q_0^*(x),\ r(x,t),\ \beta_0^*(t), \beta_1^*(t)$ are continuously differentiable functions.

In the subsequent presentation the reference external and boundary inputs $r(x,t)$, $\beta_0^*(t)$, $\beta_1^*(t)$ are assumed to persistently excite the reference model (7.171). Under this assumption it is possible to construct MRAC of the heat process (7.142) with simultaneous plant identification. For this purpose, introduce the auxiliary dynamical system

$$\begin{aligned}
\dot{\tilde{k}}^* &= -\mu_1\frac{\partial(Q-Q^*)}{\partial x}\frac{\partial Q^*}{\partial x}, \quad \tilde{k}^*(x,0) = \tilde{k}_0^*(x),\\
\dot{\tilde{q}}^* &= -\mu_2(Q-Q^*)Q^*, \quad \tilde{q}^*(x,0) = \tilde{q}_0^*(x),\\
\dot{\tilde{\rho}}^* &= -\mu_3(Q-Q^*)\dot{Q}^*, \quad \tilde{\rho}^*(x,0) = \tilde{\rho}_0^*(x)
\end{aligned} \tag{7.172}$$

where $\mu_1 > 0, \mu_2 > 0, \mu_3 > 0$ are adaptation parameters, and $\tilde{\rho}_0^*(x) \ge \rho_0 > 0$, $\tilde{k}_0^*(x) \ge k_0 > 0,\ \tilde{q}_0^*(x) \ge 0$ are smooth functions. The following control law

$$\begin{aligned}
u(x,t) &= r(x,t) - [\tilde{\rho}^*(x,t) - \rho^*(x)]\frac{\partial Q^*(x,t)}{\partial t} - \frac{\partial\{[\tilde{k}^*(x,t) - k^*(x)]\frac{\partial Q^*(x,t)}{\partial x}\}}{\partial x}\\
&\quad +[\tilde{q}^*(x,t) - q^*(x)]Q^*(x,t) - \mu_0\Delta Q^*(x,t),\\
\beta_0(t) &= \beta_0^*(t), \quad \beta_1(t) = \beta_1^*(t)
\end{aligned} \tag{7.173}$$

is proposed with a positive gain μ_0, which determines the convergence rate.

Theorem 7.11 *Consider a heat process* (7.142) *with the assumptions above, and let the reference model be given by* (7.171) *with time-periodic inputs* $r(x,t)$, $\beta_0^*(t)$, $\beta_1^*(t)$), *which generate persistent excitation of the reference model. Then the limiting relation*

$$\lim_{t\to\infty}\int_0^1 \{[\Delta Q^*(x,t)]^2 + [\rho(x) - \rho^*(x,t)]^2 + [k(x) - k^*(x,t)]^2 + [q(x) - q^*(x,t)]^2\}dx = 0 \tag{7.174}$$

holds with the adaptive control (7.173) *and parameters* $\tilde{\rho}^*(x,t)$,$\tilde{k}^*(x,t)$, $\tilde{q}^*(x,t)$ *tuned as* (7.172). *Furthermore, all signals in the closed-loop system are bounded.*

Proof Since the proof is similar to that of Theorem 7.10, only a sketch is, therefore, provided. To begin with, a unique local solution of the overall system (7.142), (7.171)–(7.173) is established to exist based on the observation that the system is question falls under (Friedman 1969, Theorem 16.2).

Next, introduce the function $\delta Q^*(x,t) = Q^*(x,t) - Q_{ss}^*(x,t)$, specified with a steady-state periodic solution $Q_{ss}^*(x,t)$ of (7.171), and the estimation errors

$$\Delta\rho^* = \rho - \tilde{\rho}^*, \quad \Delta k^* = k - \tilde{k}^*, \quad \Delta q^* = q - \tilde{q}^*. \tag{7.175}$$

Using the Lyapunov functional

$$V(t) = \frac{1}{2}\int_0^1 \Big\{\rho(x)[\delta Q^*(x,t)]^2 + \rho(x)[\Delta Q^*(x,t)]^2 + \frac{1}{\mu_1}[\Delta k^*(x,t)]^2 + \frac{1}{\mu_2}[\Delta q^*(x,t)]^2 + \frac{1}{\mu_3}[\Delta\rho^*(x,t)]^2\Big\}dx$$

where, due to (7.142), (7.171)–(7.173), (7.175), variables $\Delta Q^*, \delta Q^*, \Delta k, \Delta q, \Delta\rho$ satisfy the equations

$$\begin{aligned}
\rho(x)\frac{\partial\Delta Q^*}{\partial t} + \Delta\rho^*(x,t)\frac{\partial Q^*}{\partial t} &= \frac{\partial[k(x)\frac{\partial\Delta Q^*}{\partial x}]}{\partial x} + \frac{\partial[\Delta k^*(x,t)\frac{\partial Q^*}{\partial x}]}{\partial x} - \Delta q^*(x,t)Q^* \\
&\quad -[q(x)+\mu_0]\Delta Q^*, \quad 0<x<1, \quad t>0, \\
\frac{\partial\Delta Q^*(0,t)}{\partial x} &= \frac{\partial\Delta Q^*(1,t)}{\partial x} = 0, \quad t>0, \\
\rho(x)\frac{\partial\delta Q^*}{\partial t} &= \frac{\partial[k(x)\frac{\partial\delta Q^*}{\partial x}]}{\partial x} - q(x)\delta Q^*, \quad 0<x<1, \quad t>0, \\
\frac{\partial\delta Q^*(0,t)}{\partial x} &= \frac{\partial\delta Q^*(1,t)}{\partial x} = 0, t>0, \\
\Delta\dot{k}^* &= -\mu_1\frac{\partial\Delta Q^*}{\partial x}\frac{\partial Q^*}{\partial x}, \quad 0<x<1, \quad t>0, \\
\Delta\dot{q}^* &= -\mu_2\Delta Q^* Q^*, \quad 0<x<1, \quad t>0, \\
\Delta\dot{\rho}^* &= -\mu_3\Delta Q^*\dot{Q}^*, \quad 0<x<1, \quad t>0,
\end{aligned} \tag{7.176}$$

the well-posedness of the solutions of the resulting system (7.176) is established for all $t \geq 0$. To reproduce this conclusion it suffices to differentiate V with respect to t along the trajectories of (7.176), employing integration by parts and applying the corresponding boundary conditions. Thus, one arrives at

$$\dot{V}(t) = \int_0^1 \Delta Q^* \Big\{ -\Delta\rho^* \frac{\partial Q^*}{\partial t} + \frac{\partial[k\frac{\partial \delta Q^*}{\partial x}]}{\partial x} - q\delta Q^* +$$

$$+\frac{\partial[k\frac{\partial \Delta Q^*}{\partial x}]}{\partial x} + \frac{\partial[\Delta k^* \frac{\partial Q^*(x)}{\partial x}]}{\partial x} - q\delta Q^* - [q+\mu_0]\Delta Q^* - \Delta q Q^* \Big\} dx +$$

$$+ \int_0^1 \Big\{ \Delta k^* \frac{\partial \Delta Q^*}{\partial x}\frac{\partial Q^*}{\partial x} + \Delta q^* \Delta Q^* Q^* + \Delta\rho^* \Delta Q^* \dot{Q}^* \Big\} dx =$$

$$= -\int_0^1 k \left(\frac{\partial \delta Q^*}{\partial x}\right)^2 dx - \int_0^1 q(\delta Q^*)^2 dx -$$

$$- \int_0^1 k \left(\frac{\partial \Delta Q^*}{\partial x}\right)^2 dx - \int_0^1 (q+\mu_0)(\Delta Q^*)^2 dx \leq 0, \quad (7.177)$$

thereby ensuring the uniform boundedness $V(t) \leq V(0) < \infty$ of the Lyapunov functional for all $t \geq 0$, and consequently, the uniform L_2-boundedness of the solutions of (7.176) and their stability.

Since $Q^*(x,t) = \delta Q^*(x,t) + Q^*_{ss}(x,t)$ represents a linear combination of the state variable δQ^* of the error system (7.176) and the time-periodic function $Q_{ss}(x,t)$, the periodic invariance principle version of Theorem 4.7 becomes applicable to the periodic system (7.176). Due to this, there must be a convergence of the trajectories of system (7.176) to the maximal invariant subset of a set of solutions of (7.176) for which

$$\dot{V}(t) = -\int_0^1 \Big\{ k(x) \left[\frac{\partial \delta Q^*(x,t)}{\partial x}\right]^2 + q(x)[\delta Q^*(x,t)]^2$$

$$+ k(x) \left[\frac{\partial \Delta Q^*(x,t)}{\partial x}\right]^2 + [q(x)+\mu][\Delta Q^*(x,t)]^2 \Big\} dx \equiv 0. \quad (7.178)$$

By virtue of (7.176), it follows that

$$\delta Q^*(x,t) = 0 \quad \Delta Q^*(x,t) = 0,$$
$$\Delta k^*(x,t) = \Delta k^*(x), \; \Delta q^*(x,t) = \Delta q^*(x), \; \Delta\rho^*(x,t) = \Delta\rho^*(x),$$
$$\Delta\rho^*(x)\frac{\partial Q^*}{\partial t} = \frac{\partial[\Delta k^*(x)\frac{\partial Q^*}{\partial x}]}{\partial x} - \Delta q^*(x) Q^*, \quad 0 < x < 1, \quad t > 0 \quad (7.179)$$

on the set where $\dot{V}(t) = 0$. Since the external and boundary inputs $u(x,t)$, $\beta_0(t)$, $\beta_1(t)$ generate persistent excitation of (7.171), the parameters $k(x)$, $\rho(x)$, $q(x)$ are identifiable under $u(x,t)$, $\beta_0(t)$, $\beta_1(t)$ due to Theorem 7.9. Hence by Definition 7.9, it follows from (7.179) that

$$\Delta k^*(x) \equiv \Delta q^*(x) \equiv \Delta \rho^*(x) \equiv 0. \tag{7.180}$$

Thus, the maximal invariant subset of (7.178) is governed by (7.179) and (7.180). The desired L_2-convergence (7.154) is then concluded by applying the aforementioned extension of the invariance principle. This proves Theorem 7.11. ∎

The choice of the MRAC law (7.173) comes from the extension to the infinite-dimensional case of the Lyapunov–Krasovskii redesign methods (see, e.g., Utkin and Orlov 1990). The series–parallel/parallel configuration of the state and parameter errors (7.176) is inspired by the finite-dimensional treatment (Ioannou and Sun 1996; Landau 1979; Narendra and Annaswamy 1989). Similar to the identifier design, the proposed MRAC algorithm does not require the first-order temporal derivative of the state as well as the second-order spatial derivative of the state, and therefore it does not lead to the loss of the robustness in the presence of measurement noise and dynamical nonidealities. Meanwhile, an implementation of the first-order spatial derivative in the MRAC algorithm admits a simple realization via its regularizing replacement by the first-order difference with a sufficiently small step (Kravaris and Seinfeld 1985) or by the SM first-order spatial derivative observer output (Orlov 2000).

If some of the plant parameters are known a priori or are spatially invariant, then the corresponding Eq. (7.172) can be replaced by the relations $\tilde{k}^*(x,t) = k(x)$, $\tilde{q}^*(x,t) = q(x)$, $\tilde{\rho}^*(x,t) = \rho(x)$ or, respectively, by the corresponding ODEs (similar to (7.169)) with respect to the lumped variables $\hat{k}^*(t) = \int_0^1 \tilde{k}^*(x,t)dx$, $\hat{q}^*(t) = \int_0^1 \tilde{q}^*(x,t)dx$, $\hat{\rho}^*(t) = \int_0^1 \tilde{\rho}^*(x,t)dx$ as it has been made in the identifier design case.

7.2.2 *Identification over Boundary Sensing*

So far, on-line parameter identification was addressed within the distributed sensing framework only. The present development, inherited from Smyshlyaev et al. (2009), goes beyond such a restrictive identification framework by focusing on unstable plants with boundary actuation, that results in the closed-loop identification, and also introducing boundary sensing. This makes technique proposed applicable to a much broader range of practical problems. As a first step toward identification of general reaction–advection–diffusion systems, two benchmark plants are considered, one with an uncertain parameter g in the domain:

$$u_t(x,t) = u_{xx}(x,t) + gu(0,t), \qquad 0 < x < 1 \tag{7.181}$$

$$u_x(0,t) = 0, \tag{7.182}$$

and the other with an unknown parameter q in the boundary condition:

$$u_t(x,t) = u_{xx}(x,t), \qquad 0 < x < 1 \tag{7.183}$$

$$u_x(0,t) = -qu(0,t), \tag{7.184}$$

Both plants are actuated from the boundary:

$$u(1,t) = U(t), \tag{7.185}$$

and it is assumed that only $u(0,t)$ is measured. Both plants are of infinite relative degree and are unstable for $g > 2$ and $q > 1$, respectively. The objective is to identify the unknown parameters g and q. The adaptive identifier proposed consists of standard gradient/least squares estimators and backstepping adaptive controllers. The parameter estimates are shown to converge to the true parameters when the closed-loop system is excited by an additional constant input at the boundary.

The above plants capture significant features of the model of thermal instability in solid propellant rockets (Boskovic and Krstic 2003). However, the primary motivation to study these systems comes from Smyshlyaev and Krstic (2006), where it was shown that a general reaction–advection problem

$$v_t(x,t) = v_{xx}(x,t) + b(x)v_x(x,t) + \lambda(x)v(x,t) \tag{7.186}$$

$$v_x(0,t) = 0 \tag{7.187}$$

with output $v(0,t)$ and input $v(1,t)$ can be transformed into the form

$$u_t(x,t) = u_{xx}(x,t) + g(x)u(0,t) \tag{7.188}$$

$$u_x(0,t) = -qu(0,t) \tag{7.189}$$

while keeping input and output the same: $u(1,t) = v(1,t)$, $u(0,t) = v(0,t)$. The finite-dimensional analog of the form (7.188), (7.189) is the *observer canonical form*, in which parametric uncertainties multiply the measured output. The comparison of (7.181)–(7.184) with (7.188), (7.189) reveals that the benchmark problems presented here serve as a first step to the output feedback adaptive identification of reaction–advection systems (7.186), (7.187).

7.2.2.1 Open-Loop Adaptive Identification of Stable PDE

For transparency, open-loop stable plants are first addressed. While dealing with the open-loop identification of the plant equation (7.181), (7.182), (7.185) with an uncertain parameter in the domain, the extension for the other plant (7.183), (7.184), (7.185) with an unknown boundary parameter appears to be very similar.

Definition 7.11 The parameter g of the plant (7.181), (7.182), (7.185) is said to be identifiable from the measured output $u(0,t)$ if there exists a control input $U(t)$ such that for initial conditions $u(x,0) \in L_2(0,1)$ and $\tilde{u}(x,0) \in L_2(0,1)$ of the plant and its model

$$\tilde{u}_t(x,t) = \tilde{u}_{xx}(x,t) + \tilde{g}\tilde{u}(0,t) \quad (7.190)$$

$$\tilde{u}_x(0,t) = 0, \quad (7.191)$$

$$\tilde{u}(1,t) = U(t), \quad (7.192)$$

the identity $u(0,t) \equiv \tilde{u}(0,t)$ results in $g = \tilde{g}$.

The following result is instrumental for the subsequent identifiability analysis.

Lemma 7.2 *Let system* (7.181)–(7.182), (7.185) *with* $U(t) \equiv 0$ *be asymptotically stable. Then, whenever forced by a constant input* $U(t) = u_1 \neq 0$*, the system possesses a time-invariant steady-state solution*

$$u^{ss}(x) = u_1 \frac{2 - gx^2}{2 - g} \quad (7.193)$$

such that $u^{ss}(0) = u_0 \neq 0$.

Proof The steady-state solution has to satisfy the following ODE:

$$\frac{d^2 u^{ss}(x)}{dx^2} = -g u^{ss}(0)$$

$$\frac{d}{dx} u^{ss}(0) = 0, \ u^{ss}(1) = u_1. \quad (7.194)$$

This two-point boundary-value problem has a unique solution given by (7.193). Introducing the error variable $\omega(x,t) = u(x,t) - u^{ss}(x)$, one arrives at the system

$$\omega_t(x,t) = \omega_{xx}(x,t) + g\omega(0,t) \quad (7.195)$$

$$\omega_x(0,t) = \omega(1,t) = 0, \quad (7.196)$$

which is asymptotically stable by assumption, so that (7.193) is indeed a steady-state solution of (7.181)–(7.182), (7.185). To complete the proof it remains to note that $u^{ss}(0) \neq 0$. ■

For certain parameter values, its identifiability is shown to be enforced by any nontrivial constant input.

Theorem 7.12 *The parameter g of the system* (7.181)–(7.182), (7.185) *is identifiable from the measured output* $u(0,t)$ *with an arbitrary constant input* $U(t) = u_1 \neq 0$ *under the assumption* $g < 2$.

Proof Let us fix a constant input $U(t) = u_1 \neq 0$. Since $g < 2$, the system (7.181)–(7.182), (7.185) has asymptotically stable internal dynamics. Let us assume that the model system (7.190)–(7.192) produces the same output as the original system, i.e., $u(0,t) = \tilde{u}(0,t)$ for all $t \geq 0$. Then system (7.190)–(7.192) also has asymptotically stable internal dynamics. By Lemma 7.2, both systems (7.181)–(7.182), (7.185), and (7.190)–(7.192) have time-invariant steady-state solutions $u^{ss}(x)$ and $\tilde{u}^{ss}(x)$, respectively, and these solutions take nonzero values $u^{ss}(0) \neq 0$ and $\tilde{u}^{ss}(0) \neq 0$ at $x = 0$. Setting $\delta u(x,t) = u(x,t) - \tilde{u}(x,t)$, $\delta g = g - \tilde{g}$, one obtains

$$\delta u_t(x,t) = \delta u_{xx}(x,t) + \delta g u(0,t), \tag{7.197}$$

$$\delta u_x(0,t) = \delta u(1,t) = 0. \tag{7.198}$$

This system has a steady-state solution $\delta u^0(x) = u^{ss}(x) - \tilde{u}^{ss}(x)$ which satisfies

$$\delta u^0_{xx} + \delta g u_0 = 0, \tag{7.199}$$

$$\delta u^0_x(0) = \delta u^0(1) = 0\,, \tag{7.200}$$

where $u_0 = u^{ss}(0) \neq 0$. It follows that

$$\delta u^0(x) = \frac{1}{2}\delta g u_0(1 - x^2). \tag{7.201}$$

Since $\delta u(0,t) = 0$ by assumption, one concludes that $\delta u^0(0) = 0$, and therefore, the relation $\delta g u_0 = 0$ results from from (7.201). Since $u_0 \neq 0$ it follows that $\delta g = 0$. Hence, $g = \tilde{g}$ and the parameter g is identifiable under an arbitrary nonzero constant input $U(t) = u_1 \neq 0$. The proof is completed. ■

For the on-line identification of the parameter g (Boskovic and Krstic 2003), the following gradient update law

$$\dot{\hat{g}} = \gamma \frac{\hat{e}(0,t)v(0,t)}{1 + v^2(0,t)}\,, \tag{7.202}$$

is proposed with normalization and adaptation gain $\gamma > 0$. The above identifier utilizes the prediction error

$$\hat{e}(x,t) = u(x,t) - \hat{g}(t)v(x,t) - \eta(x,t) \tag{7.203}$$

evaluated at $x = 0$, and the filters

$$v_t(x,t) = v_{xx}(x,t) + u(0,t) \tag{7.204}$$

$$v_x(0,t) = v(1,t) = 0\,, \tag{7.205}$$

$$\eta_t(x,t) = \eta_{xx}(x,t) \tag{7.206}$$

$$\eta_x(0,t) = 0, \quad \eta(1,t) = u(1,t)\,, \tag{7.207}$$

such that the boundary-value problem

$$e_t(x,t) = e_{xx}(x,t) \tag{7.208}$$

$$e_x(0,t) = e(1,t) = 0 \tag{7.209}$$

written in terms of the error $e(x,t) = u(x,t) - gv(x,t) - \eta(x,t)$ is asymptotically stable.

Theorem 7.13 *Consider the plant* (7.181), (7.182), (7.185) *with the filters* (7.204)–(7.207) *and the update law* (7.202). *Let* $U(t) = u_1 \neq 0$. *Then for any initial conditions* $u(x,0), v(x,0), \eta(x,0) \in L_2(0,1)$, $\hat{g}(0) \in \mathbf{R}$ *and any adaptation gain* $\gamma > 0$, *the signals* $\hat{g}(t)$, $u(x,t)$, $v(x,t)$, $\eta(x,t)$ *are uniformly bounded for* $t > 0$ *and* $\lim_{t\to\infty} \hat{g}(t) = g$.

Proof Following the line of reasoning used in the proof of Lemma 7.2 one shows that under a constant input $U(t) = u_1 \neq 0$, the linear time-invariant system (7.181), (7.182), (7.185), (7.204)–(7.205) has a time-invariant steady-state solution

$$u^{ss}(x) = u_1 \frac{2 - gx^2}{2 - g}, \quad v^{ss}(x) = u_1 \frac{1 - x^2}{2 - g}. \tag{7.210}$$

It is clear that the states u, v, η of heat equations (7.181), (7.182), (7.185), (7.204)–(7.207) with bounded inputs are uniformly bounded for $t > 0$.

Representing the update law (7.202) in terms of the deviation Δg from the nominal value

$$\frac{d}{dt}\Delta g = -\gamma \frac{\hat{e}(0,t)v(0,t)}{1 + v^2(0,t)}, \tag{7.211}$$

introduce the positive definite functional

$$V(e, \Delta g) = \frac{1}{2}\int_0^1 e^2(x,t)dx + \frac{1}{2\gamma}[\Delta g(t)]^2 \tag{7.212}$$

and compute its time derivative

$$\begin{aligned}
\dot{V}(t) &= -\int_0^1 e_x^2(x,t)dx - \frac{\Delta g(t)\hat{e}(0,t)v(0,t)}{1 + v^2(0,t)} \\
&= -\int_0^1 e_x^2(x,t)dx - \frac{\hat{e}^2(0,t)}{1 + v^2(0,t)} + \frac{\hat{e}(0,t)e(0,t)}{1 + v^2(0,t)} \\
&\le -\|e_x(\cdot,t)\|^2 - \frac{\hat{e}^2(0,t)}{1 + v^2(0,t)} + \frac{|\hat{e}(0,t)|\|e_x(\cdot,t)\|}{\sqrt{1 + v^2(0,t)}} \\
&\le -\frac{1}{2}\|e_x(\cdot,t)\|^2 - \frac{1}{2}\frac{\hat{e}^2(0,t)}{1 + v^2(0,t)}.
\end{aligned} \tag{7.213}$$

along the solutions of system (7.208), (7.209), (7.211). Therefore, V is bounded and thus $\Delta g(t)$ and $\hat{g}(t)$ are bounded as well.

By taking into account the relation

$$\hat{e}(x,t) = e(x,t) + \Delta g(t)v(x,t), \tag{7.214}$$

resulting from (7.203), (7.209), the update law (7.202), rewritten in the form

$$\frac{d}{dt}\Delta g = -\gamma \frac{[e(0,t) + \Delta g(t)v(0,t)]v(0,t)}{1 + v^2(0,t)}, \tag{7.215}$$

and coupled to the system (7.208), (7.209), turns out to be asymptotically autonomous because

$$\lim_{t\to\infty} v(0,t) = v^{ss}(0) = \frac{u_1}{2-g} \neq 0. \tag{7.216}$$

By extending the Krasovskii–LaSalle invariance principle to the asymptotically autonomous parabolic system (7.208), (7.209), (7.215) in analogy to that of Theorem 4.7, there must occur a convergence of the system trajectories to the maximal invariant subset of a set of solutions of (7.208), (7.209), (7.215), for which $\dot{V}(t) = 0$. Since (7.208), (7.209) is exponentially stable, it follows that

$$e(x,t) = 0, \quad \hat{e}(0,t) = 0, \qquad \Delta g(t)v(0,t) = \hat{e}(0,t) - e(0,t) = 0. \tag{7.217}$$

Due to (7.216), the relation $\lim_{t\to\infty} \Delta g(t) = 0$ is then concluded and Theorem 7.13 is thus proved. ∎

7.2.2.2 Closed-Loop Identification: Plant with Unknown Parameter in the Domain

If the internal dynamics of (7.181), (7.182), (7.185) are no longer stable, the following boundary controller

$$u(1,t) = u_1 + \int_0^1 \hat{k}(1,\xi)(\hat{g}v(\xi,t) + \eta(\xi,t))\,d\xi, \tag{7.218}$$

$$\hat{k}(x,\xi) = \begin{cases} -\sqrt{\hat{g}}\sinh\sqrt{\hat{g}}(x-\xi), & \hat{g} \geq 0, \\ \sqrt{-\hat{g}}\sin\sqrt{-\hat{g}}(x-\xi), & \hat{g} < 0, \end{cases} \tag{7.219}$$

coupled to the filters (7.204)–(7.207) and the gradient update law (7.202), is proposed. Note that $\hat{k}(x,\xi)$ depends on time through $\hat{g}$. The above controller with $u_1 = 0$ was introduced in Smyshlyaev and Krstic (2007). As shown below, the present controller modification with a constant component $u_1 \neq 0$ makes the control signal sufficiently rich to persistently excite the system, yielding the desired parameter convergence.

Theorem 7.14 *Consider the system* (7.181), (7.182) *with the boundary controller* (7.218), (7.219) *and* $u_1 \neq 0$, *with filters* (7.204)–(7.207), *and with the update law, given by (7.202). Then for any* $\hat{g}(0) \in \mathbf{R}$ *and any initial conditions* u_0, v_0, $\eta_0 \in L_2(0,1)$, *the signals* $\hat{g}(t)$, $u(x,t)$, $v(x,t)$, $\eta(x,t)$ *are uniformly bounded for* $t > 0$ *and* $\hat{g}(t) \to g$ *as* $t \to \infty$.

Proof For later use, recall an instrumental lemma.

Lemma 7.3 (Lemma B.6 in Krstic et al. 1995) *Let* v, l_1, *and* l_2 *be real-valued functions defined on* $\mathbf{R}_+$, *and let* c *be a positive constant. If* l_1 *and* l_2 *are nonnegative integrable functions of time and* $\dot{v} \leq -cv + l_1(t)v + l_2(t)$, $\quad v(0) \geq 0$, *then* $v \in \mathscr{L}_1 \cap \mathscr{L}_\infty$.

The proof of the theorem is broken into several simple steps.

Step 1. It is straightforward to verify that estimate (7.213) of the time derivative of the positive definite functional (7.212), computed on the solutions of (7.202), (7.208), (7.209), is still in force. This ensures that

$$\frac{\hat{e}(0,t)}{\sqrt{1+v^2(0,t)}} \in \mathscr{L}_2, \quad \Delta g \in \mathscr{L}_\infty\,. \tag{7.220}$$

Moreover, since

$$\frac{\hat{e}(0,t)}{\sqrt{1+v^2(0,t)}} = \frac{e(0,t)}{\sqrt{1+v^2(0,t)}} + \Delta g \frac{v(0,t)}{\sqrt{1+v^2(0,t)}}, \tag{7.221}$$

$$\dot{\hat{g}} = \gamma \frac{\hat{e}(0,t)}{\sqrt{1+v^2(0,t)}} \frac{v(0,t)}{\sqrt{1+v^2(0,t)}}, \tag{7.222}$$

it is concluded that

$$\frac{\hat{e}(0,t)}{\sqrt{1+v^2(0,t)}} \in \mathscr{L}_2 \cap \mathscr{L}_\infty, \quad \Delta g \in \mathscr{L}_\infty, \quad \dot{\hat{g}} \in \mathscr{L}_2 \cap \mathscr{L}_\infty\,. \tag{7.223}$$

Step 2. For the transformation

$$\hat{w}(x,t) = \hat{g}v(x,t) + \eta(x,t) - \int_0^x \hat{k}(x,\xi)(\hat{g}v(\xi,t) + \eta(\xi,t))\,d\xi \tag{7.224}$$

with $\hat{k}(x,\xi)$, given by (7.219), it is established that (7.224) maps (7.181), (7.182), (7.203)–(7.207), (7.218) into the following system:

$$\hat{w}_t(x,t) = \hat{w}_{xx}(x,t) + \beta(x)\hat{e}(0,t) + \dot{\hat{g}}v(x,t) + \dot{\hat{g}} \int_0^x \alpha(x-\xi)\left(\hat{g}v(\xi,t) + \hat{w}(\xi,t)\right)\,d\xi, \tag{7.225}$$

$$\hat{w}_x(0,t) = 0, \quad \hat{w}(1,t) = u_1\,, \tag{7.226}$$

where

$$
\alpha(x)=\begin{cases}\frac{1}{\sqrt{\hat g}}\sinh\sqrt{\hat g}x, & \hat g\ge 0,\\ \frac{1}{\sqrt{-\hat g}}\sin\sqrt{-\hat g}x, & \hat g<0.\end{cases}\qquad \beta(x)=\hat k_\xi(x,0)=\begin{cases}\hat g\cosh\sqrt{\hat g}x, & \hat g\ge 0,\\ \hat g\cos\sqrt{-\hat g}x, & \hat g<0.\end{cases} \tag{7.227}
$$

To reproduce this conclusion the boundary conditions in (7.226) are verified first:

$$
\hat w_x(0,t)=-\hat g v_x(0,t)-\hat k(0,0)\hat g v(0,t)-\hat k(0,0)\eta(0,t)=0\,, \tag{7.228}
$$

$$
\hat w(1,t)=u(1,t)-\int_0^1\hat k(1,\xi)(\hat g v(\xi,t)+\eta(\xi,t))\,d\xi=u_1\,. \tag{7.229}
$$

Next, the derivatives of $\hat w$ are computed:

$$
\begin{aligned}
\hat w_{xx}(x,t)&=\hat g v_{xx}(x,t)+\eta_{xx}(x,t)+\hat g^2 v(x,t)-\hat g\eta(x,t)\\
&\quad+\hat g^2\int_0^x\alpha(x-\xi)(\hat g v(\xi,t)+\eta(\xi,t))\,d\xi,
\end{aligned} \tag{7.230}
$$

$$
\begin{aligned}
\hat w_t(x,t)&=\hat g v_t(x,t)+\eta_t(x,t)-\hat g u(0,t)+\beta(x)u(0,t)+\hat g^2 v(x,t)+\hat g\eta(x,t)\\
&\quad-\beta(x)(\hat g v(0,t)+\eta(0,t))+\dot{\hat g}v(x,t)+\dot{\hat g}\int_0^x\hat g\,\alpha(x-\xi)v(\xi,t)\,d\xi\\
&\quad+\frac{\dot{\hat g}}{2\hat g}\int_0^x[(x-\xi)\beta(x-\xi)+\hat g\alpha(x-\xi)](\hat g v(\xi,t)+\eta(\xi,t))\,d\xi,
\end{aligned} \tag{7.231}
$$

where α and β are defined in (7.227) and the integration by parts is invoked twice. Subtracting (7.230) from (7.231) and using the transformation

$$
\hat g v(x,t)+\eta(x,t)=\hat w(x,t)-\hat g\int_0^x(x-\xi)\hat w(\xi,t)\,d\xi\,, \tag{7.232}
$$

inverse to (7.224), yield

$$
\begin{aligned}
\hat w_t(x,t)&=\hat w_{xx}(x,t)+\beta(x)\hat e(0,t)+\dot{\hat g}v(x,t)+\dot{\hat g}\int_0^x\hat g\alpha(x-\xi)v(\xi,t)\,d\xi\\
&+\dot{\hat g}\int_0^x[(x-\xi)\beta(x-\xi)+\hat g\alpha(x-\xi)]\left(\hat w(\xi,t)-\hat g\int_0^\xi(\xi-y)\hat w(y,t)\,dy\right)d\xi.
\end{aligned} \tag{7.233}
$$

Finally, changing the order of integration in the double integral (last term in (7.233)), computing the internal integral, and gathering all the terms together, result in (7.225).

Step 3. By employing $u(0,t)=\hat w(0,t)+\hat e(0,t)$, the filter equations (7.204), (7.205) are represented in the form

$$v_t(x,t) = v_{xx}(x,t) + \hat{w}(0,t) + \hat{e}(0,t) \tag{7.234}$$
$$v_x(0,t) = v(1,t) = 0\,. \tag{7.235}$$

The resulting interconnected systems (7.225)–(7.226) and (7.234)–(7.235) for $\hat{w}$ and v are driven by three external signals, namely, a constant u_1 and the inputs $\hat{e}(0,t)$, $\dot{\hat{g}}(t)$ of class (7.223).

Step 4. The next goal is to demonstrate that v-system and $\hat{w}$-system asymptotically approach the limit points

$$\hat{w}^{lp}(x) = u_1, \qquad v^{lp}(x) = u_1 \frac{1-x^2}{2}\,. \tag{7.236}$$

For this purpose, introduce the error variables $\bar{w} = \hat{w} - \hat{w}^{lp}$, $\bar{v} = v - v^{lp}$. The equations for $\bar{w}$ and $\bar{v}$ are

$$\begin{aligned} \bar{w}_t(x,t) &= \bar{w}_{xx}(x,t) + \beta(x)\hat{e}(0,t) + \dot{\hat{g}}\bar{v}(x,t) \\ &\quad + \dot{\hat{g}} \int_0^x \alpha(x-\xi)\left(\hat{g}\bar{v}(\xi,t) + \bar{w}(\xi,t)\right) d\xi + \dot{\hat{g}}\beta(x)\frac{u_1}{2\hat{g}} \end{aligned} \tag{7.237}$$
$$\bar{w}_x(0,t) = \bar{w}(1,t) = 0\,, \tag{7.238}$$

and

$$\bar{v}_t(x,t) = \bar{v}_{xx}(x,t) + \bar{w}(0,t) + \hat{e}(0,t) \tag{7.239}$$
$$\bar{v}_x(0,t) = \bar{v}(1,t) = 0\,. \tag{7.240}$$

Let us along the solutions of the boundary-value problem (7.239), (7.240), consider the Lyapunov–Krasovskii functional

$$V_1 = \frac{1}{2}\int_0^1 \bar{v}^2(x)\,dx + \frac{1}{2}\int_0^1 \bar{v}_x^2(x)\,dx\,. \tag{7.241}$$

Using the inequalities

$$\|w(x)\| \le 2\|w_x\|, \quad |w(0)| \le \|w_x\|, \quad w^2(x) \le 4\|w\|\|w_x\|\,, \tag{7.242}$$

which result from Wirtinger's, Poincare's, and Agmon's inequalities of Sect. 3.1 for any function $w \in H^1(0,1)$ such that $w(1) = 0$, yields

$$\begin{aligned} \dot{V}_1 &= -\int_0^1 \bar{v}_x^2\,dx + (\bar{w}(0) + \hat{e}(0))\int_0^1 \bar{v}\,dx - \int_0^1 \bar{v}_{xx}^2\,dx - (\bar{w}(0) + \hat{e}(0))\int_0^1 \bar{v}_{xx}\,dx \\ &\le -\|\bar{v}_x\|^2 + \frac{1}{8}\|\bar{v}\|^2 + 4\frac{\hat{e}^2(0)}{1+v^2(0)}(1+v^2(0)) \end{aligned}$$

$$+4\|\bar{w}_x\|^2 - \|\bar{v}_{xx}\|^2 + \frac{1}{2}\|\bar{v}_{xx}\|^2 + \|\bar{w}_x\|^2 + \frac{\hat{e}^2(0)}{1+v^2(0)}(1+v^2(0))$$
$$\leq -\frac{1}{2}\|\bar{v}_x\|^2 - \frac{1}{2}\|\bar{v}_{xx}\|^2 + 5\|\bar{w}_x\|^2 + 5\frac{\hat{e}^2(0)}{1+v^2(0)}\left(1+2\|\bar{v}_x\|^2 + \frac{u_1^2}{2}\right)$$
$$\leq -\frac{1}{2}\|\bar{v}_x\|^2 - \frac{1}{2}\|\bar{v}_{xx}\|^2 + 5\|\bar{w}_x\|^2 + l_1\|\bar{v}_x\|^2 + l_1, \tag{7.243}$$

where for ease of reference, the dependence on time is dropped, and l_1 stands for a generic time function of class $\mathscr{L}_1 \cap \mathscr{L}_\infty$.

For the $\bar{w}$–system, another Lyapunov–Krasovskii functional

$$V_2 = \frac{1}{2}\int_0^1 \bar{w}^2\, dx\,, \tag{7.244}$$

is involved to estimate its time derivative

$$\dot{V}_2 = -\int_0^1 \bar{w}_x^2\, dx + \hat{e}(0)\int_0^1 \beta\bar{w}\, dx + \dot{\hat{g}}\int_0^1 \bar{w}\bar{v}\, dx + \dot{\hat{g}}\int_0^1 \bar{w}(x)v^{lp}(x)\, dx$$
$$+\dot{\hat{g}}\int_0^1 \bar{w}(x)\int_0^x \alpha(x-y)(\hat{g}v^{lp}(y) + u_1 + \hat{g}\bar{v}(y) + \bar{w}(y))\, dy\, dx \tag{7.245}$$

on the solutions of the boundary-value problem (7.237), (7.238). Note that Δg is bounded and, therefore, $\hat{g}$ is also bounded; let us denote this bound by g_0. The functions α and β are also bounded, let us denote these bounds by α_0 and β_0. By applying Wirtinger's, Poincare's, and Agmon's inequalities in the form of (7.242) and employing the well-known inequality

$$2\phi\psi \leq \varepsilon\psi^2 + \varepsilon^{-1}\phi, \tag{7.246}$$

which is valid for arbitrary reals ϕ, ψ, and for arbitrary $\varepsilon > 0$, the following estimate

$$\dot{V}_2 \leq -\|\bar{w}_x\|^2 + \frac{\varepsilon}{2}\|\bar{w}\|^2 + \frac{|\dot{\hat{g}}|^2}{8\varepsilon}u_1^2\cosh^2(\sqrt{g_0}) + \varepsilon\|\bar{w}\|^2 + \frac{|\dot{\hat{g}}|^2(1+\alpha_0 g_0)^2}{2\varepsilon}\|\bar{v}\|^2$$
$$+\frac{\beta_0^2}{2\varepsilon}\frac{\hat{e}^2(0)}{1+v^2(0)}\left(1+\|\bar{v}_x\|^2 + u_1^2\right) + \varepsilon\|\bar{w}\|^2 + \frac{|\dot{\hat{g}}|^2\alpha_0^2}{2\varepsilon}\|\bar{w}\|^2$$
$$\leq -(1-10\varepsilon)\|\bar{w}_x\|^2 + l_1\|\bar{w}\|^2 + l_1\|\bar{v}_x\|^2 + l_1 \tag{7.247}$$

is obtained. Setting $\varepsilon = 1/40$ and using the Lyapunov–Krasovskii functional $V = V_2 + (1/20)V_1$, it follows from (7.243) and (7.247), coupled together, that

$$\dot{V} \leq -\frac{1}{2}\|\bar{w}_x\|^2 - \frac{1}{40}\|\bar{v}_x\|^2 - \frac{1}{40}\|\bar{v}_{xx}\|^2 + l_1\|\bar{w}\|^2 + l_1\|\bar{v}_x\|^2 + l_1$$
$$\leq -\frac{1}{4}V + l_1 V + l_1\,, \tag{7.248}$$

and by Lemma 7.3, the uniform boundedness and square integrability of $\|\bar{w}\|$, $\|\bar{v}\|$, $\|\bar{v}_x\|$ are established. Using these properties, one computes

$$\frac{1}{2}\frac{d}{dt}\|\bar{w}_x\|^2 \leq -\|\bar{w}_{xx}\|^2 + \beta_0|\hat{e}(0)|\|\bar{w}_{xx}\| + |\dot{\hat{g}}|\|\bar{w}_{xx}\|((1+\alpha_0 g_0)\|\bar{v}\| + \alpha_0\|\bar{w}\|)$$
$$+ |\dot{\hat{g}}|\|\bar{w}_{xx}\||u_1|\frac{\cosh^2(\sqrt{g_0})}{2} \leq -\frac{1}{8}\|\bar{w}_x\|^2 + l_1, \tag{7.249}$$

so that by Lemma 7.3, $\|\bar{w}_x\| \in \mathscr{L}_2 \cap \mathscr{L}_\infty$. Using the fact that $\|\bar{v}_x\|$, $\|\bar{w}_x\|$ are uniformly bounded, it is clear that

$$\left|\frac{d}{dt}(\|\bar{v}\|^2 + \|\bar{w}\|^2)\right| < \infty. \tag{7.250}$$

By Barbalat's Lemma 2.2, the convergence $\|\bar{w}\|^2 + |\bar{v}\|^2 \to 0$ is in force. Moreover, by the third inequality in (7.242), both $\bar{v}(x,t)$ and $\bar{w}(x,t)$ are uniformly bounded.

Step 5. In order to show the uniform boundedness of $\eta(x,t)$ and $u(x,t)$, it suffices to express η in terms of v and $\hat{w}$ by means of the inverse transformation (7.232). Since $v(x,t)$ and $\hat{w}(x,t)$ are uniformly bounded, it is seen from (7.232) that $\eta(x,t)$ is also uniformly bounded. Finally, the uniform boundedness of $u(x,t)$ is obtained from the relationship $u = e + gv + \eta$.

Step 6. It has been shown that

$$\lim_{t\to\infty} v(0,t) = v^{lp}(0) = \frac{u_1}{2} \neq 0. \tag{7.251}$$

Thus, the update law (7.202), being rewritten in the form (7.215) and coupled to the parabolic system (7.208), (7.209), turns out to be asymptotically autonomous. To complete the proof it remains to apply the Krasovskii–LaSalle invariance principle, properly extended to the asymptotically autonomous system (7.208), (7.209), (7.215) in analogy to that of Theorem 4.7, and to note that due to (7.251), the convergence of the solution to the maximal invariant set $\Delta g(t)v(0,t) = 0$ results in $\lim_{t\to\infty}\Delta g(t) = 0$. Theorem 7.14 is thus proved. ■

7.2.2.3 Closed-Loop Identification: Plant with Unknown Parameter in the Boundary Condition

The internal dynamics of the boundary-value problem (7.183)–(7.185) with the unknown boundary parameter q are unstable for $q > 1$. Following the line of reasoning of Sect. 7.2.2.1, one can establish the identifiability of (7.183)–(7.185) for $q < 1$ with constant nonzero input. The present study skips this calculation and starts directly with the unstable case.

When q is known a priori, the transformation

$$w(x,t) = u(x,t) + \int_0^x q e^{q(x-\xi)} u(\xi,t)\, d\xi \tag{7.252}$$

along with the feedback

$$U(t) = -\int_0^1 q e^{q(1-\xi)} u(\xi,t)\, d\xi \tag{7.253}$$

was used in Smyshlyaev and Krstic (2007) to map (7.183)–(7.185) into the target system

$$w_t(x,t) = w_{xx}(x,t)\,, \tag{7.254}$$

$$w_x(0,t) = w(1,t) = 0\,. \tag{7.255}$$

The adaptive control synthesis to be developed relies on the least-square identifier, proposed below, and it is based on the input and output filters

$$v_t(x,t) = v_{xx}(x,t)\,, \tag{7.256}$$

$$v_x(0,t) = -u(0,t)\,,\ v(1,t) = 0\,, \tag{7.257}$$

$$\eta_t(x,t) = \eta_{xx}(x,t)\,, \tag{7.258}$$

$$\eta_x(0,t) = 0\,, \qquad \eta(1,t) = u(1,t)\,. \tag{7.259}$$

Clearly, the error $e = u - qv - \eta$ satisfies the exponentially stable parabolic equations (7.208)–(7.209). Let the prediction error be defined as

$$\hat{e}(x,t) = u(x,t) - \hat{q}v(x,t) - \eta(x,t)\,. \tag{7.260}$$

Then the least squares update law, given by

$$\dot{\hat{q}}(t) = \gamma(t)\frac{\hat{e}(0,t)v(0,t)}{1+\gamma(t)v^2(0,t)}\,, \tag{7.261}$$

$$\dot{\gamma}(t) = -\frac{\gamma^2(t)v^2(0,t)}{1+\gamma(t)v^2(0,t)}\,, \quad \gamma(0) > 0\,, \tag{7.262}$$

comes with attractive features, which are summarized in the next lemma.

Lemma 7.4 *Identifier* (7.256)–(7.262) *guarantees the properties:*

1. $0 < \gamma(t) < \infty$, $|\dot{\gamma}(t)| < \infty$ *for all* $t \geq 0$.
2. $\Delta q(t)$ *is bounded.*
3. $\frac{\hat{e}(0)}{\sqrt{1+v^2(0,t)}} \in \mathcal{L}_2 \cap \mathcal{L}_\infty$ *and* $\dot{\hat{q}} \in \mathcal{L}_2 \cap \mathcal{L}_\infty$.
4. *There exist* γ_∞, q_∞ *such that* $\lim_{t\to\infty} \gamma(t) = \gamma_\infty$, $\lim_{t\to\infty} q(t) = q_\infty$.

Proof The assertions of the lemma are proved step by step.

1. Rewriting (7.262) as

$$\frac{d}{dt}(\gamma(t)^{-1}) = \frac{\gamma(t)^{-1}v^2(0,t)}{\gamma(t)^{-1} + v^2(0,t)}, \tag{7.263}$$

one can see that $\gamma(t)^{-1} \geq \gamma(0)^{-1} > 0$. Therefore, $\gamma(t)$ is bounded and positive. Then, it follows from (7.262) that $|\dot{\gamma}(t)| \leq |\gamma(t)| < \infty$ for all $t \geq 0$.

2. Consider the Lyapunov–Krasovskii functional

$$V = \frac{1}{2}\|e\|^2 + \frac{1}{2\gamma(t)}\Delta q^2, \tag{7.264}$$

where $\Delta q = q - \hat{q}$. Clearly,

$$\begin{aligned}\dot{V} &= -\|e_x\|^2 + \frac{1}{2}\frac{\Delta q^2 v^2(0)}{1+\gamma v^2(0)} - \frac{\Delta q v(0)\hat{e}(0)}{1+\gamma v^2(0)} = -\|e_x\|^2 + \frac{1}{2}\frac{\Delta q v(0)}{1+\gamma v^2(0)} \\ &\times[\Delta q v(0) - 2\hat{e}(0)] = -\|e_x\|^2 + \frac{1}{2}\frac{(\hat{e}(0)-e(0))(-e(0)-\hat{e}(0))}{1+\gamma v^2(0)} \\ &= -\|e_x\|^2 + \frac{1}{2}\frac{e^2(0)-\hat{e}^2(0)}{1+\gamma(t)v^2(0)} \leq -\frac{1}{2}\|e_x\|^2 - \frac{1}{2}\frac{\hat{e}^2(0)}{1+\gamma(t)v^2(0)}.\end{aligned} \tag{7.265}$$

Therefore, V is bounded which in turn ensures that Δq is bounded.

3. Integrating (7.265) in time, one gets $\frac{\hat{e}(0)}{\sqrt{1+\gamma(t)v^2(0,t)}} \in \mathscr{L}_2$. Apart from this,

$$\frac{\hat{e}(0,t)}{\sqrt{1+v^2(0,t)}} = \frac{\hat{e}(0,t)}{\sqrt{1+\gamma(t)v^2(0,t)}}\frac{\sqrt{1+\gamma(t)v^2(0,t)}}{\sqrt{1+v^2(0,t)}} \leq \frac{\hat{e}(0,t)\sqrt{1+\gamma(0)}}{\sqrt{1+\gamma(t)v^2(0,t)}} \in \mathscr{L}_2$$

and

$$\frac{\hat{e}(0,t)}{\sqrt{1+v^2(0,t)}} = \frac{e(0,t)}{\sqrt{1+v^2(0,t)}} + \frac{\Delta q v(0,t)}{\sqrt{1+v^2(0,t)}} < \infty. \tag{7.266}$$

In addition, it follows from (7.261) that

$$\dot{\hat{q}} = \frac{\hat{e}(0)}{\sqrt{1+\gamma(t)v^2(0,t)}}\frac{\gamma(t)v(0,t)}{\sqrt{1+\gamma(t)v^2(0,t)}} \leq \frac{\sqrt{\gamma(t)}\hat{e}(0)}{\sqrt{1+\gamma(t)v^2(0,t)}} \in \mathscr{L}_2 \cap \mathscr{L}_\infty.$$

4. Since $\gamma(t)$ is monotonically decreasing and is bounded from below, it possesses a limit $\lim_{t\to\infty}\gamma(t) = \gamma_\infty$. Let us rewrite (7.261) as

$$\frac{d}{dt}\Delta q(t) = -\gamma\frac{v(0,t)(e(0,t)+\Delta q(t)v(0,t))}{1+\gamma(t)v^2(0,t)} = -\gamma\frac{v(0,t)e(0,t)}{1+\gamma(t)v^2(0,t)} + \frac{\Delta q(t)}{\gamma(t)}\dot{\gamma}(t).$$

The solution to this ODE is

$$\Delta q(t) = \frac{\Delta q(0)}{\gamma(0)}\gamma(t) - \gamma(t)\int_0^t \frac{v(0,\tau)e(0,\tau)}{1+\gamma(\tau)v^2(0,\tau)}\,d\tau\,. \tag{7.267}$$

Therefore,

$$\lim_{t\to\infty}\hat{q}(t) = q - \frac{\Delta q(0)}{\gamma(0)}\gamma_\infty + \gamma_\infty\int_0^\infty \frac{v(0,\tau)e(0,\tau)}{1+\gamma(\tau)v^2(0,\tau)}\,d\tau = q_\infty\,, \tag{7.268}$$

because the integral in (7.268) is readily shown to converge. ■

Based on the identifier proposed, the adaptive control synthesis is as follows.

Theorem 7.15 *Consider the system* (7.183)–(7.184), *driven by the controller*

$$u(1,t) = u_1 - \int_0^1 \hat{q}e^{\hat{q}(1-\xi)}(\hat{q}v(\xi,t)+\eta(\xi,t))\,d\xi\,, \tag{7.269}$$

accompanied with the update law (7.261)–(7.262) *and filters* (7.256)–(7.259). *Then for any* $\hat{q}(0)$ *and any initial conditions* u_0, v_0, $\eta_0 \in L_2(0,1)$, *the signals* $\hat{q}(t)$, $u(x,t)$, $v(x,t)$, $\eta(x,t)$ *are uniformly bounded and* $\hat{q}(t)\to q$ *as* $t\to\infty$.

Proof The proof is divided into several steps.

Step 1. Following (7.252), the transformation

$$\hat{w}(x,t) = \hat{q}v(x,t)+\eta(x,t)+\int_0^x \hat{q}e^{\hat{q}(x-\xi)}(\hat{q}v(\xi,t)+\eta(\xi,t))\,d\xi\,, \tag{7.270}$$

is introduced with the estimate $\hat{q}v(x,t)+\eta(x,t)$ of the state u. Let us show that this transformation maps (7.183)–(7.184), (7.269) into the following system:

$$\hat{w}_t(x,t) = \hat{w}_{xx}(x,t)+\hat{q}^2e^{\hat{q}x}\hat{e}(0,t)+\dot{\hat{q}}v + \dot{\hat{q}}\int_0^x e^{\hat{q}(x-\xi)}(\hat{q}v(\xi,t)+\hat{w}(\xi,t))\,d\xi\,, \tag{7.271}$$

$$\hat{w}_x(0,t) = -\hat{q}\hat{e}(0,t)\,,\quad \hat{w}(1,t) = u_1\,. \tag{7.272}$$

For this purpose, the boundary conditions in (7.272) are verified first:

$$\hat{w}_x(0,t) = -\hat{q}(u(0,t)-\hat{q}v(0,t)-\eta(0,t)) = -\hat{q}\hat{e}(0,t)\,, \tag{7.273}$$

$$\hat{w}(1,t) = u(1,t)+\int_0^1 \hat{q}e^{\hat{q}(x-\xi)}(\hat{q}v(\xi,t)+\eta(\xi,t))\,d\xi = u_1\,. \tag{7.274}$$

Next, the derivatives of $\hat{w}$ are computed:

$$\hat{w}_{xx}(x,t) = \hat{q}v_{xx}(x,t)+\eta_{xx}(x,t)+\hat{q}^2v_x(x,t)+\hat{q}\eta_x(x,t)+\hat{q}^3v(x,t) + \hat{q}^2\eta(x,t)+\int_0^x \hat{q}^3e^{\hat{q}(x-\xi)}(\hat{q}v(\xi,t)+\eta(\xi,t))\,d\xi, \tag{7.275}$$

$$\hat{w}_t(x,t) = \hat{q}v_t(x,t) + \eta_t(x,t) + \hat{q}^2 v_x(x,t) - \hat{q}^2 e^{\hat{q}x} v_x(0,t) + \hat{q}^3 v(x,t)$$
$$+ \hat{q}^2\eta(x,t) + \hat{q}\eta_x(x,t) \qquad - \hat{q}^3 e^{\hat{q}x} v(0,t) - \hat{q}^2 e^{\hat{q}x}\eta(0,t) + \dot{\hat{q}}v(x,t)$$
$$+ \int_0^x \hat{q}^3 e^{\hat{q}(x-\xi)}(\hat{q}v(\xi,t) + \eta(\xi,t))\,d\xi + \dot{\hat{q}}\int_0^x \hat{q}e^{\hat{q}(x-\xi)}v(\xi,t)\,d\xi$$
$$+ \dot{\hat{q}}\int_0^x (1+\hat{q}(x-\xi))e^{\hat{q}(x-\xi)}(\hat{q}v(\xi,t) + \eta(\xi,t))\,d\xi, \qquad (7.276)$$

where the integration by parts is applied twice. Subtracting (7.275) from (7.276) and using the inverse transformation

$$\hat{q}v(x,t) + \eta(x,t) = \hat{w}(x,t) - \hat{q}\int_0^x \hat{w}(y,t)\,dy \qquad (7.277)$$

yield

$$\hat{w}_t(x,t) = \hat{w}_{xx}(x,t) + \hat{q}^2 e^{\hat{q}x}\hat{e}(0,t) + \dot{\hat{q}}v(x,t) + \dot{\hat{q}}\int_0^x \hat{q}e^{\hat{q}(x-\xi)}v(\xi,t)\,d\xi$$
$$+ \dot{\hat{q}}\int_0^x (1+\hat{q}(x-\xi))e^{\hat{q}(x-\xi)}\left(\hat{w}(\xi,t) - \hat{q}\int_0^\xi \hat{w}(y,t)\,dy\right)d\xi. \qquad (7.278)$$

Finally, changing the order of integration in the double integral (last term in (7.278)) and computing the internal integral result in (7.271).

Step 2. Taking into account that $u(0,t) = \hat{w}(0,t) + \hat{e}(0,t)$, the v-filter is represented in the form

$$v_t(x,t) = v_{xx}(x,t)\,, \qquad (7.279)$$
$$v_x(0,t) = -\hat{w}(0,t) - \hat{e}(0,t)\,, \qquad (7.280)$$
$$v(1,t) = 0\,. \qquad (7.281)$$

The interconnected systems for $\hat{w}$ and v, excited by the constant boundary input u_1 and by the signal $\hat{e}(0,t)$, are coming with the properties, established in Lemma 7.4.

Step 3. The next goal is to demonstrate that these systems asymptotically approach the limit points

$$\hat{w}^{lp}(x) = u_1, \qquad v^{lp}(x) = u_1(1-x)\,. \qquad (7.282)$$

Introducing the error variables $\bar{w} = \hat{w} - \hat{w}^{lp}$, $\bar{v} = v - v^{lp}$ yields

$$\bar{w}_t(x,t) = \bar{w}_{xx}(x,t) + \hat{q}^2 e^{\hat{q}x}\hat{e}(0,t)$$
$$+ \dot{\hat{q}}\left[\bar{v} + u_1 e^{\hat{q}x} + \int_0^x e^{\hat{q}(x-\xi)}(\hat{q}\bar{v}(\xi,t) + \bar{w}(\xi,t))\,d\xi\right] \qquad (7.283)$$
$$\bar{w}_x(0,t) = -\hat{q}\hat{e}(0,t)\,, \quad \bar{w}(1,t) = 0\,, \qquad (7.284)$$

$$\bar{v}_t(x,t) = \bar{v}_{xx}(x,t) \tag{7.285}$$

$$\bar{v}_x(0,t) = -\bar{w}(0,t) - \hat{e}(0,t)\,, \quad \bar{v}(1,t) = 0\,. \tag{7.286}$$

For the Lyapunov–Krasovskii functional

$$V = \frac{1}{2}\int_0^1 \bar{w}^2(x)\,dx + \frac{1}{2}\int_0^1 \bar{v}^2(x)\,dx\,, \tag{7.287}$$

one computes its time derivative by using (7.242) and (7.246):

$$\begin{aligned}
\dot{V} &= \hat{q}\bar{w}(0)\hat{e}(0) - \int_0^1 \bar{w}_x^2\,dx + \dot{\hat{q}}\int_0^1 \bar{w}(x)\bar{v}(x)\,dx + \hat{e}(0)\int_0^1 \hat{q}^2 e^{\hat{q}x}\bar{w}(x)\,dx \\
&\quad - \int_0^1 \bar{v}_x^2\,dx + \dot{\hat{q}}\int_0^1 \bar{w}(x)\int_0^x e^{\hat{q}(x-\xi)}[\hat{q}\bar{v}(\xi) + \bar{w}(\xi)]\,d\xi\,dx \\
&\quad + \bar{v}(0)[\bar{w}(0) + \hat{e}(0)] + \dot{\hat{q}}u_1\int_0^1 e^{\hat{q}x}\bar{w}(x)\,dx \le -\|\bar{w}_x\|^2 + |\hat{e}(0)|\Big(q_0|\bar{w}(0)| \\
&\quad + q_0^2 e^{q_0}\|\bar{w}\|\Big) + \frac{(1+q_0e^{q_0})^2|\dot{\hat{q}}|^2}{\varepsilon_1}\|\bar{v}\|^2 + \frac{e^{2q_0}|\dot{\hat{q}}|^2}{\varepsilon_1}\|\bar{w}\|^2 + \frac{\varepsilon_1}{2}\|\bar{w}\|^2 \\
&\quad - \|\bar{v}_x\|^2 + \frac{1}{2}\|\bar{v}_x\|^2 + \frac{1}{2}\|\bar{w}_x\|^2 + |\bar{v}(0)||\hat{e}(0)| + \frac{\varepsilon_1}{2}\|\bar{w}\|^2 + \frac{|\dot{\hat{q}}|^2u_1^2e^{2q_0}}{2\varepsilon_1} \\
&\le -\left(\frac{1}{2} - 4\varepsilon_1\right)\|\bar{w}_x\|^2 - \frac{1}{2}\|\bar{v}_x\|^2 + l_1\|\bar{w}\|^2 + l_1\|\bar{v}\|^2 + l_1 \\
&\quad + q_0|\hat{e}(0)||\bar{w}(0)| + q_0^2e^{q_0}|\hat{e}(0)|\|\bar{w}\| + |\bar{v}(0)||\hat{e}(0)|\,.
\end{aligned} \tag{7.288}$$

Hereinafter, q_0 and γ_0 stand for the bounds on $\hat{q}$ and γ, respectively, l_1 is a generic function of time of class $\mathscr{L}_1 \cap \mathscr{L}_\infty$, and $\varepsilon_1 > 0$ is a generic constant. In a similar manner, the last three terms of (7.288) are separately estimated by employing (7.242) and (7.246):

$$\begin{aligned}
q_0|\hat{e}(0)||\bar{w}(0)| &\le q_0|\bar{w}(0)|\frac{\hat{e}(0)}{\sqrt{1+v^2(0)}}(1 + |u_1| + |\bar{v}(0)|) \\
&\le \varepsilon_2\|\bar{w}_x\|^2 + \frac{q_0^2(1+|u_1|)^2}{4\varepsilon_2}\frac{\hat{e}^2(0)}{1+v^2(0)} \\
&\quad + 2q_0\sqrt{\|\bar{w}\|\|\bar{w}_x\|\|\bar{v}\|\|\bar{v}_x\|}\frac{|\hat{e}(0)|}{\sqrt{1+v^2(0)}} \le \varepsilon_2\|\bar{w}_x\|^2 + l_1 \\
&\quad + \frac{q_0|\hat{e}(0)|}{\sqrt{1+v^2(0)}}(\|\bar{w}\|\|\bar{w}_x\| + \|\bar{v}\|\|\bar{v}_x\|) \le \varepsilon_2\|\bar{w}_x\|^2 + l_1 \\
&\quad + \varepsilon_3\|\bar{w}_x\|^2 + \varepsilon_4\|\bar{v}_x\|^2 + q_0^2\frac{\hat{e}^2(0)}{1+v^2(0)}\left(\frac{\|\bar{v}\|^2}{4\varepsilon_3} + \frac{\|\bar{w}\|^2}{4\varepsilon_4}\right) \\
&\le \varepsilon_2\|\bar{w}_x\|^2 + \varepsilon_3\|\bar{w}_x\|^2 + \varepsilon_4\|\bar{v}_x\|^2 + l_1\|\bar{v}\|^2 + l_1\|\bar{w}\|^2 + l_1,
\end{aligned} \tag{7.289}$$

$$
\begin{aligned}
q_0^2 e^{q_0} |\hat{e}(0)| \|\bar{w}\| &\leq q_0^2 e^{q_0} \|\bar{w}\| \frac{\hat{e}(0)}{\sqrt{1+v^2(0)}}(1+|u_1|+|v(0)|) \\
&\leq \varepsilon_5 \|\bar{w}\|^2 + \frac{q_0^4 e^{2q_0}(1+|u_1|)^2}{4} \frac{\hat{e}^2(0)}{1+v^2(0)} \left(\frac{1}{\varepsilon_5} + \frac{1}{\varepsilon_6}\|\bar{w}\|^2\right) \\
&\quad + \varepsilon_6 \|\bar{v}_x\|^2 \leq \varepsilon_5 \|\bar{w}\|^2 + \varepsilon_6 \|\bar{v}_x\|^2 + l_1 \|\bar{w}\|^2 + l_1, \qquad (7.290)
\end{aligned}
$$

$$
\begin{aligned}
|\bar{v}(0)||\hat{e}(0)| &\leq \frac{|\bar{v}(0)||\hat{e}(0)|}{1+v^2(0)}(1+2u_1^2+4\|\bar{v}\|\|\bar{v}_x\|) \\
&\leq \frac{\varepsilon_7}{2}\|\bar{v}_x\|^2 + \frac{(1+2u_1^2)^2}{2\varepsilon_7} \frac{\hat{e}^2(0)}{1+v^2(0)} + \frac{\varepsilon_7}{2}\|\bar{v}_x\|^2 \\
&\quad + \frac{8}{\varepsilon_7}\left(\frac{|\bar{v}(0)||\hat{e}(0)|}{1+v^2(0)}\right)^2 \|\bar{v}\|^2 \leq \varepsilon_7 \|\bar{v}_x\|^2 + l_1\|\bar{v}\|^2 + l_1 \,. \qquad (7.291)
\end{aligned}
$$

Substituting (7.289), (7.290), and (7.291) into (7.288), one obtains

$$
\begin{aligned}
\dot{V} \leq &-\left(\frac{1}{2} - 4\varepsilon_1 - \varepsilon_2 - \varepsilon_3 - 4\varepsilon_5\right)\|\bar{w}_x\|^2 + l_1\|\bar{w}\|^2 \\
&-\left(\frac{1}{2} - \varepsilon_4 - \varepsilon_6 - \varepsilon_7\right)\|\bar{v}_x\|^2 + l_1\|\bar{v}\|^2 + l_1 \,. \qquad (7.292)
\end{aligned}
$$

Now setting $4\varepsilon_1 = \varepsilon_2 = \varepsilon_3 = 4\varepsilon_5 = 1/16$, $\varepsilon_4 = \varepsilon_6 = \varepsilon_7 = 1/12$, one arrives at

$$
\dot{V} \leq -\frac{1}{8}V + l_1 V + l_1 \qquad (7.293)
$$

and by applying Lemma 7.3, the inclusion $\|\bar{w}\|, \|\bar{v}\| \in \mathscr{L}_2 \cap \mathscr{L}_\infty$ is verified. By integrating (7.292), $\|\bar{w}_x\|, \|\bar{v}_x\| \in \mathscr{L}_2$ is additionally concluded.

Step 4. To establish the parameter convergence, note that

$$
|\hat{e}(0,t)| \leq \frac{\hat{e}(0,t)}{\sqrt{1+v(0,t)^2}}(1+\|\bar{v}_x\|+|u_1|) \in \mathscr{L}_2 \qquad (7.294)
$$

because $\|\bar{v}_x\| \in \mathscr{L}_2$ and $\frac{\hat{e}(0,t)}{\sqrt{1+v(0,t)^2}} \in \mathscr{L}_2 \cap \mathscr{L}_\infty$ in accordance with Lemma 7.4. Due to the definition of $\hat{e}$, one has

$$
\hat{e}(0,t) = e(0,t) + \Delta q(t) v(0,t) = e(0,t) + \Delta q(t)\bar{v}(0,t) + \Delta q(t) u_1 \,, \qquad (7.295)
$$

and for $u_1 \neq 0$ it follows that

$$|\Delta q(t)| \leq \frac{|\hat{e}(0,t)| + |e(0,t)| + |\Delta q(t)| \|\bar{v}_x\|}{|u_1|}. \tag{7.296}$$

Since $\Delta q(t)$ is bounded and $\hat{e}(0,t)$, $e(0,t)$, and $\|\bar{v}_x\|$ are all square integrable, it is concluded from (7.296) that $\Delta q(t)$ is square integrable. By Lemma 7.4, $\Delta q(t)$ and $\frac{d}{dt}\Delta q$ are bounded. By applying Barbalat's Lemma 2.2, the convergence $\lim_{t\to\infty} \Delta q(t) = 0$ is established.

Step 5. In spite of the L_2-boundedness demonstration, the spatially uniform boundedness is, however, considerably harder to show. The main difficulty is the presence of nonhomogeneous terms in the boundary condition (7.286) at the left end, which does not allow one to use the H_1-norm as a standard Lyapunov–Krasovskii functional. To avoid this difficulty, the following transformation

$$\breve{w}(x) = \bar{w}(x) + \hat{q}(1-x) \int_0^x \hat{e}(y)\, dy \tag{7.297}$$

$$\breve{v}(x) = \bar{v}(x) + (1-x) \int_0^x (\hat{e}(y) + \bar{w}(y))\, dy\,. \tag{7.298}$$

from $(\bar{w}, \bar{v})$ into $(\breve{w}, \breve{v})$ is introduced. The purpose of this transformation is to homogenize the boundary condition (7.286) at the left end. One can readily check that in the new variables, $\breve{w}_x(0,t) = \breve{w}(1,t) = \breve{v}_x(0,t) = \breve{v}(1,t) = 0$. The right-hand side of the resulting PDEs for $\breve{w}$ and $\breve{v}$ is quite complicated, but has a simple structure with all the terms proportional either to $\dot{\hat{q}}$, $\hat{e}(0)$, or Δq, all of which are square integrable and bounded as shown before. After such a crucial step, the rest of the proof follows the proof for the g-case. One should first demonstrate the boundedness of $\|\breve{w}_x\|$ and $\|\breve{v}_x\|$ with the Lyapunov–Krasovskii functional

$$V = \frac{1}{2}\left(\|\breve{w}\|^2 + \|\breve{v}\|^2 + \|\breve{w}_x\|^2 + \|\breve{v}_x\|^2\right). \tag{7.299}$$

Then by Agmon's inequality Lemma 3.4, this ensures the uniform boundedness of $\breve{w}(x,t)$ and $\breve{v}(x,t)$. Then transformations (7.297), (7.298), coupled to the substitution $\hat{e} = e + \Delta q \bar{v} + \Delta q u_1(1-x)$, allow one to conclude that $\bar{w}(x,t)$ and $\bar{v}(x,t)$ are uniformly bounded as well. Moreover, from the invertible transformation (7.270), it follows that $\eta(x,t)$ is also uniformly bounded. Finally, since $u = e + qv + \eta$, the uniform boundedness of $u(x,t)$ is verified. Theorem 7.15 is thus completely proved. ■

7.2.3 Simulation Results

To illustrate the proposed identification scheme, the plant (7.181), (7.182) is viewed with the unknown parameter $g = 4$. It is worth recalling that the open-loop system is unstable for this value of the parameter. The initial estimate of the parameter is

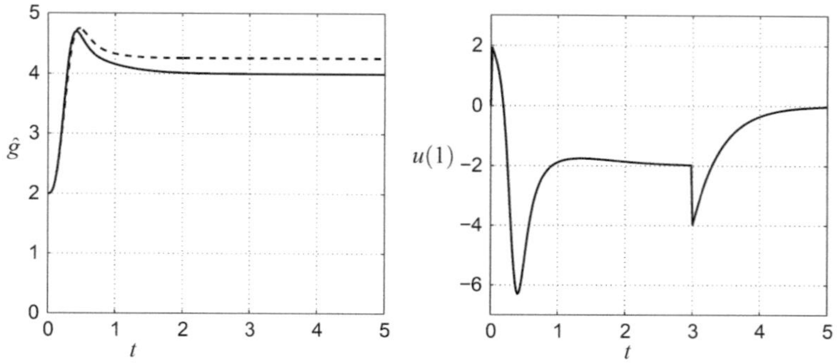

Fig. 7.7 Left: convergence of the parameter estimate $\hat{g}$ to the true value $g = 4$ with (solid) and without (dashed) additional constant boundary input. Right: the control effort with the additional constant input, turned off at $t = 3$. © 2008 *WILEY*. Reprinted, with permission, from Smyshlyaev et al. (2009)

Fig. 7.8 The closed-loop state $u(x, t)$. © 2008 *WILEY*. Reprinted, with permission, from Smyshlyaev et al. (2009)

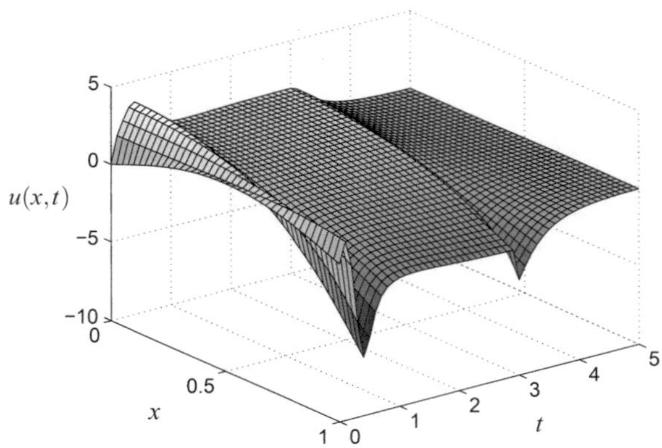

$\hat{g}(0) = 2$. A constant boundary input $u_1 = 2$ is added to the feedback. The results of the closed-loop simulations are presented in Figs. 7.7 and 7.8. In Fig. 7.7 the evolution of the parameter estimate is shown in comparison with the case $u_1 = 0$. It is seen from the figure that the unknown parameter is successfully identified. The additional constant input is turned off at $t = 3$ to achieve regulation to zero. In Fig. 7.8, the closed-loop state is shown to be stabilized.

7.3 Concluding Remarks

Lyapunov–Razumikhin and Lyapunov–Krasovskii redesigns are developed for linear TDS and parabolic DPS, respectively. Sufficient conditions of the plant parameters to be identifiable are given in terms of the plant structure and its state to be persistently excited. Input generators of the persistent excitation for both classes of infinite-dimensional systems are explicitly constructed independently of the underlying system and its initial conditions.

For TDS, the complete treatment is made when the state feedback redesign is available, and it is confined to SISO TDS for the output feedback redesign. An extension to MIMO TDS represents a challenging problem, calling for further investigation.

For DPS, the present investigation captures parabolic PDEs with a scalar spatial variable only. The persistency of excitation property of the input can be extended to the case of the multidimensional spatial variable; however, the construction of the generator of persistent excitation becomes more complicated if the corresponding Sturm–Liouville problem has roots with multiplicity. In this case, persistent excitation cannot be generated by time-invariant input, however all time-varying inputs with linearly independent Fourier coefficients appear to generate persistent excitation of the underlying parabolic PDE. In the case of the state feedback redesign, the adaptive identification and redesign of linear hyperbolic PDEs are addressed in Bentsman and Orlov (2001), Orlov and Bentsman (1995).

On-line identification schemes are additionally developed for the output feedback redesign of two unstable parabolic systems with unknown spatially independent parameters and with boundary sensing and actuation. Further research should be focused on extending these results to reaction–advection–diffusion PDEs with spatially varying parametric uncertainties. A promising approach to such a challenging problem is to approximate the unknown spatially varying parameter, e.g., by truncating its Fourier series, and consider a problem of simultaneous identification of a finite set of the unknown Fourier coefficients. There are no fundamental obstacles in using the developed redesign for that problem. However, one would need to establish the persistency of excitation condition on the boundary input, since a constant input is no longer sufficient for the identification of more than one parameter. Once the persistency of excitation is established on the boundary input, one can choose sufficiently many Fourier coefficients in the properly truncated Fourier series to identify an unknown spatially varying parameter with any prescribed accuracy.

As a matter of fact, the adaptive redesign presented do not admit an easy generalization to nonlinear TDS and DPS that constitutes a formidable open problem.

References

Alt HW, Hoffmann KH, Sprekels J (1984) A numerical procedure to solve certain identification problems. Int Ser Numer Math 68:11–43

Åström KJ, Wittenmark B (1989) Adaptive control. Addison-Wesley, Massachusetts

Banks BT, Kunish K (1989) Estimation techniques for distributed parameter systems. Birkhäuser, Boston

Baumeister J, Scondo W (1987) Asymptotic embedding methods for parameter estimation. In: IEEE conference on decision and control, pp 170–174

Baumeister J, Demetriou MA, Rosen IG, Scondo W (1997) On-line parameter estimation for infinite-dimensional dynamical systems. SIAM J Control Optim 35:678–713

Belkoura L, Orlov Y (2002) Identifiability analysis of linear delay-differential systems. IMA J Math Control Inf 19:73–81

Bentsman J, Orlov Y (2001) Reduced spatial order model reference adaptive control of spatially-varying distributed parameter systems of parabolic and hyperbolic types. Int J Adapt Control Signal Process 15:679–696

Boskovic DM, Krstic M (2003) Stabilization of a solid propellant rocket instability by state feedback. Int J Robust Nonlinear Control 13:483–495

Courdesses M, Polis MG, Amouroux M (1981) On the identifiability of parameters in a class of parabolic distributed systems. IEEE Trans Auto Control AC-26:474–477

Demetriou MA, Rosen IG (1994a) Adaptive identification of second order distributed parameter systems. Inverse Probl 10:261–294

Demetriou MA, Rosen IG (1994b) On the persistence of excitation in the adaptive identification of distributed parameter systems. IEEE Trans Auto Control AC-39:1117–1123

Demetriou MA, Rosen IG (1995) Adaptive parameter estimation for degenerate parabolic systems. J. Math. Anal. Applicat. 189:815–847

Filippov AF (1988) Differential equations with discontinuous right-hand sides. Kluwer Academic Publisher, Dordrecht

Friedman A (1969) Partial differential equations. Holt, Reinhart, and Winston, New York

Gomez O, Orlov Y, Kolmanovsky IV (2007) On-line identification of SISO linear time-invariant delay systems from output measurements. Automatica 43: 2060–2069

Goodwin GC, Sin KS (1984) Adaptive filtering prediction and control. Prentice Hall, Englewood Cliffs

Gu K, Kharitonov V, Chen J (2003) Stability of time delay systems. Birkhäuser, Boston

Haddock J, Terjeki J (1983) Lyapunov-Razumikhin functions and an invariance principle for functional differential equations. J Differ Equ 48:95–121

Hale J (1971) Functional differential equations. Springer, New York

Henry D (1981) Geometric theory of semilinear parabolic equations. Lecture notes in mathematics. Springer, Berlin

Hoffmann KH, Sprekels J (1984/1985) On the identification of coefficients of elliptic problems by asymptotic regularization. Numer Funct Anal Optim 7:157–177

Hong KS, Bentsman J (1994a) Direct adaptive control of parabolic systems: algorithm synthesis and convergence and stability analysis. IEEE Trans Auto Control AC-39:2018–2033

Hong KS, Bentsman J (1994b) Application of averaging method for integro-differential equations to model reference adaptive control of parabolic systems. Automatica 30:1415–1420

Ioannou PA, Sun J (1996) Robust adaptive control. Dover, New York

Khalil H (2002) Nonlinear systems, 3rd edn. Prentice Hall, Upper Saddle River

Kitamura S, Nakagiri S (1977) Identifiability of spatially-varying and constant parameters in distributed systems of parabolic type. SIAM J Control Optim 15:785–802

Kolmanovskii VB, Nosov VR (1986) Stability of functional differential equations. Academic, New York

Kravaris C, Seinfeld JH (1985) Identification of parameters in distributed parameters systems by regularization. SIAM J Control Optim 23:217–241

Krstic H, Kokotovic PV, Kanellakopoulos I (1995) Nonlinear and adaptive control design. Wiley, New York

Kwakernaak H, Sivan R (1991) Modern signals and systems. Prentice-Hall, Englewood Cliffs

Landau YD (1979) Adaptive control - the model reference approach. Marcel Dekker, New York

Miller RK, Michel AN (1990) An invariance theorem with applications to adaptive control. IEEE Trans Auto Control 35:744–748

Morse AS (1976) Ring models for delay-differential systems. Automatica 12:529–531

Nakagiri S (1983) Identifiability of linear systems in Hilbert space. SIAM J Control Optim 21:501–530

Nakagiri S, Yamamoto M (1995) Unique identification of coefficient matrices, time delay and initial function of functional differential equations. J Math Syst 5(3):323–344

Narendra KS, Annaswamy A (1989) Stable adaptive systems. Prentice Hall, Englewood Cliffs
Omatu S, Seinfeld JM (1989) Distributed parameter systems. Clarendon Press, Oxford
Orlov Y (2000) Sliding mode observer-based synthesis of state derivative-free model reference adaptive control of distributed parameter systems. ASME J Dyn Syst Meas Control 122(4):725–731
Orlov Y, Belkoura L, Richard JP, Dambrine M (2002) On identifiability of linear time-delay systems. IEEE Trans Auto Control 47:1319–1324
Orlov Y, Belkoura L, Richard JP, Dambrine M (2003) Adaptive identification of linear time-delay systems. Int J Robust Nonlinear Control 13:857–872
Orlov Y, Bentsman J (1995) Model reference adaptive control (MRAC) of heat processes with simultaneous plant identification. In: IEEE conference on decision and control, pp 1165–1170
Orlov Y, Bentsman J (2000) Adaptive distributed parameter systems identification with enforceable identifiability conditions and reduced-order spatial differentiation. IEEE Trans Auto Control 45:203–216
Orlov Y, Kolmanovsky IV, Gomez O (2009) Adaptive identification of linear time-delay systems: from theory toward application to engine transient fuel identification. International Journal of Adaptive Control and Signal Processing 23(2): 150–165
Pierce A (1979) Unique identification of eigenvalues and coefficients in a parabolic problem. SIAM J Control Optim 17:494–499
Rouche N, Habets P, Laloy M (1977) Stability theory by Lyapunov's direct method. Springer, New York
Sastry SS, Bodson M (1989) Adaptive control: stability, convergence and robustness. Prentice Hall, Englewood Cliffs
Smyshlyaev A, Krstic M (2006) Output-feedback adaptive control for parabolic PDEs with spatially-varying coefficients. In: IEEE conference on decision and control, pp 3099–3104
Smyshlyaev A, Krstic M (2007) Adaptive boundary control for unstable parabolic PDEs–Part III: output-feedback examples with swapping identifiers. Automatica 43:1557–1564
Smyshlyaev A, Orlov Y, Krstic M (2009) Adaptive identification of two unstable PDEs with boundary sensing and actuation. International Journal of Adaptive Control and Signal Processing 23: 131–149
Solo V, Bentsman J (1993) Adaptive control of parabolic systems with spatially varying parameters: an averaging analysis In: IEEE conference on decision and control, pp 2435–2437
Sontag ED (1976) Linear systems over commutative rings: a survey. Ricerche di Automatica 7(1):1–34
Utkin VI, Orlov YV (1990) Theory of infinite-dimensional sliding mode control systems. Nauka, Moscow
Verduyn Lunel SM (2001) Parameter identifiability of differential delay equations. Adapt Control Signal Process 15(6):655–678
Zemanian AH (1965) Distribution theory and transform analysis. McGrow-Hill, New York

Chapter 8
Control Applications

Several engineering applications are addressed to make the monograph appropriately complete. Specific CLFs are constructed for the periodic bipedal gait synthesis under unilateral constraints and for the energy control of a nonlinear sine-Gordon model with complex behavior such as solitons, kinks, antikinks, and breathers. The sine-Gordon model describes a continuum of oscillators, governed by a nonlinear hyperbolic PDE. It constitutes a paradigmatic model for infinite-dimensional nonlinear dynamics, e.g., in nonlinear optics (propagation of an optical pulse in fiber waveguide) and in mechanics (transition from a static state to a dynamic frictional mode). The exposition of the chapter focuses on the control algorithms rather than their technical implementation, thus avoiding inconsistency with the general theoretical line of the monograph.

8.1 Synthesis of Mechanical Systems Under Unilateral Constraints

Lyapunov function constructions, relying on proximal solutions of Hamilton–Jacobi PDE/PDIs (see Sect. 5.5), form a powerful method of nonsmooth robust control synthesis in the presence of external disturbances. Further on, the capability of this method is illustrated for hybrid mechanical systems, operating under uncertainty conditions. Such systems are governed by continuous-time differential equations and difference equations, accompanied with a rule of switching between them, which is defined according to output and/or time constraints. Hybrid systems have attracted a lot of research interest (see, e.g., Goebel et al. 2009; Hamed and Grizzle 2013; Naldi and Sanfelice 2013; Nesic et al. 2013 and references therein) due to the wide variety of their applications and due to the need of special analysis tools for this type of system. Robust control of hybrid mechanical systems to be presented was developed in Montano et al. (2016) and successfully applied in Montano et al. (2016) to biped gait stabilization.

Y. Orlov, *Nonsmooth Lyapunov Analysis in Finite and Infinite Dimensions*, Communications and Control Engineering,
https://doi.org/10.1007/978-3-030-37625-3_8

8.1.1 Robust Tracking Problem and Hybrid Error Dynamics

A hybrid mechanical system of interest is composed of free-motion phases, governed by

$$\mathbf{D}(\mathbf{q})\ddot{\mathbf{q}} + \mathbf{H}(\mathbf{q}, \dot{\mathbf{q}}) = \mathbf{D}_{\tau}\boldsymbol{\tau} + \mathbf{w}_1 \tag{8.1}$$

beyond a unilateral time-invariant constraint $F_0(\mathbf{q}) = \mathbf{0}$ where

$$F_0(\mathbf{q}) > \mathbf{0}, \tag{8.2}$$

whereas these free-motion phases are separated by transition phases according to the restitution rule

$$\mathbf{q}(t_i^+) = \mathbf{q}(t_i^-) \tag{8.3}$$

$$\dot{\mathbf{q}}(t_i^+) = \boldsymbol{\phi}(\mathbf{q}(t_i))\dot{\mathbf{q}}(t_i^-) + \boldsymbol{\omega}^0(\mathbf{q}(t_i), \dot{\mathbf{q}}(t_i^-), t)\mathbf{w}_\mathbf{i}^\mathbf{d} \tag{8.4}$$

when the state trajectory hits the surface

$$F_0(\mathbf{q}(t_i)) = 0, \quad i = 1, 2, \ldots. \tag{8.5}$$

Hereinafter, the bold font is for the vector notation, typically used in mechanics, $\mathbf{q}, \dot{\mathbf{q}} \in \mathbb{R}^n$ are generalized position and velocity vectors, the control input $\boldsymbol{\tau} \in \mathbb{R}^n$ is a vector of external torques, $\mathbf{w}_1 \in \mathbb{R}^n$ is an external disturbance, $\mathbf{w}_\mathbf{i}^\mathbf{d}$, $i = 1, 2, \ldots$ are discrete perturbations of the velocity restitution rule (8.4) at a priori *unknown* time instants t_i; $\boldsymbol{\phi}(\mathbf{q}) \in \mathbb{R}^{n\times n}$ is a position-dependent restitution matrix; $\mathbf{H}(\mathbf{q}, \dot{\mathbf{q}}) \in \mathbb{R}^n$ is the vector of Coriolis, centrifugal, and gravitational torques, the inertia matrix $\mathbf{D}(\mathbf{q})$ and the actuation matrix $\mathbf{D}_\tau$ are of appropriate dimensions such that $\mathbf{D}(\mathbf{q})$ is symmetric and positive definite, and $\mathbf{D}_\tau$ is invertible and is composed of zero and unit values (thus considering only fully actuated mechanical systems); the scalar function $F_0(\mathbf{q})$ relies on the unilateral constraint, imposed on the robot. As a matter of fact, $\mathbf{D}(\mathbf{q})$, $\mathbf{H}(\mathbf{q}, \dot{\mathbf{q}})$, $\boldsymbol{\phi}(\mathbf{q})$, and $\boldsymbol{\omega}^0(\mathbf{q}, \dot{\mathbf{q}}, t)$ are smooth functions in their arguments.

In what follows, the research is confined to the tracking control problem where the output to be controlled is given in terms of the state deviation from a reference trajectory $\mathbf{q}^\mathbf{r}(t)$ and it is composed of the continuous-time component

$$\mathbf{z} = \begin{bmatrix} \mathbf{0} \\ \rho_p(\mathbf{q} - \mathbf{q}^\mathbf{r}) \\ \rho_v(\dot{\mathbf{q}} - \dot{\mathbf{q}}^\mathbf{r}) \end{bmatrix} + \begin{bmatrix} \mathbf{1} \\ \mathbf{0} \\ \mathbf{0} \end{bmatrix} \mathbf{u} \tag{8.6}$$

with positive weight coefficients ρ_p, ρ_v, and of its discrete counterpart

$$\mathbf{z}_\mathbf{i}^\mathbf{d} = \dot{\mathbf{q}}(t_i^+) - \dot{\mathbf{q}}^\mathbf{r}(t_i^+), \tag{8.7}$$

whereas the available measurement

$$\mathbf{y} = \mathbf{q} - \mathbf{q^r} + \mathbf{w_0} \tag{8.8}$$

is affected by the measurement error $\mathbf{w_0}(t)$.

The $\mathscr{H}_\infty$-tracking control problem for mechanical systems subject to unilateral constraints on the position is formally stated as follows. Given a mechanical system (8.1)–(8.5) a desired trajectory $\mathbf{q^r}(t)$ to track, and a real number $\tau > 0$, it is required to find (if any) a causal output feedback (or state feedback) controller such that the undisturbed closed-loop system is uniformly asymptotically stable around $\mathbf{q^r}(t)$ and its $\mathscr{L}_2$-gain is locally less than γ.

A causal dynamic output feedback controller is of the same structure

$$\begin{aligned} &\dot{\boldsymbol{\xi}}_1 = \boldsymbol{\xi}_2, \quad \dot{\boldsymbol{\xi}}_2 = \boldsymbol{\eta}(\boldsymbol{\xi}_1, \boldsymbol{\xi}_2, \mathbf{y}, t) \\ &\boldsymbol{\xi}_1(t_j^+) = \boldsymbol{\xi}_1(t_j^-), \quad \boldsymbol{\xi}_2(t_j^+) = \boldsymbol{\nu}(\boldsymbol{\xi}_1(t_j), \boldsymbol{\xi}_2(t_j^-), t_j) \\ &\mathbf{u} = \boldsymbol{\theta}(\boldsymbol{\xi}, t) \end{aligned} \tag{8.9}$$

as that of the plant. It comes with the internal state $\boldsymbol{\xi} = (\boldsymbol{\xi}_1, \boldsymbol{\xi}_2)^T \in \mathbb{R}^{2s}$, with the time instants $t = t_j, \ j = 1, 2, \ldots$ (not necessarily coinciding with the collision time instants in the plant equations (8.1)–(8.5)), and with uniformly bounded in t functions $\boldsymbol{\eta}(\boldsymbol{\xi}, \mathbf{y}, t)$, $\boldsymbol{\nu}(\boldsymbol{\xi}, t)$, and $\boldsymbol{\theta}(\boldsymbol{\xi}, t)$ of class C^1 such that $\boldsymbol{\eta}(\mathbf{0}, \mathbf{0}, t) = \mathbf{0}$, $\boldsymbol{\nu}(\mathbf{0}, t) = \mathbf{0}$, and $\boldsymbol{\theta}(\mathbf{0}, t) = \mathbf{0}$ for all t.

By Definition 5.1, the $\mathscr{L}_2$-gain is locally less than γ iff the inequality

$$\begin{aligned} \int_{t_0}^{T} \|\mathbf{z}(t)\|^2 \mathrm{d}t + \sum_{i=1}^{N_T} \|\mathbf{z_i^d}\|^2 < \gamma^2 \left[\int_{t_0}^{T} \|\mathbf{w}(t)\|^2 \mathrm{d}t + \sum_{i=1}^{N_T} \|\mathbf{w_i^d}\|^2 \right] \\ + \sum_{k=0}^{N_T} \beta_k(\mathbf{x}(t_k^-), \boldsymbol{\xi}(t_k^-), t_k) \end{aligned} \tag{8.10}$$

holds for all T and all piecewise continuous functions $\mathbf{w_1}$, $\mathbf{w_d}$, and $\mathbf{w_0}$ for which the state trajectory of the closed-loop system starting in a neighborhood of the initial point $(\mathbf{q}(0), \ \dot{\mathbf{q}}(0)) = (\mathbf{q^r}(0), \ \dot{\mathbf{q}}^r(0))$ remains in a neighborhood of the desired trajectory $\mathbf{q^r}(t)$ for all $t \in [0, T]$.

In order to accomplish the above task, the following assumptions are made

$$\mathbf{q_d}(t_i) \in \mathbf{F_0}(\mathbf{q}) = 0, \quad i = 1, 2, \ldots \tag{8.11}$$

$$\dot{q}_{d_k} \neq 0 \, \text{for almost all} \ t \tag{8.12}$$

and for all components $q_k, \ k = 1, \ldots, n$ of the reference vector $\mathbf{q_d}$. The reference trajectory $\mathbf{q^r}(t)$ to be tracked is a periodic or uniformly bounded trajectory subject to an impact that occurs when the reference trajectory achieves the surface $F_0(\mathbf{q^r}) = 0$.

The restitution law during this impact phase is given by

$$\dot{\mathbf{q}}^{\mathbf{r}}(t_i^+) = \boldsymbol{\phi}(\mathbf{q}^{\mathbf{r}}(t_i))\dot{\mathbf{q}}^{\mathbf{r}}(t_i^-), \quad i = 1, 2, \ldots. \tag{8.13}$$

This trajectory may be constructed off-line with a priori known impact instants t_i, $i = 1, 2, \ldots$ or it may be generated by an impact reference model, possessing a stable limit cycle, similar to the constrained Van der Pol oscillator of Sect. 2.4.

In terms of the state deviation vector $\mathbf{x} = (\mathbf{x}_1, \mathbf{x}_2)^T$ where $\mathbf{x}_1(t) = \mathbf{q}(t) - \mathbf{q}^{\mathbf{r}}(t)$ and $\mathbf{x}_2(t) = \dot{\mathbf{q}}(t) - \dot{\mathbf{q}}^{\mathbf{r}}(t)$, the state equations (8.1)–(8.5), (8.6)–(8.8) are represented in the form

$$\begin{aligned} \dot{\mathbf{x}}_1 &= \mathbf{x}_2 \\ \dot{\mathbf{x}}_2 &= \mathbf{D}^{-1}(\mathbf{x}_1 + \mathbf{q}^{\mathbf{r}})[-\mathbf{H}(\mathbf{x}_1 + \mathbf{q}^{\mathbf{r}}, \mathbf{x}_2 + \dot{\mathbf{q}}^{\mathbf{r}}) + \mathbf{D}_\tau \boldsymbol{\tau} + \mathbf{w}_1] - \ddot{\mathbf{q}}^{\mathbf{r}} \end{aligned} \tag{8.14}$$

of the error system.

The transitions occur in the error dynamics according to the following scenarios.

(T1) The reference trajectory reaches its predefined impact-time instant $t = t^k$, $k = 1, 2, \ldots$ when it hits the unilateral constraint whereas the plant remains beyond this constraint, i.e., $F_0(\mathbf{q}^{\mathbf{r}}(t^k)) = 0$, $F_0(\mathbf{x}_1(t^k) + \mathbf{q}^{\mathbf{r}}(t^k)) \neq 0$;

(T2) The plant hits the unilateral constraint at $t = t^j$, $j = 1, 2, \ldots$ while the reference trajectory is beyond this constraint, i.e., $F_0(\mathbf{q}^{\mathbf{r}}(t^j)) \neq 0$, $F_0(\mathbf{x}_1(t^j) + \mathbf{q}^{\mathbf{r}}(t^j)) = 0$;

(T3) Both the reference trajectory and the plant hit the unilateral constraint at the same time instant $t = t^l$, $l = 1, 2, \ldots$ (what can deliberately be enforced by modifying the prespecified reference trajectory online), i.e., $F_0(\mathbf{q}^{\mathbf{r}}(t^l)) = 0$, $F_0(\mathbf{x}_1(t^l) + \mathbf{q}^{\mathbf{r}}(t^l)) = 0$.

Transition errors are then represented as follows.

Scenario T1:

$$\begin{aligned} \mathbf{x}_1(t^{k+}) &= \mathbf{x}_1(t^{k-}) \\ \mathbf{x}_2(t^{k+}) &= \boldsymbol{\mu}^1(\mathbf{x}(t^{k-}), t^k) + \mathbf{w}_{\mathbf{k}}^{\mathbf{d}}, \end{aligned} \tag{8.15}$$

provided that $F_0(\mathbf{q}^{\mathbf{r}}(t^k)) = 0$ and $F_0(\mathbf{x}_1(t^k) + \mathbf{q}^{\mathbf{r}}(t^k)) \neq 0$, $k = 1, 2, \ldots$;
Scenario T2:

$$\begin{aligned} \mathbf{x}_1(t^{j+}) &= \mathbf{x}_1(t^{j-}) \\ \mathbf{x}_2(t^{j+}) &= \boldsymbol{\mu}^2(\mathbf{x}(t^{j-}), t^j) + \mathbf{w}_{\mathbf{j}}^{\mathbf{d}}, \end{aligned} \tag{8.16}$$

provided that $F_0(\mathbf{q}^{\mathbf{r}}(t^j)) \neq 0$ and $F_0(\mathbf{x}_1(t^j) + \mathbf{q}^{\mathbf{r}}(t^j)) = 0$, $j = 1, 2, \ldots$;

Scenario T3:

$$\mathbf{x_1}(t^{l+}) = \mathbf{x_1}(t^{l-})$$
$$\mathbf{x_2}(t^{l+}) = \boldsymbol{\mu}^3(\mathbf{x}(t^{l-}), t^l) + \mathbf{w_l^d}, \ l = 1, 2, \dots \quad (8.17)$$

provided that $F_0(\mathbf{q^r}(t^l)) = 0$ and $F_0(\mathbf{x_1}(t^l) + \mathbf{q^r}(t^l)) = 0, \ l = 1, 2, \dots$

where $\mathbf{w_k^d}$, $\mathbf{w_j^d}$, $\mathbf{w_l^d}$ are discrete perturbations, counting for restitution inadequacies, and functions $\boldsymbol{\mu}^1$, $\boldsymbol{\mu}^2$, and $\boldsymbol{\mu}^3$ are given by

$$\boldsymbol{\mu}^1(\mathbf{x}, t) = \mathbf{x_2} + [\mathbf{I} - \boldsymbol{\phi}(\mathbf{q^r}(t))]\dot{\mathbf{q}}^{\mathbf{r}}(t^-)$$
$$\boldsymbol{\mu}^2(\mathbf{x}, t) = \boldsymbol{\phi}(\mathbf{x_1} + \mathbf{q^r}(t))[\mathbf{x_2} + \dot{\mathbf{q}}^{\mathbf{r}}(t^-)] - \dot{\mathbf{q}}^{\mathbf{r}}(t^-)$$
$$\boldsymbol{\mu}^3(\mathbf{x}, t) = \boldsymbol{\phi}(\mathbf{x_1} + \mathbf{q^r}(t))[\mathbf{x_2} + \dot{\mathbf{q}}^{\mathbf{r}}(t^-)] - \boldsymbol{\phi}(\mathbf{q^r}(t))\dot{\mathbf{q}}^{\mathbf{r}}(t^-). \quad (8.18)$$

In terms of the error vector $\mathbf{x}$, the unilateral constraint $F_0(\mathbf{x_1})$ becomes time-varying and it is modified to

$$F(\mathbf{x}, t) = F_0(\mathbf{x_1} + \mathbf{q^r}(t)), \quad (8.19)$$

whereas the restitution rule (8.4) takes the form

$$\dot{\mathbf{q}}(t_i^+) = \boldsymbol{\mu_0}(\mathbf{x}(t_i), t_i) + \boldsymbol{\omega}(\mathbf{x}(t_i^-, t_i)\mathbf{w_i^d}, \ i = 1, 2, \dots \quad (8.20)$$

specified with $\boldsymbol{\omega}(\mathbf{x}(t_i^-, t_i) = \boldsymbol{\omega}^0(\mathbf{x_1}(t_i) + \mathbf{q^r}(t_i), \mathbf{x_2}(t_i^-) + \dot{\mathbf{q}}^{\mathbf{r}}(t_i^-), t)$ and

$$\boldsymbol{\mu_0}(\mathbf{x},t) = \begin{cases} \boldsymbol{\mu}^1(\mathbf{x}, t) \text{ if } F_0(\mathbf{q^r}(t)) = 0, & F_0(\mathbf{x_1} + \mathbf{q^r}) \neq 0 \\ \boldsymbol{\mu}^2(\mathbf{x}, t) \text{ if } F_0(\mathbf{q^r}(t)) \neq 0, & F_0(\mathbf{x_1} + \mathbf{q^r}) = 0 \\ \boldsymbol{\mu}^3(\mathbf{x}, t) \text{ if } F_0(\mathbf{q^r}(t)) = 0, & F_0(\mathbf{x_1} + \mathbf{q^r}) = 0. \end{cases} \quad (8.21)$$

With a prespecified feedback design, the error system falls under a canonical form, used in the nonlinear/nonsmooth $\mathscr{H}_\infty$-synthesis.

8.1.2 Pre-feedback Design and $\mathscr{H}_\infty$ Synthesis

In the case where only the generalized positions of the mechanical system are available for measurements, the pre-feedback design

$$\boldsymbol{\tau} = \mathbf{D}_\tau{}^{-1}[\mathbf{D}(\mathbf{q^r})\ddot{\mathbf{q}}^{\mathbf{r}} + \mathbf{H}(\mathbf{q^r}, \dot{\mathbf{q}}^{\mathbf{r}}) + \mathbf{u}] \quad (8.22)$$

computes the Coriolis, centrifugal, and gravitational torques on the reference trajectories rather than those occurring in the plant. Thus, the position feedback controller to be constructed consists of a disturbance attenuator $\mathbf{u}$, internally stabilizing the biped around the desired trajectory, and the remainder, which is responsible for the compensation of the reference trajectory and the torques, associated with this trajectory.

Substituting the position pre-feedback (8.22) into (8.14) yields the impact-free error dynamics in the form

$$\dot{\mathbf{x}}_1 = \mathbf{x}_2$$
$$\dot{\mathbf{x}}_2 = \mathbf{D}^{-1}(\mathbf{x}_1 + \mathbf{q}^r)[-\mathbf{H}(\mathbf{x}_1 + \mathbf{q}^r, \mathbf{x}_2 + \dot{\mathbf{q}}^r) + \mathbf{D}(\mathbf{q}^r)\ddot{\mathbf{q}}^r + \mathbf{H}(\mathbf{q}^r, \dot{\mathbf{q}}^r) + \mathbf{u} + \mathbf{w}_1] - \ddot{\mathbf{q}}^r. \tag{8.23}$$

When specified with (8.19)–(8.21) and $\mathbf{x}^T = [\mathbf{x}_1^T, \mathbf{x}_2^T]$,

$$\mathbf{f}(\mathbf{x}, t) = \begin{bmatrix} \mathbf{x}_2 \\ \mathbf{D}^{-1}(\mathbf{x}_1 + \mathbf{q}^r)[-\mathbf{H}(\mathbf{x}_1 + \mathbf{q}^r, \mathbf{x}_2 + \dot{\mathbf{q}}^r) + \mathbf{D}(\mathbf{q}^r)\ddot{\mathbf{q}}^r + \mathbf{H}(\mathbf{q}^r, \dot{\mathbf{q}}^r)] - \ddot{\mathbf{q}}^r \end{bmatrix}$$

$$\mathbf{g}_1(\mathbf{x}, t) = \begin{bmatrix} \mathbf{0} & \mathbf{0} \\ \mathbf{0} & \mathbf{D}^{-1}(\mathbf{x}_1 + \mathbf{q}^r) \end{bmatrix}, \ \mathbf{h}_1(\mathbf{x}) = \begin{bmatrix} \mathbf{0} \\ \rho_p \mathbf{x}_1 \\ \rho_v \mathbf{x}_2 \end{bmatrix},$$

$$\mathbf{g}_2(\mathbf{x}, t) = \begin{bmatrix} \mathbf{0} \\ \mathbf{D}^{-1}(\mathbf{x}_1 + \mathbf{q}^r) \end{bmatrix}, \ \mathbf{k}_{12}(\mathbf{x}) = \begin{bmatrix} \mathbf{1} \\ \mathbf{0} \\ \mathbf{0} \end{bmatrix},$$

$$\mathbf{h}_2(\mathbf{x}) = \begin{bmatrix} \mathbf{x}_1 & \mathbf{0} \end{bmatrix}, \ \mathbf{k}_{21}(\mathbf{x}) = \begin{bmatrix} \mathbf{1} & \mathbf{0} \end{bmatrix}, \ \mathbf{w} = [\mathbf{w}_0^T \mathbf{w}_1^T]^T, \tag{8.24}$$

these motion-free error dynamics are governed by a generic system

$$\dot{\mathbf{x}} = \mathbf{f}(\mathbf{x}, t) + \mathbf{g}_1(\mathbf{x}, t)\mathbf{w} + \mathbf{g}_2(\mathbf{x}, t)\mathbf{u}, \tag{8.25}$$

$$\mathbf{z} = \mathbf{h}_1(\mathbf{x}_1, \mathbf{x}_2, t) + \mathbf{k}_{12}(\mathbf{x}_1, \mathbf{x}_2, t)\mathbf{u}, \tag{8.26}$$

$$\mathbf{y} = \mathbf{h}_2(\mathbf{x}_1, \mathbf{x}_2, t) + \mathbf{k}_{21}(\mathbf{x}_1, \mathbf{x}_2, t)\mathbf{w}. \tag{8.27}$$

In accordance with (8.20), the motion-free dynamics (8.25)–(8.27) are accompanied by the restitution rule

$$\mathbf{x}(t_i^+) = \boldsymbol{\mu}(\mathbf{x}(t_i^-), t_i) + \boldsymbol{\Omega}(\mathbf{x}(t_i^-), t_i)\mathbf{w}_i^d, \ i = 1, 2, \ldots, \tag{8.28}$$

imposed at the impact-time instants, when the constraint function (8.19) is nullified, and specified with

$$\boldsymbol{\mu}^T(\mathbf{x}, t) = [\mathbf{x}_1^T, \boldsymbol{\mu}_0^T(\mathbf{x}, t)], \ \boldsymbol{\Omega}^T(\mathbf{x}, t) = [\mathbf{0}, \omega(\mathbf{x}, t)]. \tag{8.29}$$

By straightforward inspection, the standard assumptions

$$\begin{gathered} \mathbf{f}(\mathbf{0}, t) \equiv 0, \ \mathbf{h}_1(\mathbf{0}, t) \equiv 0, \ \mathbf{h}_2(\mathbf{0}, t) \equiv 0, \\ \mathbf{h}_1{}^T\mathbf{k}_{12} \equiv \mathbf{0}, \quad \mathbf{k}_{12}{}^T\mathbf{k}_{12} \equiv \mathbf{I}, \quad \mathbf{k}_{21}\mathbf{g}_1{}^T \equiv \mathbf{0}, \quad \mathbf{k}_{21}\mathbf{k}_{21}{}^T \equiv \mathbf{I}, \end{gathered} \tag{8.30}$$

typically used in the nonlinear/nonsmooth $\mathscr{H}_\infty$-synthesis (Isidori and Astolfi 1992; Orlov and Aguilar 2014), are verified to hold true.

8.1.2.1 State-Space Solution

Let $B_\delta^{2n} \in \mathbb{R}^{2n}$ be a ball of radius $\delta > 0$, centered around the origin. Given $\gamma > 0$, a solution to the $\mathscr{H}_\infty$-problem in question is derived under the hypotheses, specified below in a domain $(\mathbf{x}, \boldsymbol{\xi}) \in B_\delta^{2n}$, $t \in \mathbb{R}$ of interest:

(H1) The norm of the matrix function $\boldsymbol{\omega}$ is upper bounded by $\frac{\sqrt{2}}{2}\gamma$, i.e.,

$$\|\boldsymbol{\omega}(\mathbf{x}, t)\| \le \frac{\sqrt{2}}{2}\gamma; \tag{8.31}$$

(H2) The Hamilton–Jacobi–Isaacs inequality

$$\frac{\partial V}{\partial t} + \frac{\partial V}{\partial \mathbf{x}}(\mathbf{f}(\mathbf{x}, t) + \mathbf{g}_1(\mathbf{x}, t)\boldsymbol{\alpha}_1 + \mathbf{g}_2(\mathbf{x}, t)\boldsymbol{\alpha}_2) + \mathbf{h}_1{}^T\mathbf{h}_1 + \boldsymbol{\alpha}_2{}^T\boldsymbol{\alpha}_2 - \gamma^2\boldsymbol{\alpha}_1{}^T\boldsymbol{\alpha}_1 < 0, \tag{8.32}$$

specified with

$$\boldsymbol{\alpha}_1(\mathbf{x}, t) = \frac{1}{2\gamma^2}\mathbf{g}_1^T(\mathbf{x}, t)\left(\frac{\partial V}{\partial \mathbf{x}}\right)^T, \quad \boldsymbol{\alpha}_2(\mathbf{x}, t) = -\frac{1}{2}\mathbf{g}_2^T(\mathbf{x}, t)\left(\frac{\partial V}{\partial \mathbf{x}}\right)^T,$$

possesses a smooth, positive definite solution $V(\mathbf{x}, t)$;

(H3) There exist a continuous uniformly bounded function $G(t)$ such that the Hamilton–Jacobi–Isaacs inequality

$$\frac{\partial W}{\partial t} + \left(\frac{\partial W}{\partial \mathbf{x}} \; \frac{\partial W}{\partial \boldsymbol{\xi}}\right)\mathbf{f}_\mathbf{e}(\mathbf{x}, \boldsymbol{\xi}, t) + \mathbf{h}_\mathbf{e}{}^T\mathbf{h}_\mathbf{e} - \gamma^2\boldsymbol{\psi}^T\boldsymbol{\psi} < 0, \boldsymbol{\xi}) \tag{8.33}$$

specified with

$$\mathbf{f}_\mathbf{e}(\mathbf{x}, \boldsymbol{\xi}, t) = \begin{pmatrix} \mathbf{f}(\mathbf{x}, t) + \mathbf{g}_1(\mathbf{x}, t)\boldsymbol{\alpha}_1(\mathbf{x}, t) + \mathbf{g}_2(\mathbf{x}, t)\boldsymbol{\alpha}_2(\boldsymbol{\xi}, t) \\ \mathbf{f}(\boldsymbol{\xi}, t) + \mathbf{g}_1(\boldsymbol{\xi}, t)\boldsymbol{\alpha}_1(\boldsymbol{\xi}, t) + \mathbf{g}_2(\boldsymbol{\xi}, t)\boldsymbol{\alpha}_2(\boldsymbol{\xi}, t) + \mathbf{G}(t)(\mathbf{h}_2(\mathbf{x}, t) - \mathbf{h}_2(\boldsymbol{\xi}, t)) \end{pmatrix}$$

$$\mathbf{h}_\mathbf{e}(\mathbf{x}, \boldsymbol{\xi}, t) = \boldsymbol{\alpha}_2(\mathbf{x}, t) - \boldsymbol{\alpha}_2(\boldsymbol{\xi}, t),$$

$$\boldsymbol{\psi}(\mathbf{x}, \boldsymbol{\xi}, t) = \frac{1}{2\gamma^2}\mathbf{g}_\mathbf{e}{}^T(\mathbf{x}, t)\begin{pmatrix} \left(\frac{\partial W}{\partial \mathbf{x}}\right)^T \\ \left(\frac{\partial W}{\partial \boldsymbol{\xi}}\right)^T \end{pmatrix},$$

$$\mathbf{g}_e(\mathbf{x}, t) = \begin{pmatrix} \mathbf{g}_1(\mathbf{x}, t) \\ \mathbf{G}(t)\mathbf{k}_{21}(\mathbf{x}, t) \end{pmatrix},$$

possesses a smooth, positive semi-definite solution $W(\mathbf{x}, \boldsymbol{\xi}, t)$ with $W(\mathbf{0}, \boldsymbol{\xi}, t)$ being positive definite;

(H4) Hypotheses (H2) and (H3) are satisfied with the functions $V(\mathbf{x}, t)$ and $W(\mathbf{x}, \boldsymbol{\xi}, t)$ which decrease along the direction $\boldsymbol{\mu}$ in the sense that the inequalities

$$V(\mathbf{x}, t) \geq V(\boldsymbol{\mu}(\mathbf{x}, t), t), \tag{8.34}$$

$$W(\mathbf{x}, \boldsymbol{\xi}, t) \geq W(\boldsymbol{\mu}(\mathbf{x}, t), \boldsymbol{\mu}(\boldsymbol{\xi}, t), t) \tag{8.35}$$

hold in the domains of V and W.

The following result is in force.

Theorem 8.1 *Given $\gamma > 0$, suppose that Hypotheses (H1)–(H3) are satisfied for system* (8.25)–(8.28) *in a domain $\mathbf{x} \in B_\delta^{2n}$, $\boldsymbol{\xi} \in B_\delta^{2n}$, $t \in \mathbb{R}$ with functions $V(x, t)$ and $W(x, \xi, t)$. Then, the closed-loop system* (8.25)–(8.28), *driven by the dynamic controller*

$$\begin{aligned} \dot{\boldsymbol{\xi}} = \mathbf{f}(\boldsymbol{\xi}, t) + \mathbf{g}_1(\boldsymbol{\xi}, t)\boldsymbol{\alpha}_1(\boldsymbol{\xi}, t) + \mathbf{g}_2(\boldsymbol{\xi}, t)\boldsymbol{\alpha}_2(\boldsymbol{\xi}, t) + \mathbf{G}(t)(\mathbf{y}(\mathbf{x}, t) - \mathbf{h}_2(\boldsymbol{\xi}, t)) \\ \boldsymbol{\xi}_1(t_i^+) = \boldsymbol{\xi}_1(t_i^-),\ \boldsymbol{\xi}_2(t_i^+) = \boldsymbol{\mu}_0(\boldsymbol{\xi}_1(t_i), \boldsymbol{\xi}_2(t_i^-), t_i) \\ \mathbf{u} = \boldsymbol{\alpha}_2(\boldsymbol{\xi}, t), \end{aligned} \tag{8.36}$$

locally possesses a $\mathscr{L}_2$-gain less than γ. Once Hypothesis (H4) is satisfied as well, the function

$$U(\mathbf{x}, \boldsymbol{\xi}, t) = V(\mathbf{x}, t) + W(\mathbf{x}, \boldsymbol{\xi}, t) \tag{8.37}$$

constitutes a Lyapunov function of the disturbance-free closed-loop system (8.25)–(8.28), (8.36), *the uniform asymptotic stability of which is thus additionally guaranteed.*

Proof Since the proof follows the same line of reasoning, used in the proof of Theorem 5.5, only a sketch is further provided. To begin with, let us differentiate function (8.37) along the disturbed closed-loop system (8.25)–(8.28), (8.36) and estimate its derivative between collision time instants:

$$\begin{aligned} \frac{dU}{dt} < -\|\mathbf{z}(t)\|^2 + \gamma^2\|\mathbf{w}\|^2 - \gamma^2\|\mathbf{w} - \boldsymbol{\alpha}_1(\mathbf{x}, t) - \boldsymbol{\psi}(\mathbf{x}, \boldsymbol{\xi}, t)\|^2, \\ t \in (t_k, t_{k+1}),\ k = 0, 1, \ldots. \end{aligned} \tag{8.38}$$

Then integrating (8.38) from t_k to t_{k+1}, $k = 0, 1, \ldots,$ yields

$$\int_{t_k}^{t_{k+1}} [\gamma^2\|\mathbf{w}\|^2 - \|\mathbf{z}(t)\|^2]\mathrm{d}t > \int_{t_k}^{t_{k+1}} \frac{\mathrm{d}U(\mathbf{x}(t), \boldsymbol{\xi}(t), t)}{\mathrm{d}t}\mathrm{d}t \\ +\gamma^2 \int_{t_k}^{t_{k+1}} \|\mathbf{w}(t) - \boldsymbol{\alpha}_1(\mathbf{x}(t), t) - \boldsymbol{\psi}(\mathbf{x}(t), \boldsymbol{\xi}(t), t)\|^2\mathrm{d}t. \tag{8.39}$$

Taking (8.37) into account and skipping the positive term in the right-hand side of (8.39), it follows that

$$\int_{t_0}^{T} (\gamma^2\|\mathbf{w}\|^2 - \|\mathbf{z}(t)\|^2)\mathrm{d}t > U(\mathbf{x}(T), \boldsymbol{\xi}(T), T) + \sum_{i=1}^{N_T}[V(\mathbf{x}(t_i^-), t_i) - V(\mathbf{x}(t_i^+), t_i)] \\ +\sum_{i=1}^{N_T}[W(\mathbf{x}(t_i^-), \boldsymbol{\xi}(t_i^-), t_i) - W(\mathbf{x}(t_i^+), \boldsymbol{\xi}(t_i^+), t_i)] - U(\mathbf{x}(t_0), \boldsymbol{\xi}(t_0), t_0). \tag{8.40}$$

Since the functions V and W are smooth by Hypotheses (H2) and (H3), the following relations

$$|V(\mathbf{x}(t_i^-), t_i) - V(\mathbf{x}(t_i^+), t_i)| \le L_i^V\|\mathbf{x}(t_i^-) - \mathbf{x}(t_i^+)\| \le L_i^V[\|\mathbf{x}(t_i^-)\| + \|\mathbf{x}(t_i^+)\|] \\ |W(\mathbf{x}(t_i^-), \boldsymbol{\xi}(t_i^-), t_i) - W(\mathbf{x}(t_i^+), \boldsymbol{\xi}(t_i^-), t_i)| \le L_i^W[\|\mathbf{x}(t_i^-) - \mathbf{x}(t_i^+)\| \\ +\|\boldsymbol{\xi}(t_i^-) - \boldsymbol{\xi}(t_i^+)\|] \le L_i^W[\|\mathbf{x}(t_i^-)\| + \|\mathbf{x}(t_i^+)\| + \|\boldsymbol{\xi}(t_i^-)\| + \|\boldsymbol{\xi}(t_i^+)\|] \tag{8.41}$$

hold true with $L_i^V > 0$ and $L_i^W > 0$ being local Lipschitz constants of V and W in the domain $B_\delta^{2n} \in \mathbb{R}^{2n}$. Relations (8.40) and (8.41), coupled together, result in the inequality

$$\int_{t_0}^{T} (\gamma^2\|\mathbf{w}\|^2 - \|\mathbf{z}(t)\|^2)\mathrm{d}t > -\sum_{i=1}^{N_T}[2(L_i^V + L_i^W)\|\mathbf{x}(t_i^-)\| + 2L_i^W\|\boldsymbol{\xi}(t_i^-)\|] \\ -U(\mathbf{x}(t_0), \boldsymbol{\xi}(t_0), t_0), \tag{8.42}$$

thus being verified in the domain $B_\delta^{2n} \in \mathbb{R}^{2n}$. Apart from this, inequality

$$\sum_{i=1}^{N_T} \|\mathbf{z}_\mathbf{i}^\mathbf{d}\|^2 = \sum_{i=1}^{N_T} \|\mathbf{x}_2(t_i^+)\|^2 \le 2\sum_{i=1}^{N_T} \|\boldsymbol{\mu}_0(\mathbf{x}(t_i^-), t_i)\|^2 \\ +2\sum_{i=1}^{N_T} \|\boldsymbol{\omega}(\mathbf{x}(t_i^-), t_i)\mathbf{w}_\mathbf{i}^\mathbf{d}\|^2 \le 2\sum_{i=1}^{N_T} \|\boldsymbol{\mu}_0(\mathbf{x}(t_i^-), t_i)\|^2 + \gamma^2\sum_{i=1}^{N_T} \|\mathbf{w}_\mathbf{i}^\mathbf{d}\|^2 \tag{8.43}$$

is ensured by Hypothesis (H1). Thus, combining (8.42) and (8.43), one derives

$$\int_{t_0}^{T} \|\mathbf{z}(t)\|^2 \mathrm{d}t + \sum_{i=1}^{N_T} \|\mathbf{z}_\mathbf{i}^\mathbf{d}\|^2 < U(\mathbf{x}(t_0), \boldsymbol{\xi}(t_0), t_0)$$
$$+\gamma^2 \left[\int_{t_0}^{T} \|\mathbf{w}(t)\|^2 \mathrm{d}t + \sum_{i=1}^{N_T} \|\mathbf{w}_\mathbf{i}^\mathbf{d}\|^2 \right] + 2 \sum_{i=1}^{N_T} \|\boldsymbol{\mu}_\mathbf{0}(\mathbf{x}(t_i^-), t_i)\|^2 \tag{8.44}$$
$$+ \sum_{i=1}^{N_T} [(2L_i^V + 2L_i^W)\|\mathbf{x}(t_i^-)\| + 2L_i^W \|\boldsymbol{\xi}(t_i^-)\|],$$

i.e., the disturbance attenuation inequality (8.10) is established with the positive definite functions

$$\beta_0(\mathbf{x}(t_0), \boldsymbol{\xi}(t_0), t_0) = U(\mathbf{x}(t_0), \boldsymbol{\xi}(t_0), t_0),$$
$$\beta_i(\mathbf{x}(t_i), \boldsymbol{\xi}(t_i), t_i) = (2L_i^V + 2L_i^W)\|\mathbf{x}(t_i^-)\| + 2L_i^W \|\boldsymbol{\xi}(t_i^-)\| + 2\|\boldsymbol{\mu}_\mathbf{0}(\mathbf{x}(t_i^-), t_i)\|^2,$$
$$i = 1, \ldots, N. \tag{8.45}$$

To complete the proof, it remains to establish the asymptotic stability of the undisturbed version of the closed-loop system (8.25)–(8.28), (8.36). Indeed, if coupled to Hypothesis (H4), the negative definiteness (8.38) of the time derivative of the Lyapunov function (8.37) between the collision time instants ensures that Theorem 4.9 is applicable to the undisturbed version of the closed-loop system (8.25)–(8.28), (8.36). By applying Theorem 4.9, the required asymptotic stability is thus validated. Theorem 8.1 is proved. ■

It should be noted that the influence of the discrete disturbance w_d, acting at the transition phase (8.4), cannot be attenuated by an admissible non-impulsive control input. Due to this, the best disturbance attenuation level γ of Theorem 8.1 is limited by Hypothesis (H1).

8.1.2.2 Local Output-Feedback Synthesis

To circumvent the difficulty of solving the Hamilton–Jacobi–Isaacs PDIs (8.32), (8.33) their solutions are proposed to be approximated by those to the corresponding Riccati equations that appear in solving the $\mathscr{H}_\infty$ control problem for the linearized system which is given by Isidori and Astolfi (1992)

$$\begin{aligned} \dot{\mathbf{x}} &= \mathbf{A}(t)\mathbf{x} + \mathbf{B_1}(t)\mathbf{w} + \mathbf{B_2}(t)\mathbf{u}, \\ \mathbf{z} &= \mathbf{C_1}(t)\mathbf{x} + \mathbf{D_{12}}(t)\mathbf{u}, \\ \mathbf{y} &= \mathbf{C_2}(t)\mathbf{x} + \mathbf{D_{21}}(t)\mathbf{w}, \end{aligned} \tag{8.46}$$

within impact-free time intervals (t_{i-1}, t_i) where t_0 is the initial time instant and $t_i,\ i = 1, 2, \ldots$ are the collision time instants whereas

$$\mathbf{A}(t) = \left.\frac{\partial \mathbf{f}}{\partial \mathbf{x}}\right|_{\mathbf{x}=\mathbf{0}}, \ \mathbf{B}_1(t) = \mathbf{g}_1(0,t), \mathbf{B}_2(t) = \mathbf{g}_2(0,t), \ \mathbf{C}_1(t) = \left.\frac{\partial \mathbf{h}_1}{\partial \mathbf{x}}\right|_{\mathbf{x}=\mathbf{0}}$$
$$\mathbf{C}_2(t) = \left.\frac{\partial \mathbf{h}_2}{\partial \mathbf{x}}\right|_{\mathbf{x}=\mathbf{0}}, \ \mathbf{D}_{12}(t) = \mathbf{k}_{12}(0,t), \ \mathbf{D}_{21}(t) = \mathbf{k}_{21}(0,t) \tag{8.47}$$

By the time-varying strict bounded real lemma (Orlov and Aguilar 2014, p. 46), the following conditions are necessary and sufficient for the linear $\mathscr{H}_\infty$ control problem (8.46) to possess a solution: given $\gamma > 0$,

(C1) There exists a positive constant ε_0 such that the differential Riccati equation

$$\begin{aligned} -\dot{\mathbf{P}}_\varepsilon(t) &= \mathbf{P}_\varepsilon(t)\mathbf{A}(t) + \mathbf{A}^T(t)\mathbf{P}_\varepsilon(t) + \mathbf{C}_1{}^T(t)\mathbf{C}_1(t) \\ &+\mathbf{P}_\varepsilon(t)\left[\frac{1}{\gamma^2}\mathbf{B}_1\mathbf{B}_1{}^T - \mathbf{B}_2\mathbf{B}_2{}^T\right](t)\mathbf{P}_\varepsilon(t) + \varepsilon\mathbf{I} \end{aligned} \tag{8.48}$$

has a uniformly bounded symmetric positive definite solution $\mathbf{P}_\varepsilon(t)$ for each $\varepsilon \in (0, \varepsilon_0)$;

(C2) While being coupled to (8.48), the differential Riccati equation

$$\begin{aligned} \dot{\mathbf{Z}}_\varepsilon(t) &= \mathbf{A}_\varepsilon(t)\mathbf{Z}_\varepsilon(t) + \mathbf{Z}_\varepsilon(t)\mathbf{A}_\varepsilon^{\mathbf{T}}(t) + \mathbf{B}_1(t)\mathbf{B}_1{}^T(t) \\ &+\mathbf{Z}_\varepsilon(t)\left[\frac{1}{\gamma^2}\mathbf{P}_\varepsilon\mathbf{B}_2\mathbf{B}_2{}^T\mathbf{P}_\varepsilon - \mathbf{C}_2{}^T\mathbf{C}_2\right](t)\mathbf{Z}_\varepsilon(t) + \varepsilon\mathbf{I} \end{aligned} \tag{8.49}$$

has a uniformly bounded symmetric positive definite solution $\mathbf{Z}_\varepsilon(t)$ with $\mathbf{A}_\varepsilon(t) = \mathbf{A}(t) + \frac{1}{\gamma^2}\mathbf{B}_1(t)\mathbf{B}_1{}^T(t)\mathbf{P}_\varepsilon(t)$.

These conditions, if coupled to a certain monotonicity condition, prove to be sufficient as well for a local solution to exist for the nonlinear $\mathscr{H}_\infty$ control problem under unilateral constraints.

Theorem 8.2 *Let conditions (C1) and (C2) be satisfied with some* $\gamma > 0$. *Then Hypotheses (H2) and (H3) hold locally around the equilibrium* $(\mathbf{x}, \boldsymbol{\xi}) = (0, 0)$ *of the nonlinear system* (8.25)–(8.28) *with*

$$V(\mathbf{x}, t) = \mathbf{x}^T\mathbf{P}_\varepsilon(t)\mathbf{x}, \tag{8.50}$$
$$W(\mathbf{x}, \boldsymbol{\xi}, t) = \gamma^2(\mathbf{x} - \boldsymbol{\xi})^T\mathbf{Z}_\varepsilon^{-1}(t)(\mathbf{x} - \boldsymbol{\xi}), \tag{8.51}$$
$$G(t) = \mathbf{Z}_\varepsilon(t)\mathbf{C}_2{}^T(t), \tag{8.52}$$

and the closed-loop system driven by the output feedback

$$\begin{aligned} \dot{\boldsymbol{\xi}} &= \mathbf{f}(\boldsymbol{\xi}, t) + \mathbf{G}(t)[\mathbf{y} - \mathbf{h}_2(\boldsymbol{\xi}, t)] + \left[\tfrac{1}{\gamma^2}\mathbf{g}_1(\boldsymbol{\xi}, t)\mathbf{g}_1{}^T(\boldsymbol{\xi}, t) - \mathbf{g}_2(\boldsymbol{\xi}, t)\mathbf{g}_2{}^T(\boldsymbol{\xi}, t)\right]\mathbf{P}_\varepsilon(t)\boldsymbol{\xi} \\ \boldsymbol{\xi}_1(t_i^+) &= \boldsymbol{\xi}_1(t_i^-), \ \boldsymbol{\xi}_2(t_i^+) = \boldsymbol{\mu}_0(\boldsymbol{\xi}_1(t_i), \boldsymbol{\xi}_2(t_i^-), t_i) \\ \mathbf{u} &= -\mathbf{g}_2(\boldsymbol{\xi}, t)^T\mathbf{P}_\varepsilon(t)\boldsymbol{\xi} \end{aligned} \tag{8.53}$$

locally possesses a $\mathscr{L}_2$-gain less than γ provided that Hypothesis (H1) holds as well. Moreover, the disturbance-free closed-loop system (8.25)–(8.28), (8.53) *is uniformly asymptotically stable provided that Hypothesis (H4) is satisfied with the quadratic functions* (8.50) *and* (8.51).

Proof Due to Orlov and Aguilar (2014, Theorem 24), Hypotheses (H2) and (H3) locally hold with (8.50)–(8.52). Then by applying Theorem 8.1, the validity of Theorem 8.2 is concluded. ■

8.1.2.3 Local State Feedback Synthesis

In the full information case, where the perfect state measurement is available, Theorem 8.2 is simplified to the static feedback design

$$\mathbf{u} = -\mathbf{g}_2(\mathbf{x}, t)^T \mathbf{P}_\varepsilon(t)\mathbf{x} \tag{8.54}$$

and it is summarized as follows.

Theorem 8.3 *Let Hypothesis (H1) and Condition (C1) be satisfied with some $\gamma > 0$. Then the closed-loop system* (8.25)–(8.28), *driven by the state feedback* (8.54), *locally possesses $\mathscr{L}_2$-gain less than γ. Moreover, the disturbance-free closed-loop system* (8.25)–(8.28), (8.54) *is uniformly asymptotically stable provided that the function $V(\mathbf{x}, t) = \mathbf{x}^T \mathbf{P}_\varepsilon(t)\mathbf{x}$ locally satisfies inequality* (8.34).

Proof Due to Orlov and Aguilar (2014, Theorem 23), Hypothesis (H2) locally holds with $V(\mathbf{x}, t)$, given by (8.50). Thus, the closed-loop system (8.25)–(8.28), (8.54) fails under the conditions of Theorem 5.5. By applying Theorem 5.5, the validity of Theorem 8.3 is then concluded. ■

8.1.2.4 On Synthesis of Autonomous and Periodic Systems

For autonomous systems, all functions (8.24) and (8.46) are time-independent, and the differential Riccati equations (8.48) and (8.49) degenerate to the algebraic Riccati equations with $\dot{\mathbf{P}}_\varepsilon(t) = 0$ and $\dot{\mathbf{Z}}_\varepsilon(t) = 0$. For periodic systems, all functions (8.24) and (8.46) are time-periodic, and Theorem 8.2 admits a time-periodic synthesis (8.53) which is based on appropriate periodic solutions $\mathbf{P}_\varepsilon(t)$ and $\mathbf{Z}_\varepsilon(t)$ to the periodic differential Riccati equations (8.48) and (8.49).

8.1.3 $\mathscr{H}_\infty$-Control of Mass–Spring–Barrier System

The objective of this section is to illustrate the effectiveness of the developed synthesis with a simple example that captures essential features of the general treatment under unilateral constraints.

A simple testbed of interest is depicted in Fig. 8.1 where m represents the mass, k the spring constant, b the damping constant, τ is the applied control force, and q represents the position. For the free-motion dynamics ($q > 0$), the plant equation reads

$$\ddot{q} = -\frac{k}{m}q - \frac{b}{m}\dot{q} + \frac{1}{m}\tau + \frac{1}{m}w_1 \tag{8.55}$$

whereas for the transition phase ($q = 0$), the restitution rule is given by

$$q^+ = q^-, \quad \dot{q}^+ = -e\dot{q}^- + w_i^d, \quad e \in [0, 1]. \tag{8.56}$$

For brevity, the notation q^+ (q^-) stands for the postimpact (pre-impact) values $q(t_i^+)$ ($q(t_i^-)$) at impact instants t_1, $i = 1, 2, \ldots$. The variables w_1 and w_i^d are introduced to account for non-modeled external forces and model inadequacies such as friction and restitution uncertainties.

In order to address position feedback tracking of a reference trajectory $q^r(t)$, the state error variables

$$x_1 = q - q^r, \; x_2 = \dot{q} - \dot{q}^r \tag{8.57}$$

and the position measurement

$$y = x_1 + w_0 \tag{8.58}$$

are involved where w_0 stands for the measurement noise. Inspired from (8.22), the pre-feedback control law

$$\tau = m\ddot{q}^r + b\dot{q}^r + kq^r + u \tag{8.59}$$

is composed of a controller u to be designed and the rest being a trajectory compensator. Then, setting $\mathbf{x} = (x_1, x_2)^T$, $\mathbf{w} = (w_0, w_1)^T$, and rewriting the system (8.55)–(8.59) in terms of the tracking error variables, one derives:

Free-motion phase error dynamics

$$\dot{\mathbf{x}} = \underbrace{\begin{bmatrix} 0 & 1 \\ -\frac{k}{m} & -\frac{b}{m} \end{bmatrix}}_{\mathbf{A}} \mathbf{x} + \underbrace{\begin{bmatrix} 0 & 0 \\ 0 & \frac{1}{m} \end{bmatrix}}_{\mathbf{B}_1} \mathbf{w} + \underbrace{\begin{bmatrix} 0 \\ \frac{1}{m} \end{bmatrix}}_{\mathbf{B}_2} u \tag{8.60}$$

within the constraint $F_0(\mathbf{x}, t) = x_1 + q^r(t) > 0$ and transition phase error system

$$\mathbf{x}^+ = \begin{bmatrix} x_1^- \\ \mu_0(\mathbf{x}, t) \end{bmatrix} + \begin{bmatrix} 0 \\ 1 \end{bmatrix} w_i^d \tag{8.61}$$

on the constraint surface $F_0(\mathbf{x}, t) = x_1 + q^r(t) = 0$ where

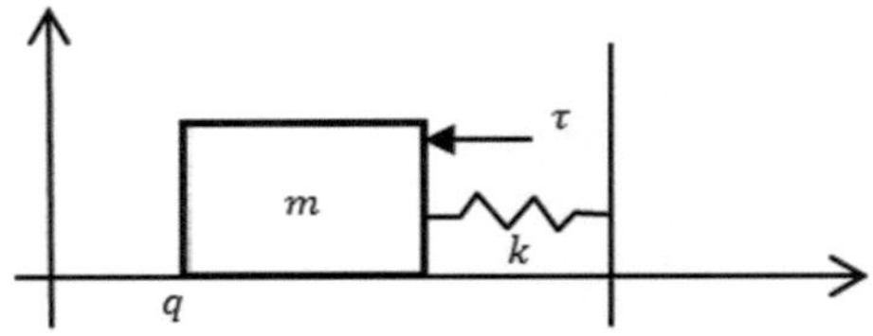

Fig. 8.1 Mass–spring–barrier system

$$\mu_0(\mathbf{x}, t) = \begin{cases} x_2 + (1+e)\dot{q}^r & \text{if } F_0(q^r(t)) = 0, \ \ F_0(x_1 + q^r) \neq 0 \\ -e(x_2 + \dot{q}^r) - \dot{q}^r & \text{if } F_0(q^r(t)) \neq 0, \ \ F_0(x_1 + q^r) = 0 \\ -ex_2 & \text{if } F_0(q^r(t)) = 0, \ \ F_0(x_1 + q^r) = 0 \end{cases} \tag{8.62}$$

is obtained by specifying (8.18)–(8.21) to the present case.

In terms of the tracking errors, the variables to be controlled are specified in the form

$$\mathbf{z} = \underbrace{\begin{bmatrix} 0 & 0 \\ \rho_p & 0 \\ 0 & \rho_v \end{bmatrix}}_{\mathbf{C}_1} \mathbf{x} + \underbrace{\begin{bmatrix} 1 \\ 0 \\ 0 \end{bmatrix}}_{\mathbf{D}_{12}} u \tag{8.63}$$

$$z_i^d = -ex_2^- + w_i^d, \tag{8.64}$$

complying with (8.30).

8.1.3.1 State Feedback Regulation

To begin with, the tracking of the mass–spring–barrier system is treated in a particular case under the perfect knowledge of the state vector $\mathbf{q} = (q, \dot{q})^T$ with the trivial reference trajectory degenerated to the origin $q^r = 0, \dot{q}^r = 0, \ddot{q}^r = 0$ while the pre-feedback (8.59) is simplified to the form $\tau = u$ with no trajectory compensation. Just in the case, the robust regulation to the origin is synthesized according to Theorem 8.3, the applicability of which to the MSDB error system (8.60)–(8.64) is verified as follows.

To verify Condition (C1), the generic time-invariant terms $\mathbf{A}, \ \mathbf{B}_1, \ \mathbf{B}_2, \ \mathbf{C}_1$ in the Riccati equation (8.48) are specified from (8.60) to (8.64). A constant positive semi-definite solution of the corresponding time-invariant Eq. (8.48) subject to $\varepsilon = 0$ is then obtained by iterating on γ in order to approach the infimal achievable level $\gamma_{min} \approx 1.01$. The value $\gamma = 2$ is subsequently selected to avoid an undesired high-gain controller design that would appear for a value of γ close to the infimal $\gamma_{min} \approx 1.01$. With $\gamma = 2$, the value $\varepsilon = 0.01$ is obtained so that the corresponding perturbed Riccati equation (8.48) possesses a constant positive definite solution, given by

$$\mathbf{P}_\varepsilon = \begin{bmatrix} 4.9542 & 0.0504 \\ 0.0504 & 4.9542 \end{bmatrix}. \tag{8.65}$$

Inequality (8.31) of Hypothesis (H1) is straightforwardly verified for $\omega = 1$ and for the corresponding value $\gamma = 2$.

Finally, it remains to verify the last condition (8.34) of Theorem 8.3. For this purpose, it suffices to note that only Scenario (T3) is in force for the state feedback regulation, and therefore in the disturbance-free case, $x_2^+ = -ex_2^-$ by virtue of (8.62), and hence $\|\mathbf{x}^+\| \leq \|\mathbf{x}^-\|$ for an admissible restitution parameter $e \in [0, 1]$. Thus, Theorem 8.3, constituting the robust state feedback synthesis under unilateral constraints, becomes applicable to the stabilization of the testbed system around the origin, and function $V(\mathbf{x}, t) = \mathbf{x}^{\mathbf{T}}\mathbf{P}_{\varepsilon}\mathbf{x}$, specified with (8.65), is a Lyapunov function for the undisturbed system.

The performance of the closed-loop system, driven by the controller, designed according to Theorem 8.3, is numerically illustrated in the sequel. The parameters, used in the simulation, are presented in Table 8.1.

Figure 8.2a depicts the disturbance-free regulation errors, escaping to zero. From this figure, one concludes that the testbed system is actually regulated to the barrier. The monotonically decreasing evolution between and across the impacts is presented in Fig. 8.2b for the quadratic Lyapunov function (8.50), specified with the Riccati matrix (8.65). Figure 8.3a shows that while the disturbing friction force w_1 and deviation w_i^d in the restitution coefficient are added to the underlying system (see Table 8.1 for their numerical values), these disturbances are actually attenuated by the controller designed.

Table 8.1 Simulation data

Param	Value
k	10 N/m
b	1 N/m/s
m	1 kg
e	0.5
ρ_p	1
$\dot{q}(0)$	−0.2 m/s
$\xi_1(0)$	0.1 m
ρ_v	1
ε	0.01
w_i^d	$0.2q_2$ m/s
w_1	$0.1q_2 + 0.1\text{sign}(q_2)$ N
$q(0)$	0.5 m
w_2	$0.1\sin(1.5t)$ m
$\xi_2(0)$	0.2 m/s

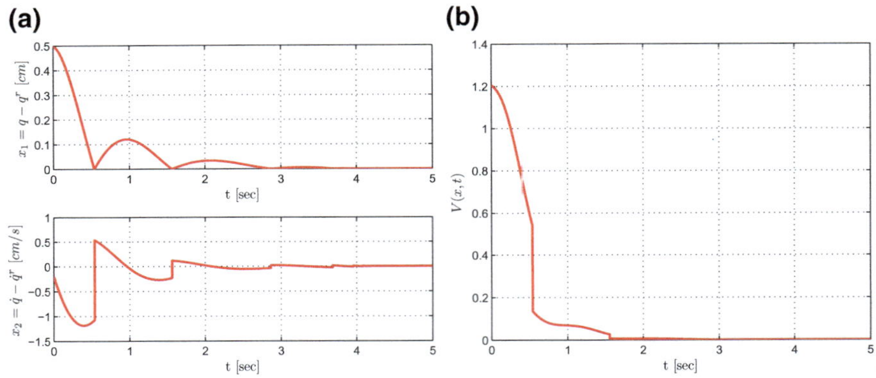

Fig. 8.2 Disturbance-free regulation: **a** position and velocity errors, **b** Lyapunov function evolution

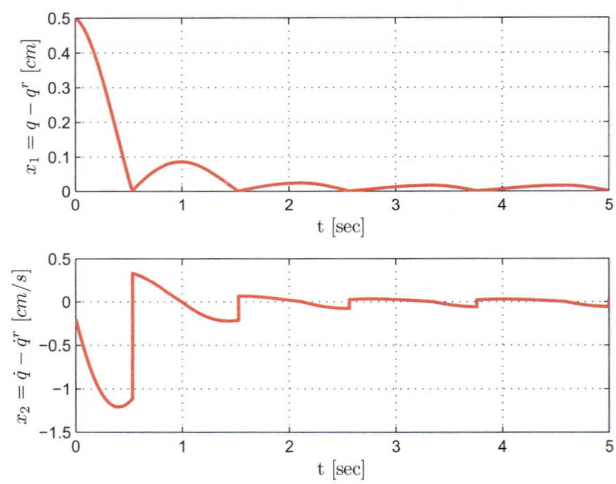

Fig. 8.3 Disturbance-corrupted regulation: position and velocity errors

8.1.3.2 Position Feedback Tracking of a Constrained Van der Pol Oscillator

In the remainder, the $\mathscr{H}_\infty$-orbitally stabilizing output feedback synthesis is developed using a hybrid version of the Van der Pol oscillator, generating a stable limit cycle to follow. Such a constrained Van der Pol oscillator is studied in Sect. 2.4.1, and it is described as follows:

Free-motion phase ($q^r > 0$)

$$\ddot{q}^r + \varepsilon \left[\left(q^{r2} + \frac{\dot{q}^{r2}}{\mu^2} \right) - \rho^2 \right] \dot{q} + \mu^2 q^r = 0. \tag{8.66}$$

Transition phase ($q^r = 0$)

$$q^r(t_i^+) = q^r(t_i^-), \quad \dot{q}^r(t_i^+) = -e\dot{q}^r(t_i^-), \tag{8.67}$$

where q^r represents the desired position and $\dot{q}^r$ the velocity, $t_i,\ i = 1, 2, \ldots$ are impact instants when the nonlinear oscillator hits the constraint $q^r = 0$. Recall that the oscillator parameters $\varepsilon, \mu, \rho > 0$ are respectively for the transient speed, frequency, and amplitude of the limit cycle, and $e \in (0, 1)$ is the restitution parameter.

As shown in Sect. 2.4.2, the constrained Van der Pol oscillator possesses a stable limit cycle, the transient speed, frequency, and amplitude which are determined by the oscillator parameters $\varepsilon, \mu, \rho > 0$, and dependent on the value of the restitution parameter $e \in (0, 1)$, the limit cycle is capable of degenerating to the asymptotically stable equilibrium, located in the origin. Due to these properties, the constrained Van der Pol oscillator is extremely suited for its use in the model reference robust control where the desired magnitude and frequency of the resulting oscillation are capable of online adaptation, thereby yielding the closed-loop system, producing its own limit cycle (particularly, being asymptotically stable) at the designer will.

In what follows, the oscillator parameters are set to $\varepsilon = \mu = \rho = 1$ and $e = 0.5$ under which the constrained oscillator generates a stable limit cycle due to the numerical analysis of Sect. 2.4.3. The position feedback synthesis is based on Theorem 8.2, which is now applied to the error dynamics (8.58), (8.60)–(8.64), driven by (8.59), to ensure robust tracking of the desired trajectory, governed by (8.66)–(8.67). By substituting the right-hand side of (8.66) into (8.59) for $\ddot{q}^r$, the pre-feedback controller (8.59), fed by the output of the impact Van der Pol reference model (8.66)–(8.67), is represented in the form

$$\tau = -m[(1 - q^{r2})\dot{q}^r - q^r] + b\dot{q}^r + kq^r + u. \tag{8.68}$$

The error restitution rule is actually given by (8.62).

The applicability of Theorem 8.2 to the present case is verified as follows. Similar to the regulation case, Conditions (C1) and (C2) are verified with the linearizing terms $\mathbf{A},\ \mathbf{B}_1,\ \mathbf{B}_2,\ \mathbf{C}_1,\ \mathbf{C}_2$, which are required to specify the Riccati equations (8.48)–(8.49) and which are identified from (8.58) to (8.64). A constant positive semidefinite solution of the corresponding (now time-invariant) system (8.48)–(8.49) subject to $\varepsilon = 0$ is then derived for sufficiently large γ and by iterating on γ the infimal achievable level $\gamma_{min} \approx 1.01$ is approached. The value $\gamma = 2$ was however selected to avoid an undesirable high-gain controller design which would appear for a value of γ close to the optimum $\gamma_{min} \approx 1.01$. With $\gamma = 2$, the corresponding perturbed Riccati equations (8.48)–(8.49) are carried out to possess constant positive definite solutions

$$\mathbf{P}_\varepsilon = \begin{bmatrix} 4.9542 & 0.0504 \\ 0.0504 & 4.9542 \end{bmatrix}, \quad \mathbf{Z}_\varepsilon = \begin{bmatrix} 0.0715 & -0.0024 \\ -0.0024 & 0.7107 \end{bmatrix}, \tag{8.69}$$

under the value $\varepsilon = 0.01$ which is obtained by iterating on ε.

After that, the value $\gamma = 2$ is straightforwardly verified to meet Hypothesis (H1) with $\omega = 1$, corresponding to the present investigation. Thus, Theorem 8.2 ensures that the underlying closed-loop system possesses $\mathscr{L}_2$-gain less than $\gamma = 2$.

Since the impact instants of the reference trajectory are not in general synchronized with the plant impact instants (unless the reference initial state coincides with that of the plant), whichever Scenario (T1)–(T3) may occur according to the adopted state error restitution rule (8.62). Therefore, Hypothesis (H4) is ruled out by the resulting synthesis which proves to be incapable to asymptotically stabilize the closed-loop system even in the disturbance-free case as is well-known from Biemond et al. (2013). Nevertheless, the proposed controller does attenuate external/restitution disturbances and measurement noise as established by Theorem 8.2 before, and while being numerically tested, the performance of the closed-loop system is observed to be acceptable.

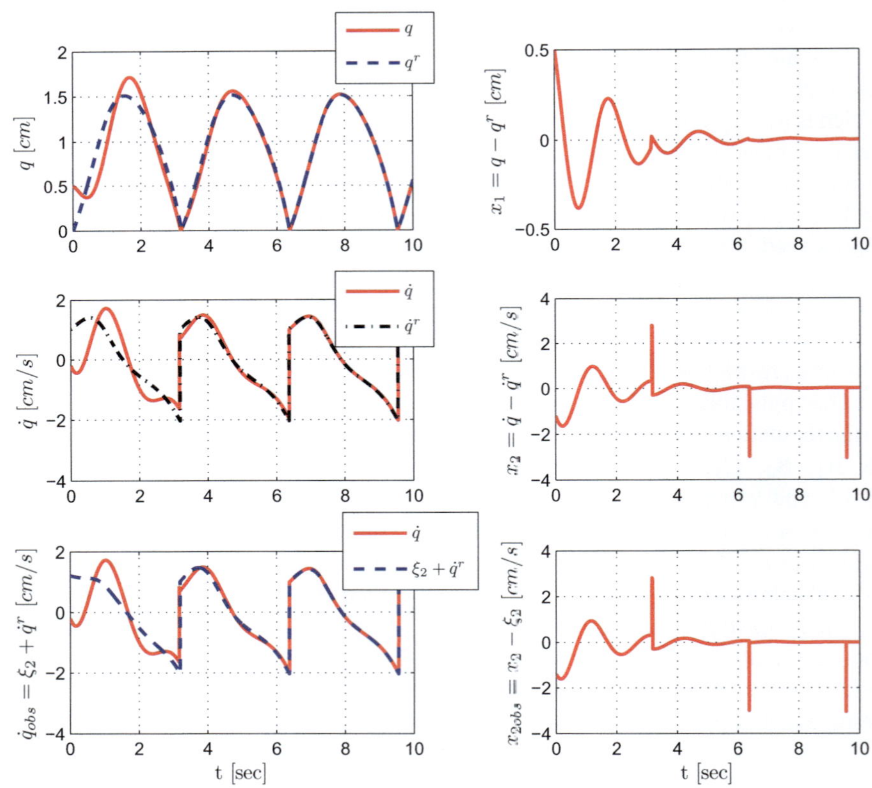

Fig. 8.4 Plots of the position and velocity tracking errors, and of velocity estimation error in the disturbance-free case

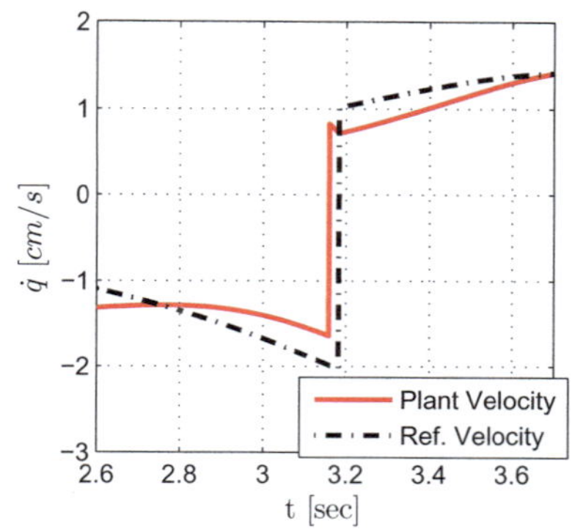

Fig. 8.5 Desynchronization of the reference trajectory with the plant trajectory in the disturbance-free case

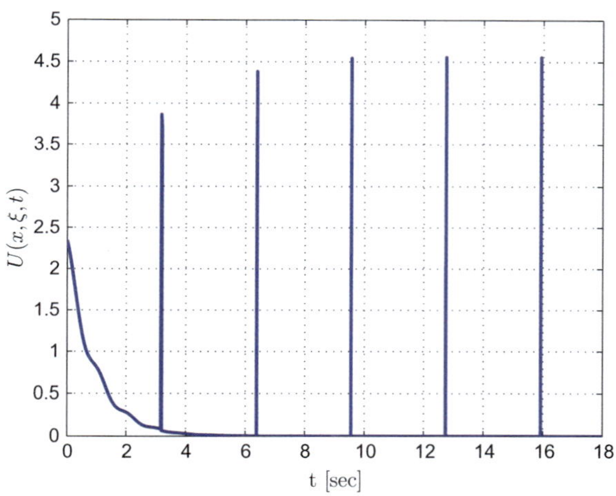

Fig. 8.6 Desynchronized tracking: evolution of the Lyapunov function in the disturbance-free case

The simulation results, shown in Figs. 8.4, 8.5, 8.6, 8.7 and 8.8, were performed under the same circumstances of Sect. 5.2.1, using the parameters from Table 8.1. The disturbance-free case is presented in Figs. 8.4, 8.5 and 8.6. These figures exhibit peaking phenomena since the plant velocity jumps do not match the reference velocity jumps (as clearly observed in Fig. 8.5), thus falling into either Scenario T1 or T2 of Sect. 8.1.1. The Lyapunov candidate function (8.37), specified with (8.50), (8.51), and (8.69), is thus monotonically decreasing just between impacts while exhibiting undesired increments at the impact-time instants (see Fig. 8.6), and the asymptotic stability proof is no longer applicable to the disturbance-free case under both Scenarios T1 and T2. Despite the discrepancy in the impact instants of the plant velocity and of the reference velocity, the $\mathscr{L}_2$-gain inequality (8.10) is still guaranteed by Theorem 8.2, and good behavior of the closed-loop system with the tracking errors, approaching zero between the impact instants, is concluded from Fig. 8.4 in the disturbance-free case. From Figs. 8.7 and 8.8, good performance is also concluded for the periodic tracking synthesis despite the added disturbances, affecting the

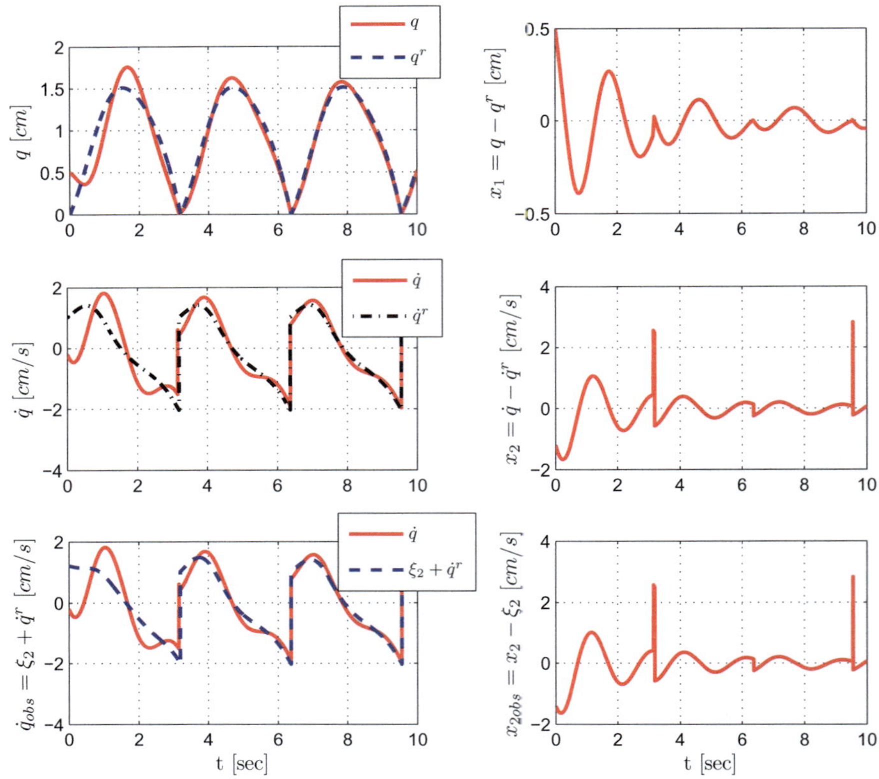

Fig. 8.7 Plots of the position and velocity tracking errors, and of the velocity estimation error for the disturbed case

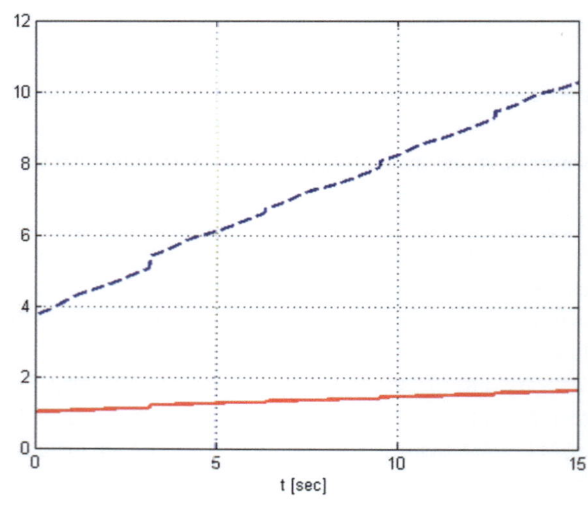

Fig. 8.8 $\mathscr{L}_2$-gain behavior without online reference model reset: $\|z\|^2_{L_2} + \|z^d\|^2_{l_2}$ (solid line) versus $\gamma^2[\|w\|^2_{L_2} + \|w^d_i\|^2_{l_2}] + \Sigma^N_{k=0}\beta_k$ (dashed line) with $\gamma = 2$

free-motion (due to friction) and transition phases (due to uncertainty in the restitution coefficient).

Finally, the estimated velocity $\dot{q}_{obs} = \xi_2 + \dot{q}^r$ and the observation error $x_{2obs} := x_2 - \xi_2$ are compared in Figs. 8.4 and 8.7 for the disturbed and undisturbed cases, respectively. One can observe that if disturbances are not applied, the filter adequately tracks the system velocity between the impact times (Fig. 8.4), whereas a reasonably small observation error persists in the disturbed case (Fig. 8.7) such that good tracking performance is achieved.

8.1.3.3 Impact Synchronization via Online Reference Model Reset

In order to suppress the peaking phenomena, depicted in Fig. 8.4 and destroying the asymptotic stability of the disturbance-free closed-loop system, the reference model is now reset online. The idea behind such a reset is in using the same hybrid Van der Pol reference model of the previous section, but instead of using its own unilateral constraint $q^r = 0$, the reset event is synchronized with the impact of the plant ($q = 0$). Thus, the restitution law (8.67) is modified to

$$q^r(t_i^+) = 0, \quad \dot{q}^r(t_i^+) = -e\dot{q}^r(t_i^-), \quad \text{iff}\, q(t_i) = 0. \tag{8.70}$$

The pre-feedback controller (8.68) and the same controller u, synthesized in Sect. 8.1.3.2, are now coupled to the Van der Pol reference model, thus modified. Hypotheses (H1)–(H3) are still in force, and it remains to show that (H4) is additionally satisfied in the present case. Since the reference trajectory is reset when the plant hits the constraint, Scenario T3 is now in order, and due to (8.62), the error transition phase is governed by $x_2^+ = -ex_2^-$. It follows that $\|x^+\| \leq \|x^-\|$ and (H4) is thus established with V and W, specified in (8.50) and (8.51), respectively. This verifies the applicability of Theorem 8.2, by virtue of which, the properly specified dynamical controller (8.53) enforces the disturbance-free mass-spring-barrier system to asymptotically track the reference trajectory while also attenuating external disturbances.

In order to demonstrate that the closed-loop system (8.55), (8.56), (8.66), (8.70), (8.53) generates an asymptotically stable limit cycle, the Poincaré analysis of Sect. 2.4.4 is revisited, using the Poincaré map

$$\tilde{G}(\zeta_k) = \zeta_{k+1} \tag{8.71}$$

associated with the Poincaré section $q = 0$, while considering the postimpact values $\zeta_k = [q_k, \dot{q}_k, \xi_k, q_k^r, \dot{q}_k^r]$ at the impact instants t_k, $k = 1, 2, \ldots$. The fixed point $\zeta^* = [0, 1.012, 0, 0, 0, 1.012]$ of the Poincaré map $\tilde{G}$ and the eigenvalues

$$eig(\nabla\tilde{G}) = [-0.1161, 0.1135, 0.0581 + 0.0204i, 0.0581 - 0.0204i, 0, 0] \tag{8.72}$$

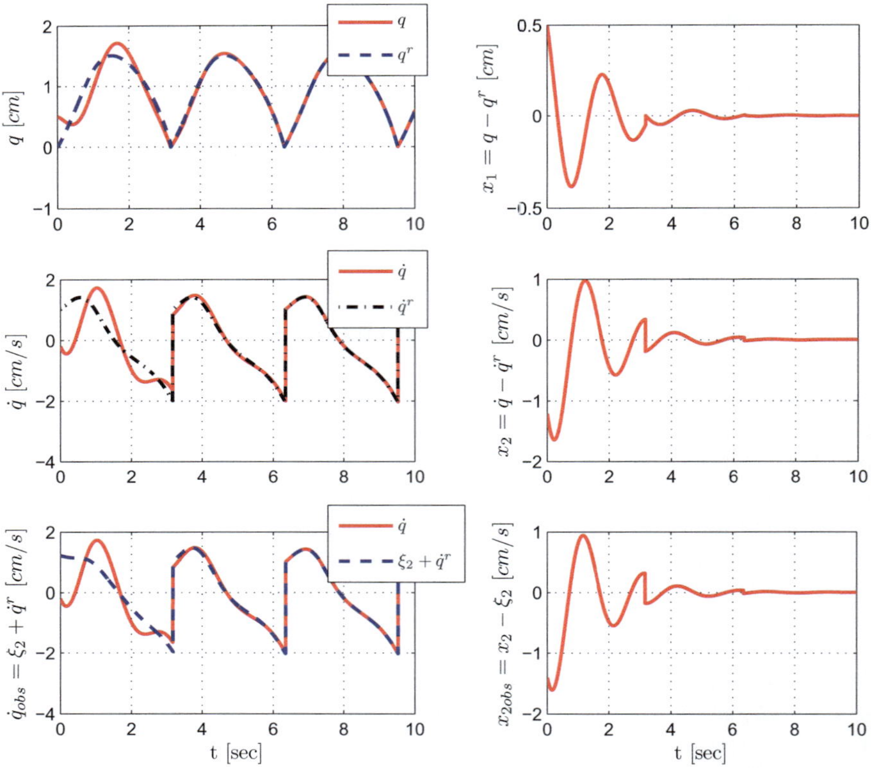

Fig. 8.9 Plots of the position and velocity tracking errors, and that of the velocity estimation error in the disturbance-free case when the online reset adaptation of the Van der Pol reference model is enforced

of the gradient $\nabla\tilde{G}$ around the fixed point are numerically computed. The asymptotic stability of a limit cycle, matching to the fixed point of the Poincaré map $\tilde{G}$, is then established by observing that eigenvalues (8.72) of the gradient $\nabla\tilde{G}$ are inside of the unit circle.

Figures 8.9, 8.10, 8.11, 8.12, and 8.13 demonstrate the numerical results performed under the same circumstances as in the previous section while the synthesized tracking controller is coupled to the Van der Pol reference model, whose online reset adaptation is synchronized with the plant impacts. It can be seen from Fig. 8.9 that in the disturbance-free case, the position, velocity and estimation errors escape to zero regardless of nonzero initial conditions. The asymptotic stability of the closed-loop system can additionally be observed from Fig. 8.10 where the plotted Lyapunov function (8.37), specified with (8.50), (8.51), and (8.69), monotonically escapes to zero. The asymptotic stability of the limit cycle, theoretically predicted by the Poincaré analysis, is illustrated in Fig. 8.11, where the plant trajectory (dashed line) converges to a periodic orbit (solid line).

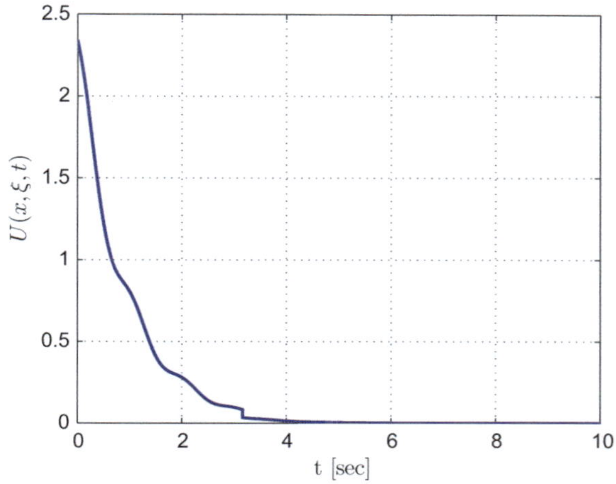

Fig. 8.10 Lyapunov function evolution in the disturbance-free case of the synchronized tracking

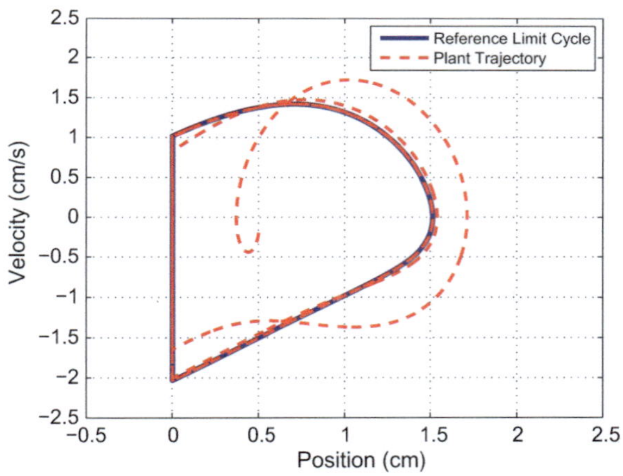

Fig. 8.11 Limit cycle of the constrained Van der Pol oscillator and closed-loop phase trajectory, approaching it: the disturbance-free case subject to synchronization

The simulations, performed in the disturbed case, are reflected in Fig. 8.12 that depicts the plots of the position and velocity tracking errors as well as the plot of the velocity estimation error. It is seen that after the transitory, the errors remain small and bounded. As seen in Fig. 8.13, this ensures that the plant trajectory evolves around the periodic orbit.

It is worth noticing that in both disturbed and undisturbed cases, the peaking effects of Fig. 8.4, matching to the desynchronized impact instants of the plant and of the reference model, disappear from the velocity tracking and velocity estimation errors of Figs. 8.9 and 8.12 where the reference model resets are synchronized with the plant impact instants. Thus, the superiority of the synthesis with the online reset adaptation is concluded.

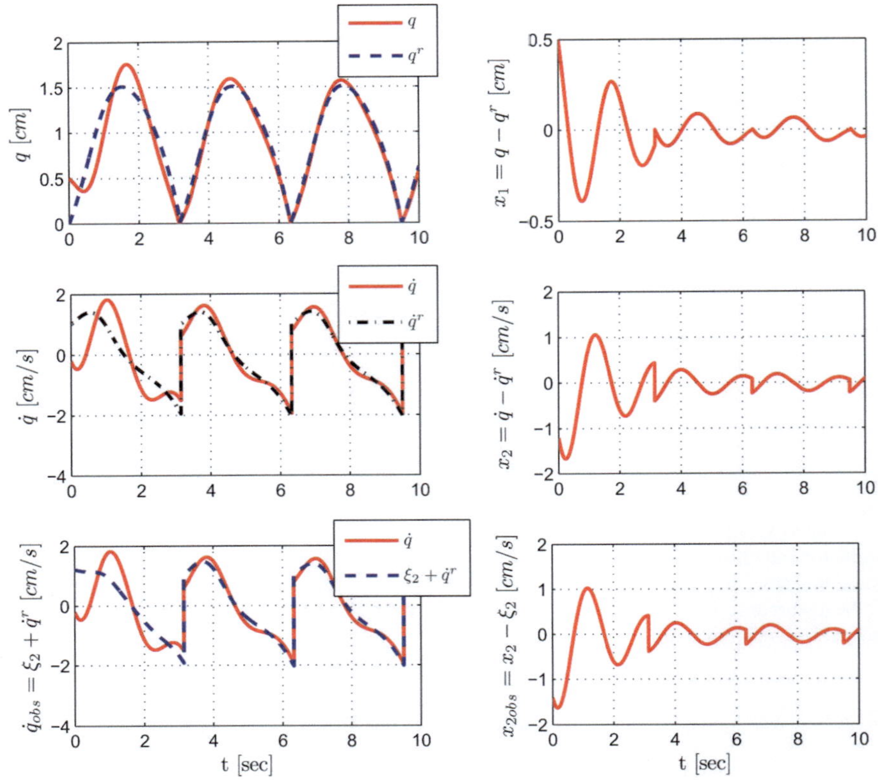

Fig. 8.12 Plots of the position and velocity tracking errors, and of the velocity estimation error in the synchronized disturbance-free case

Fig. 8.13 Limit cycle of the constrained Van der Pol oscillator and closed-loop phase trajectory, evolving around it in the presence of disturbances and synchronization

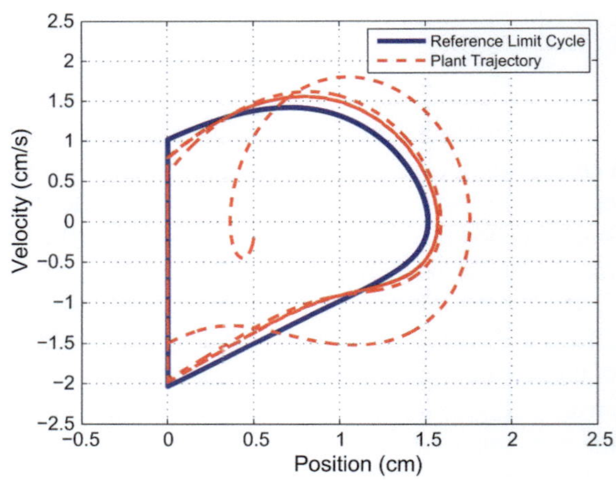

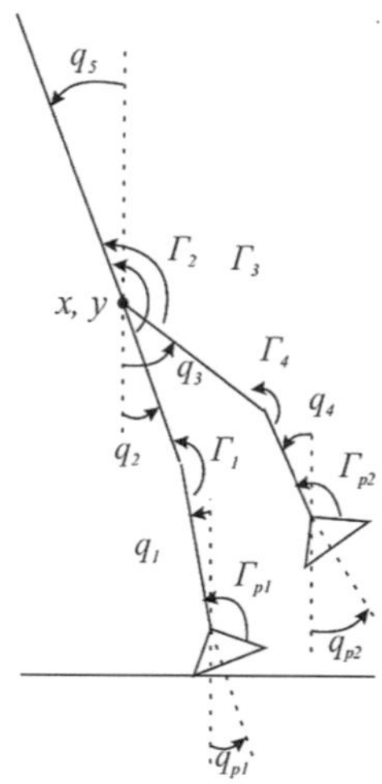

Fig. 8.14 Seven-link bipedal robot

8.1.4 $\mathscr{H}_\infty$ Tracking of a Periodical Bipedal Gait

Robust synthesis procedure, tested in the previous section on a constrained 1-DOF mechanical system, is additionally supported by the numerical study of Montano et al. (2014), made for the robust trajectory tracking of a multi-link bipedal robot. The bipedal robot considered in this section is walking on a rigid and horizontal surface. It is modeled as a planar biped of 7-DOF, which consists of a torso, hips, two legs with knees, and feet (see Fig. 8.14). The walking gait is composed of single support phases and impacts. The complete model of the biped robot consists of two parts, namely, the differential equations, describing the dynamics of the robot during the swing phase, and an impulse model of the contact event (the impact between the swing leg and the ground is modeled as a contact between two rigid bodies as in Chevallereau et al. 2003).

8.1.4.1 Dynamic Model in Single Support

The dynamic model of the biped is given by Montano et al. (2014)

$$\mathbf{D}(\mathbf{q})\ddot{\mathbf{q}} + \mathbf{H}(\mathbf{q}, \dot{\mathbf{q}}) = \mathbf{D}_\mathrm{G}\mathbf{G} + \mathbf{J}^T \left[\mathbf{R}_1\ \mathbf{R}_2\right]^T + \mathbf{w}_1 \tag{8.73}$$

with $\mathbf{J} = (\mathbf{J}_1, \mathbf{J}_2)^T$, and the constraint equations

$$\mathbf{J}_\mathrm{i}\ddot{\mathbf{q}} + \dot{\mathbf{J}}_\mathrm{i}\dot{\mathbf{q}} = \mathbf{0}, \text{ for } \mathrm{i} = 1, 2 \tag{8.74}$$

with the 9×1-vector $\mathbf{q} = (q_1, q_2, q_3, q_4, q_5, q_{p_1}, q_{p_2}, x_H, y_H)^T$ of generalized coordinates, with the symmetric, positive definite 9×9 inertia matrix $\mathbf{D}$, with

9×9 constant matrix $\mathbf{D_G}$, composed of 0 and 1, and with the 6×1 vector $\mathbf{G} = (G_1, G_2, G_3, G_4, G_{p_1}, G_{p_2})^T$ of joint torques (see Fig. 8.14). The term $\mathbf{H}(\mathbf{q}, \dot{\mathbf{q}})$ is the 9×1 vector of the centrifugal, Coriolis, and gravity forces; $\mathbf{R_1}$ and $\mathbf{R_2}$ represent the reaction forces on foot 1 and foot 2, respectively, whereas $\mathbf{J_1}$ and $\mathbf{J_2}$ are 3×9 Jacobian matrices converting these efforts to the corresponding joint torques.

In the single support phase, considering a flat foot contact of the stance foot with the ground (i.e., there is no take off, no rotation, and no sliding during this phase), there exists the orthogonal matrix $\mathbf{J}^{\perp}$ (6×9), such that left multiplying it by (8.73) yields

$$\mathbf{J}^{\perp}\mathbf{D}(\mathbf{q})\ddot{\mathbf{q}} + \mathbf{J}^{\perp}\mathbf{H}(\mathbf{q}, \dot{\mathbf{q}}) = \mathbf{J}^{\perp}\mathbf{D_G}\mathbf{G} + \mathbf{J}^{\perp}\mathbf{w_1} \tag{8.75}$$

where the product $\mathbf{J}^{\perp}\mathbf{D_G}$ is invertible, so the dynamic model, reduced to six equations, becomes fully actuated and it is written as (8.1).

8.1.4.2 Impact Model

Now, assuming a flat foot contact, the double support phase is instantaneous and it can be modeled through passive impact equations (Formalskii 2009), i.e., impulsive torques are applied in the interlink joints. An impact appears at a time $t = T$, for this time the swing leg touches the ground. The impact is assumed (Tlalolini et al. 2010) to be passive, absolutely inelastic, and that the legs do not slip. Given these conditions, the ground reactions become impulsive forces, governed by Dirac delta-functions $\mathbf{R_j} = \mathbf{I_{R_j}}\delta(t - T),\ j = 1, 2$. Here $\mathbf{I_{R_j}}$ is the vector of the magnitudes of the impulsive reaction in the swing leg j.

Impact equations can be obtained through integration of the matrix motion equation (8.73) for the infinitesimal time from $T - 0$ to $T + 0$ of instantaneous impact. The torques supplied by the actuators at the joints, the centrifugal, Coriolis, and gravity forces have finite values, thus they do not influence on impact.

The impact enforces the complete surface of the foot sole, touching the ground. This means that the velocity of the swing foot, impacting the ground, is zero after impacts. Indeed, after an impact, the right foot (previous stance foot) takes off the ground, so the vertical component of the velocity of the taking-off foot just after an impact must be directed upward and the impulsive ground reaction in this foot equals zeros $\mathbf{I_{R_1}} = \mathbf{0}$.

Thus, the impact dynamic model is

$$\mathbf{D}(\mathbf{q})\Delta\dot{\mathbf{q}} = \mathbf{J_2}^T\mathbf{I_{R_2}}, \quad \mathbf{J_2}\dot{\mathbf{q}}^+ = \mathbf{0} \tag{8.76}$$

where $\Delta\dot{\mathbf{q}}$ is the variation of the velocity at an impact-time instant, $\Delta\dot{\mathbf{q}} = \dot{\mathbf{q}}^+ - \dot{\mathbf{q}}^-$, where $\dot{\mathbf{q}}^-$ is the velocity of the robot before an impact and $\dot{\mathbf{q}}^+$ is the velocity after the impact; $\mathbf{q}$ is the position at the impact. These equations form a system of the linear equations

$$\mathbf{I_{R_2}} = -(\mathbf{J_2}\mathbf{D}^{-1}\mathbf{J_2}^T)^{-1}\mathbf{J_2}\dot{\mathbf{q}}^- \tag{8.77}$$

$$\dot{\mathbf{q}}^{+} = [-\mathbf{D}^{-1}\mathbf{J}_2{}^{T}(\mathbf{J}_2\mathbf{D}^{-1}\mathbf{J}_2{}^{T})^{-1}\mathbf{J}_2 + \mathbf{I}]\dot{\mathbf{q}}^{-}, \tag{8.78}$$

which determine the impulse forces $\mathbf{I}_{\mathbf{R}_2}$ and the velocity restitution after impact. Provided that the configuration of the biped does not change after an impact, the biped restitution Eq. (8.78) proves to be a specific case of (8.4), where $\mathbf{w_d}$ accounts for modeling errors. Thus, the overall bipedal dynamics with impacts can be written in the form (8.1)–(8.4), thereby ensuring the theory, developed in Sect. 8.1 so far, to be applicable to the bipedal gait as well.

8.1.4.3 State Feedback Synthesis

The walking gait, composed of single support phases and impacts, is prespecified with a desired trajectory $\mathbf{q_{1d}}(t)$ and $\mathbf{q_{2d}}(t)$ to track. Such a trajectory, satisfying the conditions of contact, is constructed using an off-line optimization (Haq et al. 2012; Rengifo et al. 2011). The objective of the control is that each joint angle follows its own reference trajectory. The reference walking minimizes the integral of the norm of the torque vector for a given distance. The walking velocity is selected to be 0.5 m/s. The duration of one step is 0.53 s. Since the impact is instantaneous and passive, the control law is defined only during the single support phase. The $\mathscr{H}_\infty$ state feedback is then synthesized by applying Theorem 8.3, in light of Sect. 8.1.2.4, directly to the above biped model with a prespecified periodic trajectory to track.

8.1.4.4 Simulation Results

The biped model (8.73), coupled to the velocity restitution (8.48) at the impact instants, is used to illustrate numerical results on a stable walking gait by achieving robust tracking via the $\mathscr{H}_\infty$-controller designed in Sect. 8.1.2. The biped parameters are drawn from Haq et al. (2012), and the controller parameters, chosen to respect (8.48), are $\gamma = 1.45$, $\varepsilon = 0.01$, $\rho_p = 500$, and $\rho_v = 1$. Hypothesis H4 of Theorem 8.3 is verified numerically.

The robustness of the tracking state feedback, resulting from Theorem 8.3, is verified by introducing a resultant disturbance force $F_{xw} = 80$ N in the horizontal plane. Such a force is used for the duration of 0.07 s to simulate a disturbance effect. The effect of F_{xw} represents a disturbance in the continuous phase of the dynamics (8.73) as it starts from 0.8 s in the first cycle of the biped which belongs to the continuous phase of the trajectory. The contact with the ground is stated as a linear complementary constraint problem and solved with a constrained optimization (Rengifo et al. 2011).

The robustness of the tracking state feedback, resulting from Theorem 8.3, is verified by introducing a resultant disturbance force $F_{xw} = 80$ N in the horizontal plane. Such a force is used for the duration of 0.07 s to simulate a disturbance effect. The effect of F_{xw} represents a disturbance in the continuous phase of the dynamics (8.73) as it starts from 0.8 s in the first cycle of the biped which belongs

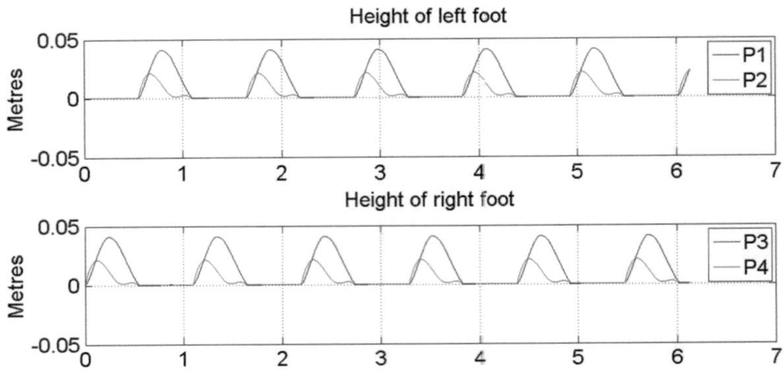

Fig. 8.15 Bipedal gait: feet height (Reproduced with permission. This figure was published in Montano et al. 2014, Copyright Elsevier (2014))

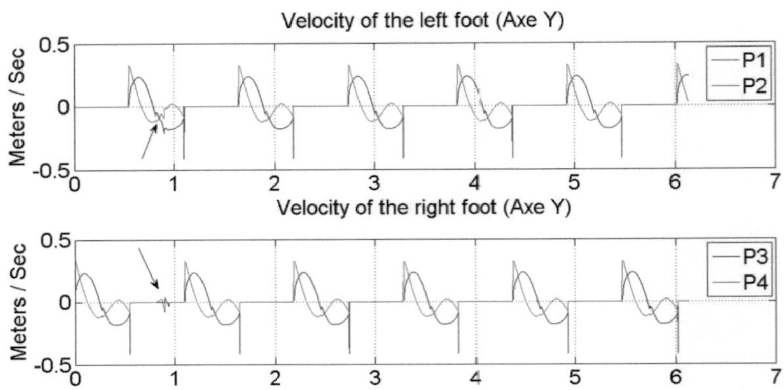

Fig. 8.16 Bipedal gait: vertical feet velocity

to the continuous phase of the trajectory. The contact with the ground is stated as a linear complementary constraint problem and solved with a constrained optimization (Rengifo et al. 2011).

Figure 8.15 shows the heights of the feet for six consecutive steps. Legends "P1" and "P3" represent the "toe" of the right and left foot, respectively. Similarly, "P2" and "P4" represent the "heel" of the right and left foot, respectively. The corresponding foot velocities in the vertical direction are depicted in Fig. 8.16. The disturbance attenuation is presented in Fig. 8.16, where the disturbance effect is pointed out by arrows. As readily seen, the biped returns to its desired gait after the disturbance disappears.

8.2 Energy Control of Continuum of Oscillators

Since recently, a number of energy-based control methods were proposed for finite-dimensional systems to particularly drive oscillatory modes (Åström and Furuta 2000; Fradkov 1996; Shiriaev and Fradkov 2000; Spong 1995) where specific values of the oscillation parameters (amplitude, frequency, and phase) may not be important, whereas some integral characteristic of the oscillation intensity, e.g., energy may be of interest. While focusing on stabilization of complex sets (non-convex sets, nonsmooth manifolds, etc.), energy control goes beyond conventional regulation and tracking objectives, thus forming a specific subarea of the control theory.

Energy control is motivated by numerous applications in physics and engineering such as control of pendulum-like models (Fantoni et al. 2000; Garofalo and Ott 2017; Spong 1995; Xin and Kaneda 2005), biped locomotion (Kajita et al. 1992), control of vibration units (Tang and Zuo 2012), and power systems (Wang et al. 2003) to name a few. Energy harvesting (Siang et al. 2018; Leong et al. 2018), deployment of tethered systems (Andrievsky and Guzenko 2014; Nikpoorparizi et al. 2018), and quantum control (Bonnard et al. 2011; Mantile 2008) are among modern applications which are relevant to energy control as well.

Distributed parameter models are capable of providing a more adequate description of real-world systems that suggest new challenges for solving energy control problems. Although there is a vast literature on the control of oscillatory systems in the infinite-dimensional setting, the majority dealt with the regulation and tracking problems (see, e.g., Lasiecka and Triggiani 1992, 2000; Krstic and Smyshlyaev 2008 and references therein).

In the sequel, the energy control is addressed for the sine-Gordon PDE, governing the dynamics of continuum of coupled oscillators, that has become an important model in nonlinear physics (Cuevas-Maraver et al. 2014). It is therefore not surprising that there were attempts to control the dynamics of sine-Gordon models. In the first publications (Kobayashi 2003; Petcu and Temam 2004; Kobayashi 2004), however only stabilization problem was studied. The energy control problem, capturing the equilibrium stabilization in particular, was initially formulated in Fradkov (2007) where the speed-gradient-based control algorithm was then proposed.

Distributed and spatially invariant energy control algorithms were rigorously studied in Orlov et al. (2017a, b) for the sine-Gordon model with a scalar spatial variable in the presence of the full information on the state variables. Both algorithms revealed, however, certain limitations. Being hardly feasible in practice, the distributed control algorithm called for further investigation on its implementation whereas the spatially invariant algorithm suffered from the generation of a nontrivial invariant manifold, possessing parasitic dynamics of distinct energy levels rather than the one to be imposed on the closed-loop system. The need for the full-state information was another shortcoming of the above works.

The present investigation is complementary to the above works and it follows Orlov et al. (2019) where spatially sampled sensors and actuators, yet feasible in practice, were admitted and placed within small spatial plant sub-domains. The output feedback synthesis to be developed along with the Luenberger-type observer design forms the core of this section.

The development is based on the speed gradient method, earlier addressed in Sect. 5.4, and it is actually well-recognized (Fradkov 1979) to successfully treat energy control problems in practice. The speed gradient method calls for the feedback design, resulting in the closed-loop system that possesses a prespecified positive function, which decreases along the system trajectories. Since such a function is not necessarily positive definite (e.g., it might be a norm of the deviation of the system energy from the desired level), the resulting closed-loop system is not generally speaking asymptotically stable around an equilibrium but it might possess an attractive manifold, generating branching solutions. Such a manifold should thus be presumed to consist not only of the desired energy trajectories but also of those with an undesired energy level, whose branching is a potential source of losing the closed-loop robustness. To prevent a non-robust scenario, one should impose extra conditions to rule out the appearance of a critical subset in the closed-loop state space where desired and undesired solutions glue together with a subsequent bifurcation to a distinct energy level. It should be noted however that no general conditions were obtained so far with a particular exception for the energy control of a simple scalar oscillator (Shiriaev and Fradkov 2000, 2001). Dealing with spatially sampled actuators and sensors is shown to yield more opportunities for reducing the invariant manifold with parasitic dynamics of undesirable energy levels.

8.2.1 *Sine-Gordon Nonlinear PDE Model and Problem Statement*

The benchmark model of interest is a dissipation-free sine-Gordon system (of unit inertia) governed by

$$x_{tt} = \kappa x_{rr} - F_0 \sin x + u(r,t), \; t \geq 0, \; 0 \leq r \leq 1, \tag{8.79}$$

where t is the time instant, $r \in [0, 1]$ is the scalar spatial variable, $x = x(\cdot, t)$ is the instant state of the system, the parameter κ is the elasticity of the system, F_0 stands for the external force, and $u(r, t)$ is for the manipulable input. As opposed to Orlov et al. (2017a), the external force is unactuated, and F_0 represents its permanent basis level.

8.2.1.1 Sampled-in-Space Actuators

As in Sect. 6.4, the sampled-in-space actuation

$$u(r,t) = \sum_{i=1}^{m} b_i(r) u_i(t) \tag{8.80}$$

is addressed to control the sine-Gordon model (8.79) which is thus representable in the form

$$x_{tt} = \kappa x_{rr} - F_0 \sin x + \sum_{i=1}^{m} b_i(r) u_i(t), \ t \geq 0, \ 0 \leq r \leq 1. \tag{8.81}$$

Such an actuation is constituted by in-domain control channels, which are characterized by spatial distributions $b_i(r) \in H^2, \ i = 1, \ldots, m$ of manipulable control inputs $u_i(t)$, located on disjoint actuator sub-domains

$$supp \ b_i(\cdot) \subseteq [r_i, r_i + h_i] \subseteq [0, 1] \tag{8.82}$$

of some lengths $h_i > 0$.

As far as the closures $[r_i, r_i + h_i]$ of the disjoint sub-domains cover the entire domain, i.e., $\cup_{i=1}^{m}[r_i, r_i + h_i] = [0, 1]$ and the maximal length $h_{max} = max_{1 \leq i \leq m} h_i$ escapes to zero as $m \to \infty$, the sampled-in-space actuation approaches a distributed one and can be viewed as an appropriate approximation of the limiting distributed control action by means of micro-electromechanical systems (MEMS) array (Bamieh et al. 2002).

Another interpretation of the sampled-in-space actuation is inspired from the single input case where $m = 1$ and the actuator distribution $b_1(r)$ is located on the infinitesimal sub-domain $(1 - h, 1)$ with the unit average value $\int_{1-h}^{1} b_1(r)dr = 1$. This case matches to the boundary actuation $u(r, t) = \delta(r - 1)u(t)$, studied in Dolgopolik et al. (2019) for the sine-Gordon energy control under the Dirichlet boundary condition at the left end and Neumann boundary condition at the right end.

The present investigation rules out the boundary actuation and for certainty, it focuses on the sine-Gordon model with fixed ends, which is why the PDE (8.81) is subsequently accompanied with the Dirichlet boundary conditions

$$x(0, t) = 0, \ x(1, t) = 0. \tag{8.83}$$

The above system belongs to a class of nonlinear wave PDEs, and it can be viewed as a continuum model of diffusively coupled pendulums, whose instant state $x(r, t)$ at t represents the deflection angle of the oscillator, located at r. Thus interpreted, the sine-Gordon model is fully actuated, if controlled by the distributed signal $u(r, t)$, and it becomes underactuated when the finite-dimensional sampled-in-space actuation (8.80) is applied.

8.2.1.2 Sampled-in-Space Sensing

Being dual to the sampled-in-space actuation (8.80), the available sensing of the sine-Gordon boundary-value problem (8.81), (8.83) is assumed to be of the form

$$y_j(t) = \int_0^1 x(r,t)\phi_j(r)dr, \; j = 1, \ldots, l \tag{8.84}$$

$$z_k(t) = \int_0^1 x_t(r,t)\psi_k(r)dr, \; k = 1, \ldots, n. \tag{8.85}$$

Both position and velocity sensors are thus admitted to be feasible. As a matter of fact, the use of either position sensors (8.84) or velocity sensors (8.85) is formally involved into the consideration by setting $\phi_j(x) \equiv 0, \; j = 1, \ldots, l$ or $\psi_k(x) \equiv 0, \; k = 1, \ldots, n$, respectively.

The in-domain measurements (8.84) and (8.85) are characterized by the sensor spatial distributions $\phi_j(r), \psi_k(r) \in H^2$ with disjoint sensor locations

$$supp\; \phi_j(\cdot) \subseteq [r_j^\phi, r_j^\phi + h_j^\phi] \subseteq [0,1], \; j = 1, \ldots, l, \tag{8.86}$$

$$supp\; \psi_k(\cdot) \subseteq [r_k^\psi, r_k^\psi + h_k^\psi] \subseteq [0,1], \; k = 1, \ldots, n. \tag{8.87}$$

These locations are introduced in a manner, similar to the actuator sub-domains (8.82). Although the sensors are not in general collocated to the actuators, however the output feedback synthesis to be developed relies on the collocated sensing and actuation such that

$$\phi_j(r) \equiv b_{i_j}(r), \; j = 1, \ldots, l \text{ for some } i_j \in \{1, \ldots, m\}, \tag{8.88}$$

$$\psi_k(r) \equiv b_{i_k}(r), \; k = 1, \ldots, n \text{ for some } i_k \in \{1, \ldots, m\}. \tag{8.89}$$

As in the sampled-in-space actuation, the distributed sensing and the boundary sensing can be interpreted in terms of the sampled-in-space sensing (8.84), (8.85). The boundary sensing, however, is beyond the development. The interested reader may refer to Dolgopolik et al. (2019) where the state observer of the sine-Gordon model was designed over the boundary velocity measurement.

8.2.2 Control Objective

The control problem to deal with is as follows. Using the available position-velocity measurements (8.84), (8.85), it is required to design a causal dynamic output feedback (8.80) for pumping or dissipating the energy

$$E(x, x_t) = \frac{1}{2}\int_0^1 \left[x_t^2 + \kappa x_r^2 + 2F_0\big(1 - \cos x\big)\right] dr \tag{8.90}$$

of the sine-Gordon system (8.81), (8.83) to a prespecified level $E_* \geq 0$. In other words, the limiting relation

$$\lim_{t\to\infty} E(t) = E_* \tag{8.91}$$

is to hold true for the system energy $E(t) = E(x(\cdot, t), x_t(\cdot, t))$, computed on the state trajectories $x(r, t)$, $x_t(r, t)$ of the closed-loop sine-Gordon model (8.81)–(8.90).

Clearly, the energy control objective (8.91) can be reformulated in terms of the exact controllability once a specific generalized coordinate among the continuum of the oscillators is required to be regulated to the desired energy level while the others are to be stabilized around the origin. It should be noted, however, that the proposed interpretation does not trivialize the energy control problem because the exact controllability of the nonlinear sine-Gordon boundary-value problem represents a real challenge when it is underactuated.

8.2.3 Energy Control Synthesis Using State Feedack

Provided that the state feedback is available, let us introduce the *goal functional* as

$$V(t) = \frac{1}{2}\big(E(t) - E_*\big)^2. \tag{8.92}$$

Following the speed gradient design procedure of Sect. 5.4, compute the time derivative of the goal functional along the trajectories of system (8.79), (8.83). Differentiating (8.92) in time, then integrating by part the resulting equality, and employing a straightforward consequence $x_t(0, t) = x_t(1, t) = 0$ of the Dirichlet boundary condition (8.83) yield

$$\begin{aligned}\dot{V} = \frac{dV}{dt} &= \big(E(t) - E_*\big)\int_0^1 (x_t \cdot x_{tt} - \kappa x_{rr} \cdot x_t + F_0 \sin x \cdot x_t)\, dr \\ &= \big(E(t) - E_*\big)\int_0^1 u(r, t) \cdot x_t\, dr.\end{aligned} \tag{8.93}$$

Let the control action be specified in the form of a finite number of sampled-in-space actuators (8.80). Then $\dot{V}$ reads as

$$\dot{V} = \big(E(t) - E_*\big)\sum_{i=1}^{m}\left(u_i(\cdot)\int_{r_i}^{r_i+h_i} b_i(r) \cdot x_t\, dr\right). \tag{8.94}$$

As the second step of the speed-gradient procedure, one should derive the gradient $\nabla_u \dot{V} \in \mathbb{R}^m$ of the resulting expression of $\dot{V}$ with respect to the control components $u_i(t)$, $i = 1, \ldots, m$, thus arriving at

$$\nabla_u \dot{V} = \big(E - E_*\big) \left[\int\limits_{r_1}^{r_1+h_1} b_1 x_t \, dr \; \ldots \int\limits_{r_m}^{r_m+h_m} b_m x_t \, dr \right]^T . \tag{8.95}$$

The third step of the speed-gradient procedure is to pick up a certain function $\eta(x, x_t)$ which satisfies the nequality $\eta^T \nabla_u \dot{V} \geq 0$. Let it be chosen in the form

$$\eta(\cdot) = \mathrm{sign}\big(E - E_*\big) \left[\int\limits_{r_1}^{r_1+h_1} b_1 x_t \, dr \; \ldots \int\limits_{r_m}^{r_m+h_m} b_m x_t \, dr \right]^T .$$

According to the speed-gradient method, the control action $u(x, \dot{x}) = -\Gamma \eta(x, \dot{x}, t)$ is then constituted with the matrix design parameter $\Gamma = diag\{\gamma_1, \ldots, \gamma_m\}$, composed of positive entries γ_i, $i = 1, \ldots, m$. Summarizing, the sampled-in-space actuation (8.80) is specified with

$$u_i(t) = -\gamma_i \mathrm{sign}\big(E(t) - E_*\big) \int\limits_{r_i}^{r_i+h_i} b_i x_t \, dr, \quad i = 1, \ldots, m. \tag{8.96}$$

For ensuring that the time derivative (8.94) of the goal functional (8.92) is nonpositive, it suffices to substitute (8.96) into (8.93) and verify that

$$\frac{dV}{dt} = -\big|E(t) - E_*\big| \sum_{i=1}^{m} \gamma_i \Big(\int_{r_i}^{r_i+h_i} b_i(r) x_t \, dr \Big)^2 \leq 0. \tag{8.97}$$

Thus, if the sine-Gordon model is enforced by the discontinuous actuation (8.96) it has a tendency to meet the control objective (8.91).

8.2.3.1 Closed-Loop Sliding Modes

It should be pointed out that due to (8.97), the closed-loop system cannot escape from the discontinuity manifold

$$\mathscr{S} = \{(s, s_t) \in H^1 \times H^0 : E(s, s_t) - E_* = 0\} \tag{8.98}$$

where it evolves in the sliding mode. Although the finite time attractiveness of manifold (8.98) is not guaranteed, the stability analysis of the closed-loop system is applicable regardless of this feature.

To describe potential sliding modes on the discontinuity manifold (8.98) a continuous equivalent control value u_{eq}, driving the system along (8.98), is to be substituted into the plant equation (8.81) for $u(r, t)$ (see Sect. 3.6.3 for details). For the purpose of deriving the equivalent control value, let us denote

$$S(s, s_t) = E(s, s_t) - E_* = 0 \tag{8.99}$$

and differentiate the sliding mode relation $S = 0$ on the trajectories of the closed-loop system (8.79)–(8.83). The equivalent control value $\text{sign}_{eq} S(s, s_t)$ is then deduced from the resulting relation

$$\dot{S} = \frac{dE}{dt} = \int_0^1 (x_t \cdot x_{tt} + x_r \cdot x_{rt} + F_0 \sin x \cdot x_t)\, dr$$
$$= -\text{sign}_{eq} S(s, s_t) \sum_{i=1}^{m} \gamma_i \left[\int_{r_i}^{r_i+h_i} b_i(r) x_t\, dr \right]^2 = 0. \tag{8.100}$$

Regardless of whether $\int_{r_i}^{r_1+h_i} b_i(r) x_t\, dr = 0$ for all $i = 1, \ldots, m$ or $\text{sign}_{eq} S(s, s_t) = 0$, it follows that

$$u_{eq}(r, t) = -\text{sign}_{eq} S(s, s_t) \left[\sum_{i=1}^{m} \gamma_i \int_{r_i}^{r_i+h_i} b_i(r) x_t\, dr \right] = 0 \tag{8.101}$$

on the discontinuity manifold (8.98). Hence, the boundary-value problem, governing the sliding modes of the closed-loop system (8.79)–(8.83), driven by the discontinuous controller (8.96) along the prespecified energy level (8.98), is naturally given by the open-loop ($u \equiv 0$) sine-Gordon model (8.79), (8.83).

8.2.3.2 Closed-Loop Well-Posedness

The closed-loop boundary-value problem (BVP) (8.81)–(8.83), (8.96) is subsequently studied in the state space $H^1 \times H^0$. For a technical reason of ensuring a sufficiently smooth classical solution of the BVP (8.81), (8.83) to exist in an appropriate Sobolev space, an initial state

$$x(r, 0) = x^0(r) \in H^3, \; x_t(r, 0) = x^1(r) \in H^2 \tag{8.102}$$

is assumed to be of class $H^3 \times H^2$ such that the compatibility condition

$$x^0(0) = x^1(0) = 0, \; x^0(1) = x^1(1) = 0 \tag{8.103}$$

holds true.

For later use, let us introduce the manifold

$$W = \left\{ (w, w_t) \in H^1 \times H^0 : \left[E(w, w_t) - E_* \right] \left[\sum_{i=1}^{m} \left| \int_{r_i}^{r_i+h_i} b_i(r) w_t\, dr \right| \right] = 0 \right\} \tag{8.104}$$

where the time derivative (8.97) of the goal functional (8.92) is nullified. The following result is established.

Theorem 8.4 *Let the sine-Gordon model* (8.81), (8.83) *be initialized with* (8.102) *under the compatibility condition* (8.103) *and let it be driven by the state feedback* (8.96). *Then in the state space* $H^1 \times H^0$ *beyond the discontinuity manifold* (8.98), *there locally exists a unique classical solution* $x(\cdot, t), x_t(\cdot, t))$ *of the closed-loop system* (8.81)–(8.83), (8.96), (8.102) *whereas along this manifold, the solution evolves in the sliding mode, governed by the open-loop boundary-value problem* (8.79), (8.83) *with* $u \equiv 0$.

Proof It is rather technical and is given in Appendix 8A.

8.2.3.3 Stability Analysis

Once the well-posedness of the closed-loop sine-Gordon system is guaranteed, the Lyapunov analysis becomes applicable.

Theorem 8.5 *Let the conditions of Theorem 8.4 be satisfied. Then the closed-loop sine-Gordon system* (8.81), (8.83), (8.96), (8.102), (8.103) *is forward complete, its arbitrary solution* $\big(x(\cdot, t), x_t(\cdot, t)\big) \in H^1 \times H^0$ *is uniformly bounded in* $H^1 \times H^0$, *and*

$$\inf_{(w, w_t) \in W_0} \|(x(\cdot, t), x_t(\cdot, t)) - (w, w_t)\|_{H^1 \times H^0} \to 0 \quad as\; t \to \infty \tag{8.105}$$

where W_0 *is the maximal invariant subset of manifold* (8.104). *In addition, the control components* (8.96) *remain uniformly bounded for all* $t \in [0, \infty)$ *and*

$$\sup_{t \in [0, \infty)} |u_i(t))| \to 0 \quad as\; \gamma_i \to 0,\; i = 1, \ldots, m. \tag{8.106}$$

Proof It is broken into several simple steps.

1. The closed-loop system is forward complete. Indeed, the closed-loop solutions are uniformly bounded in $H^1 \times H^0$ because of an a priori goal functional estimate

$$V(t) \le V(0) \quad \forall t \in [0, \infty), \tag{8.107}$$

guaranteed by (8.97). Hence, along with the goal functional (8.90), an arbitrary local solution of the closed-loop system (8.81)–(8.83), (8.96) admits an a priori estimate in $H^1 \times H^0$, too. Thus, such a solution is continuously extendible to the entire semi-infinite time axis $[0, \infty)$ because otherwise, its norm would escape to infinity in finite time what is impossible due to (8.107).

It should be pointed out that starting from a finite time instant, the global solution may evolve in the sliding mode, governed by the open-loop ($u \equiv 0$) sine-Gordon model (8.81)–(8.83) along the discontinuity manifold (8.99) (for description of potential closed-loop sliding modes, see Sect. 8.2.3.1). The forward completeness of the closed-loop system is thus concluded.

2. The uniform boundedness of the control signal and its negligibility is established as follows. Since (8.107) ensures that the goal functional (8.90) is uniformly bounded, regardless of whichever control gain $\gamma > 0$ chosen. Thus, the control signal (8.96) is uniformly bounded, too, and convergence (8.106) holds.

3. Asymptotic convergence to the maximal invariant subset Z of manifold (8.104) is demonstrated by applying the PDE-flavored invariance principle of Theorem 4.35. The applicability of this principle to the closed-loop sine-Gordon model (8.81)–(8.83), (8.96), initialized in $H^3 \times H^2$, is verified as follows.

For the space $\mathscr{D}(A_x)$, specified by (8.139) with the graph norm (Curtain and Zwart 1995), operator (8.138) possesses a compact inverse A_x^{-1}, mapping any bounded set of $\mathscr{D}(A_x)$ into a pre-compact set of $H^1 \times H^0$. Since the closed-loop system in question determines a dynamic system on the Hilbert space $H^1 \times H^0$, all solutions of which are bounded due to (8.107), it follows that each solution of (8.81)–(8.83), (8.96) is pre-compact.

Thus, Theorem 4.35 is applicable to the closed-loop sine-Gordon model. By applying Theorem 4.35, all solutions of (8.81)–(8.83), (8.96) are guaranteed to approach the maximal invariant subset W_0 of manifold (8.104) in the sense of (8.105). This completes the proof of Theorem 8.5. ■

The key role in the reachability of the control goal is played by the m integrals in (25). If a solution of the closed-loop system comes with all m integrals, nullified along it, then there is a nontrivial set of initial conditions such that the control objective (8.91) is achieved. The larger m, the narrower the bad set of initial conditions for which the goal is not achieved.

It should be pointed out that the maximal invariant subset of manifold (8.104), the attractiveness of which is established by Theorem 8.5, consists not only of the closed-loop solutions of the desired energy level but also of the solutions with trivial cumulative velocities $\int_{r_i}^{r_i+h_i} b_i(r)x_t\,dr = 0$ over the actuator sub-domains $(r_i, r_i + h_i)$, $i = 1, \ldots, m$.

In a particular case where the desired energy level is set to the lowest value $E_* = 0$, the control objective (8.90) is simplified to the plant stabilization around the origin and it is accomplished with the speed-gradient algorithm (8.96) provided that the sine-Gordon model is detectable from the output vector

$$X(x, \dot{x}) = \left[\int_0^1 b_1(r)x_t\,dr \ \ldots \ \int_0^1 b_m(r)x_t\,dr \right]^T \tag{8.108}$$

in the following sense (cf. that of Isidori and Astolfi 1992).

The closed-loop system (8.79)–(8.83), (8.96) is said to be *detectable* from (8.108) iff the trivial output $X(x, \dot{x}) \equiv 0$ can be generated by the trivial solution $x(r, t) \equiv 0$ only. The system in question is said to be *locally detectable* if its linearization is detectable from the same output.

The verification of the PDE detectability is a challenging problem, calling for an independent investigation. Once the sine-Gordon model is detectable in the adopted sense, the maximal invariant subset of manifold (8.104) is trivialized to the origin, and the asymptotic stability of the closed-loop system is then concluded from Theorem 8.5.

Moreover, the control objective (8.91) is hoped to be accomplished for $E_* > 0$ as well. Actually, if the closed-loop system is initialized at a lower energy level than the desired one, the speed gradient algorithm enforces the actuators to pump extra energy into the system, thereby avoiding its dissipation to the origin. On the contrary, once the closed-loop system is initialized at a higher energy level it appears to attain the desired energy level earlier than it escapes to zero, and hence, earlier than the output vector (8.108) is nullified. Supporting numerical evidences are presented in Sect. 8.2.6.

8.2.4 Luenberger Observer Design

For the position and velocity estimates $\xi(r,t)$ and $\zeta(r,t)$ of the state components $x(r,t)$ and, respectively, $x_t(r,t)$ of the sine-Gordon model (8.79), (8.83), a Luenberger-type observer

$$\xi_t = \zeta + \sum_{j=1}^{l} \mu_j \phi_j(r)\Big[y_j(t) - \int_0^1 \phi_j(\rho)\xi(\rho,t)d\rho\Big] \tag{8.109}$$

$$\zeta_t = \kappa\xi_{rr} - F_0 \sin\xi + u(r,t) + \sum_{k=1}^{n} \nu_k \psi_k(r)\Big[z_k(t) - \int_0^1 \psi_k(\rho)\zeta(\rho,t)d\rho\Big] \tag{8.110}$$

is designed with positive observer gains $\mu_j,\ j = 1,\ldots,l$ and $\nu_k,\ k = 1,\ldots,n$, as well as with the same weighted functions $\phi_j(r)$ and $\psi_k(r)$ of the position sensor locations (8.86) and of the velocity sensor locations (8.87).

Following the Luenberger approach, the proposed observer mimics the structure of the estimated model, separately given for the canonical position-velocity variables x and x_t. The observer PDEs (8.109), (8.110) are therefore subject to the Dirichlet boundary conditions

$$\xi(0,t) = 0,\ \xi(1,t) = 0 \tag{8.111}$$

for the first component $\xi(r,t)$ whereas no boundary conditions are needed for the second component $\zeta(r,t)$. Indeed, no spatial derivative of $\zeta(r,t)$ enters into the observer PDEs so that whichever sufficiently smooth function $\xi(r,t)$ is prespecified in (8.109), (8.110) (e.g., to be a solution of the equivalent PDE representation (8.114), given below), the observer dynamics of the second component $\zeta(r,t)$ are governed by the ordinary differential equation (8.110) where the spatial variable r can be viewed as a parameter.

It should be pointed out that the position estimate errors

$$\Delta y_j(t) = \int_0^1 \phi_j(r)\Big[\xi(r,t) - x(r,t)\Big]dr,\ j = 1,\dots,l \tag{8.112}$$

and the velocity ones

$$\Delta z_k(t) = \int_0^1 \psi_k(r)\Big[\zeta(r,t) - x_t(r,t)\Big]dr,\ k = 1,\dots,n \tag{8.113}$$

are asymmetrically injected into the position Eq. (8.109) and, respectively, into the velocity equation (8.110). This is rather unusual because in contrast to the standard finite-dimensional Luenberger design, symmetrical position and velocity injections in the present PDE setting seem to be useless as they result in inappropriate state estimates.

The well-posedness of the boundary-value problem (8.109)–(8.111) is established by representing it in the canonical wave PDE form.

8.2.4.1 Well-Posedness of Estimation Error Dynamics

By differentiating the PDE (8.109) in time and substituting (8.110) into the resulting PDE, the boundary-value problem (8.109)–(8.111) is represented in the hyperbolic PDE form

$$\xi_{tt} = \kappa\xi_{rr} - F_0 \sin\xi + u(r,t) - \sum_{j=1}^{l} \mu_j\phi_j(r)\Delta\dot{y}_j(t) - \sum_{k=1}^{n} \nu_k\psi_k(r)\Delta z_k(t), \tag{8.114}$$

coupled to the Dirichlet boundary conditions (8.111). Although the above observer model (8.114) formally relies on the non-causal differentiation of the position estimation error (8.112), however such a differentiation remains virtual as it is not required in its original form (8.109), (8.110).

Similar to the sine-Gordon model (8.79), (8.83), the boundary-value problem (8.111), (8.114) is well-posed in an appropriate state space where the compatibility condition

$$\xi(i,0) = \xi_t(i,0) = 0,\ i = 0,1 \tag{8.115}$$

holds true for initial states $\xi(r,0)$, $\xi_t(r,0)$.

Theorem 8.6 *Let system* (8.111), (8.114) *be initialized in* $H^3 \times H^2$ *and let it be driven by* (8.96). *Moreover, let the initial functions* $\xi(r,0)$ *and* $\xi_t(r,0)$ *respect the* *compatibility conditions* (8.115). *Then in the state space* $H^1 \times H^0$, *there locally exists a unique classical solution* $(\xi(\cdot, t), \xi_t(\cdot, t))$ *of the boundary-value problem* (8.111), (8.114).

Proof It is rather technical and similar to Theorem 8.4; it is postponed to Appendix 8B.

8.2.4.2 Stability of Estimation Error Dynamics

Due to (8.79), (8.83), and (8.109)–(8.111), the estimation error dynamics $\Delta\xi(x, t) = \xi(r, t) - x(r, t)$, $\Delta\zeta(x, t) = \zeta(r, t) - x_t(r, t)$ are governed by the boundary-value problem

$$\Delta\xi_t = \Delta\zeta - \sum_{j=1}^{l} \mu_j \phi_j(r) \Delta y_j(t),$$

$$\Delta\zeta_t = \kappa \Delta\xi_{rr} - F_0 \sigma(\Delta\xi, x) - \sum_{k=1}^{n} \nu_k \psi_k(r) \Delta z_k(t),$$

$$\Delta\xi(0, t) = 0, \ \Delta\xi(1, t) = 0, \tag{8.116}$$

where

$$\sigma(\Delta\xi, x) = \sin(\Delta\xi + x) - \sin x. \tag{8.117}$$

The canonical wave PDE form of the estimation error dynamics is then given by

$$\Delta\xi_{tt} = \kappa \Delta\xi_{rr} - F_0 \sigma(\Delta\xi, x) - \sum_{j=1}^{l} \mu_j \phi_j(r) \Delta\dot{y}_j(t) - \sum_{k=1}^{n} \nu_k \psi_k(r) \Delta z_k(t). \tag{8.118}$$

Let us, along with the estimation error dynamics (8.116), consider its linearized version, which is obtained by substituting

$$\sigma(\Delta\xi, x) = \Delta\xi \tag{8.119}$$

into (8.116) for $\sigma(\Delta\xi, x)$. The linear system, thus obtained, proves to be exponentially stable provided that it is detectable from the output vector $\left[Y^T(x, \dot{x}) \;\; Z^T(x, \dot{x})\right]^T$ composed of

$$Y^T(x, \dot{x}) = \left[\int_1^1 \phi_1(r)\, x\, dr \ \ldots \int_0^1 \phi_l(r)\, x\, dr\right],$$

$$Z^T(x, \dot{x}) = \left[\int_0^1 \psi_1(r) x_t\, dr \ \ldots \int_0^1 \psi_n(r) x_t\, dr\right]. \tag{8.120}$$

Theorem 8.7 *Consider system* (8.116), *specified with* (8.119) *and initialized within* $H^3 \times H^2$, *i.e.,* $\Delta\xi(\cdot, 0) \in H^3$, $\Delta\zeta(\cdot, 0) \in H^2$. *Let it be locally detectable from the output vector* (8.120). *Then the resulting linear dynamics* (8.116), (8.119) *are exponentially stable in the state space* $H^1 \times H^0$ *for arbitrary positive gains* $\mu_1, \ldots, \mu_l$ *and* $\nu_1, \ldots, \nu_n$.

Proof The proof follows the same line of reasoning as that of Theorem 8.5 and it is based on the Lyapunov functional

$$V_0(t) = \frac{1}{2}\int_0^1 \Big[F_0(\Delta\xi)^2 + \kappa(\Delta\xi_r)^2 + (\Delta\xi_t)^2\Big]dr, \tag{8.121}$$

viewed on the linearized observer error dynamics (8.116), which are specified with (8.119). It is clear that the time derivative of the Lyapunov functional (8.121) is negative semi-definite along the solutions of (8.116), (8.119). Indeed, by additionally employing the canonical PDE form (8.118), one derives that

$$\begin{aligned}
\dot{V}_0 &= \int_0^1 \Big\{F_0\Delta\xi\,\Delta\dot{\xi} + \kappa\Delta\xi_r\Delta\dot{\xi}_r + \Delta\dot{\xi}\big[\kappa\Delta\xi_{rr} \\
&\quad -F_0\Delta\xi - \sum_{j=1}^{l}\mu_j\phi_j(r)\Delta\dot{y}_j(t) - \sum_{k=1}^{n}\nu_k\psi_k(r)\Delta z_k(t)\big]\Big\}dr \\
&= -\sum_{j=1}^{l}\mu_j\Delta\dot{y}_j(t)\int_0^1 \Delta\dot{\xi}\phi_j(r)dr - \sum_{k=1}^{n}\nu_k\Delta z_k(t)\int_0^1 \Delta\dot{\xi}\psi_k(r)dr \\
&= -\sum_{j=1}^{l}\mu_j\Big[\Delta\dot{y}_j(t)\Big]^2 - \sum_{k=1}^{n}\nu_k\Big[\Delta z_k(t)\Big]^2 \le 0.
\end{aligned} \tag{8.122}$$

As in the proof of Theorem 8.5, the applicability of the invariance principle is established for the system in question. Since by the detectability condition of Theorem 8.7, the maximal invariant set of the manifold, where the Lyapunov time derivative (8.122) is nullified, consists of the system equilibrium only; the proof of Theorem 8.7 is then completed by applying the invariance principle. ■

It is clear that Theorem 8.7 allows one to utilize the proposed boundary-value problem (8.109)–(8.111) to locally asymptotically estimate the state components $x(r, t)$ and $x_t(r, t)$ of the nonlinear sine-Gordon model (8.79), (8.83). In the next section, such a Luenberger-type observer is coupled to the earlier developed energy control law to constitute the output feedback synthesis.

8.2.5 Energy Control Synthesis Using Output Feedback

In order to synthesize an output feedback, the state feedback law (8.80), (8.96) is modified to

$$u_i(t) = -\gamma_i \operatorname{sign}\left[\hat{E}(t) - E_*\right] \int_{r_i}^{r_i+h_i} b_i \xi_t(r,t)\, dr, \quad i = 1, \ldots, m. \tag{8.123}$$

Hereinafter, the instantaneous energy value

$$\hat{E}(t) = \frac{1}{2} \int_0^1 \left\{\xi_t^2(r,t) + k\xi_r^2(r,t) + 2F_0\left[1 - \cos\xi(r,t)\right]\right\} dr \tag{8.124}$$

is estimated according to (8.90), based on the observer outputs (8.109), (8.110), whereas the velocity estimate

$$\xi_t(r,t) = \zeta(r,t) - \sum_{j=1}^{l} \mu_j \phi_j(r) \Delta y_j(t) \tag{8.125}$$

is computed with the observer state component $\zeta(r,t)$, governed by (8.110).

The spatial derivatives in (8.124) are evaluated from the observer model where the finite differences can be used to reproduce a copy of the plant. Alternatively, the first-order spatial derivative could be evaluated from an appropriate sliding mode spatial derivative observer, running in parallel, with no explicit use of finite differences as proposed in Orlov (2000).

Provided that the available sensing collocate to the admissible sampled-in-space actuation, the proposed output feedback synthesis attains the energy objective, similar to the state feedback. Just in the case of the valid collocation relations (8.88), (8.89) between the system measurements (8.84), (8.85) and actuator (8.80), the Luenberger observer (8.109), (8.110) is specified to

$$\begin{aligned}
\xi_t &= \zeta + \sum_{j=1}^{l} \mu_j b_{i_j}(r)\left[y_j(t) - \int_{r_{i_j}}^{r_{i_j}+h_{i_j}} b_{i_j}(\rho)\xi(\rho,t)d\rho\right],\\
\zeta_t &= \kappa\xi_{rr} - F_0 \sin\xi + u(r,t) + \sum_{k=1}^{n} \nu_k b_{i_k}(r)\left[z_k(t) - \int_{r_{i_k}}^{r_{i_k}+h_{i_k}} b_{i_k}(\rho)\zeta(\rho,t)d\rho\right],
\end{aligned} \tag{8.126}$$

whereas its canonic PDE form (8.114) is given by

$$\xi_{tt} = \kappa \Delta\xi_{rr} - F_0 \sin\xi + u(r,t) - \sum_{j=1}^{l} \mu_j b_{i_j}(r)\Delta\dot{y}_j(t) - \sum_{k=1}^{n} \nu_k b_{i_k}(r)\Delta z_k(t). \tag{8.127}$$

By coupling Theorems 8.4–8.7 together, the following result is anticipated to hold true.

Theorem 8.8 *Consider the sine-Gordon model* (8.79), (8.83) *and the Luenberger-wise observer* (8.109), (8.110), *both initialized in* $H^3 \times H^2$ *near a desired energy level* $E_* \geq 0$ *under the compatibility conditions* (8.103), (8.115) *on their initial states. Let the sine-Gordon model be driven by the output feedback* (8.80), (8.123) *with the estimated energy* (8.124), *coupled to the velocity estimate* (8.125), *and with the Luenberger observer* (8.126) *over the collocated state measurements* (8.84)–(8.89). *Then the resulting closed-loop system satisfies the limiting relation* (8.105), *and the output feedback components* (8.123) *remain uniformly bounded in* $H^1 \times H^0$ *for all* $t \in [0, \infty)$, *and their maximal magnitudes become negligible in the sense of* (8.106).

Proof It is composed of several steps. Since these steps are similar to the corresponding steps of Theorems 8.4–8.7, only a sketch is provided.

1. Well-posedness of the closed-loop system. The observer-based closed-loop system (8.79), (8.83), (8.109), (8.110), driven by the output feedback (8.80), (8.123), (8.124) and viewed in the state space $H^1 \times H^0 \times H^1 \times H^0$, belongs to the same class of Hilbert space-valued dynamic systems (locally Lipschitz continuous beyond a regular discontinuity manifold) as its state feedback counterpart (8.79)–(8.83), (8.96), (8.102). Hence, the line of reasoning, used in the proof of Theorem 8.4 remains applicable in the present case as well. Thus, following the same line of reasoning, one establishes the local existence of a unique solution $(x, x_t, \xi, \zeta) \in H^1 \times H^0 \times H^1 \times H^0$ of the closed-loop system in question.

2. Goal functional and its time derivative. A local goal functional of the closed-loop system is determined by

$$V(t) = \frac{1}{2}\Big[\hat{E}(t) - E_*\Big]^2 + K_0 \frac{1}{2}\int_0^1 \Big[F_0(\Delta\xi)^2 + \kappa(\Delta\xi_r)^2 + (\Delta\xi_t)^2\Big] dr \tag{8.128}$$

with sufficiently large $K_0 > 0$ to subsequently be specified.

Note that by virtue of (8.126), (8.127), the estimation error dynamics (8.116) and (8.118) are represented in the form

$$
\begin{aligned}
\Delta\xi_t &= \Delta\zeta - \sum_{j=1}^{l} \mu_j b_{i_j}(r)\Delta y_j(t),\\
\Delta\zeta_t &= \kappa\Delta\xi_{rr} - F_0\sigma(\Delta\xi, x) - \sum_{k=1}^{n} \nu_k b_{i_k}(r)\Delta z_k(t),\\
\Delta\xi(0,t) &= 0,\ \Delta\xi(1,t) = 0,
\end{aligned} \tag{8.129}
$$

$$
\Delta\xi_{tt} = \kappa\Delta\xi_{rr} - F_0\sigma(\Delta\xi, x) - \sum_{j=1}^{l} \mu_j b_{i_j}(r)\Delta\dot{y}_j(t) - \sum_{k=1}^{n} \nu_k b_{i_k}(r)\Delta z_k(t). \tag{8.130}
$$

The time derivative of the goal functional (8.128) on the solutions of the closed-loop system, initialized near the origin and written in terms of the observer boundary-value problem (8.111), (8.127) and the estimation error dynamics (8.129), (8.130), is then locally estimated by (cf. that of (8.97) and (8.122))

$$
\begin{aligned}
\dot{V} = &-\big|\hat{E}(t) - E_*\big| \sum_{i=1}^{m} \gamma_i \Big[\int_{r_i}^{r_i+h_i} b_i(r)\xi_t\, dr\Big]^2\\
&- \big[\hat{E}(t) - E_*\big] \sum_{j=1}^{l} \mu_j \int_{r_{i_j}}^{r_{i_j}+h_{i_j}} b_{i_j}(r)\xi_t\, dr\, \Delta\dot{y}_j(t)\\
&- \big[\hat{E}(t) - E_*\big] \sum_{k=1}^{n} \nu_k \int_{r_{i_k}}^{r_{i_k}+h_{i_k}} b_{i_k}(r)\xi_t\, dr\, \Delta z_k(t)\\
&- K_0 \sum_{j=1}^{l} \mu_j \Big[\Delta\dot{y}_j(t)\Big]^2 - K_0 \sum_{k=1}^{n} \nu_k \Big[\Delta z_k(t)\Big]^2.
\end{aligned} \tag{8.131}
$$

Let us now apply the well-known inequality $2\alpha\beta \le \varepsilon\alpha^2 + \varepsilon^{-1}\beta^2$ on arbitrary $\alpha, \beta \in \mathbb{R}$ and $\varepsilon > 0$ to the second and third lines of (8.131) with

$$
\alpha = \int_{r_{i_j}}^{r_{i_j}+h_{i_j}} b_{i_j}(r)\xi_t\, dr,\ \beta = \Delta\dot{y}_j(t)
$$

and respectively with

$$
\alpha = \int_{r_{i_k}}^{r_{i_k}+h_{i_k}} b_{i_k}(r)\xi_t\, dr,\ \beta = \Delta z_k(t).
$$

The local negative semi-definiteness

$$\dot{V}(t) \leq -\left|\hat{E}(t) - E_*\right|\left\{\sum_{i=1}^{m}\gamma_i\left[\int_{r_i}^{r_i+h_i} b_i(r)\xi_t\,dr\right]^2 - \varepsilon\sum_{j=1}^{l}\mu_j\left[\int_{r_{i_j}}^{r_{i_j}+h_{i_j}} b_{i_j}(r)\xi_t\,dr\right]^2\right.$$

$$\left.-\varepsilon\sum_{k=1}^{n}\nu_k\left[\int_{r_{i_k}}^{r_{i_k}+h_{i_k}} b_{i_k}(r)\xi_t\,dr\right]^2\right\} - \left[K_0 - \varepsilon^{-1}\left|\hat{E}(t) - E_*\right|\right]\sum_{j=1}^{l}\mu_j\left[\Delta\dot{y}_j(t)\right]^2$$

$$-\left[K_0 - \varepsilon^{-1}\left|\hat{E}(t) - E_*\right|\right]\sum_{k=1}^{n}\nu_k\left[\Delta z_k(t)\right]^2 \leq 0 \tag{8.132}$$

of the time derivative of the goal functional is then concluded provided that

$$\varepsilon < \frac{\min_i\{\gamma_i\}}{\max_j\{\mu_j\} + \max_k\{\mu_k\}} \text{ and } K_0 > \frac{\left|\hat{E}(0) - E_*\right|}{\varepsilon}. \tag{8.133}$$

3. Uniform boundedness of the closed-loop solutions in local. By fixing sufficiently small initial conditions of the underlying sine-Gordon system and letting the parameters ε, $K_0 > 0$ to obey (8.132), the inequality $\dot{V}(t) \leq 0$ is shown to hold true. As in the proof of Theorem 8.5, this inequality locally ensures the uniform boundedness of the closed-loop solutions and the uniform negligibility of the control signals $u_i(t)$, $i = 1, \ldots, m$ as γ_i escapes to zero.

4. Asymptotic convergence. The applicability of the invariance principle is verified by following the line of reasoning used in the proof of Theorem 8.5. Applying the invariance principle to the closed-loop system in question verifies that its solutions locally approach the maximal invariant subset of manifold (8.104) where the time derivative (8.132) of the goal functional (8.128) is nullified. This completes the proof of Theorem 8.8. ∎

8.2.6 Numerical Study

The effectiveness of the energy control synthesis proposed is supported by the numerical study, reproduced from Orlov et al. (2019), where the PDE (8.79) is discretized in the spatial variable $r \in \mathbb{R}$ by uniformly splitting the segment [0, 1] into N subintervals. The *discretization step* δ is thus set to $\delta = N^{-1}$.

At the discretization nodes $r_i = i \cdot \delta$, $i = 1, \ldots, N-1$, the first- and second-order spatial derivatives of $x(r, t)$ are approximated by the finite differences of the first and second order, respectively. The boundary values $x(r_0, t) = x(r_1, t) = 0$ and $x(r_N, t) = x(r_{N-1}, t) = 0$ are specified according to the Dirichlet boundary conditions (8.83). The resulting system of $N - 1$ ordinary differential equations (ODEs) of the second order are then numerically solved over a time interval $[0, T]$ by applying the medium order variable step Runge–Kutta method (Dormand and Prince 1980), performed with the standard MATLAB routine *ode45*.

A similar procedure is applied to numerically solve observer Eq. (8.109) with respect to variables $\xi(t,r)$, $\zeta(t,r)$ under the Dirichlet boundary conditions (8.111). The first-order differences are used to approximate the spatial derivatives $x_r(t,r_i)$ and $\xi_r(t,r_i)$, $i=0,1,\ldots,N$ for the computation of the system energy $E(x,x_t)$ and for the estimated energy $\hat{E}(x,x_t)$, respectively. To calculate the values of definite integrals in (8.90), (8.96), the MATLAB standard routine *trapz* of the trapezoidal numerical integration is employed.

In the simulations of the closed-loop boundary-value problem (8.79)–(8.83), (8.96), the number of actuators $m=10$ was prespecified whereas the control gains were set to $\gamma_i=\gamma_0/h_i$, $h_i=m^{-1}$, $i=1,\ldots,m$ with $\gamma_0=5$ and the desired energy level was set to $E_*=5$. The initial states were prespecified in the form

$$x(0,r)=A\big[1-\cos(2\pi r)\big]^7,\quad x_t(0,r)=0 \tag{8.134}$$

with a certain "magnitude" parameter A.

A reasonably high number $N=2500$ was selected for the PDEs (8.79) to discretize the spatial variable r and duration of the computation time T was confined to 15.

The spatial domain [0, 1] was uniformly partitioned into $m=10$ sub-domains $[r_i, r_i+h_i]$ of lengths $h_i=m^{-1}$, $i=1,\ldots,m$ so that $r_i=(i-1)m^{-1}$. Within each sub-domain the corresponding actuator distribution $b_i(r)$ was specified as

$$b_i(r)=\begin{cases}1, & \text{if }\ r_i+0.2m^{-1}\le r\le r_i+0.8m^{-1},\\ 0, & \text{otherwise,}\end{cases} \tag{8.135}$$

i.e., the first and the last (mth) actuators were located in the distance 0.002 from the left and right boundaries, respectively, whereas the neighboring actuators possessed a slot of the length 0.04 between them.

Throughout the simulations, five position sensors were admitted only. The domain [0, 1] was uniformly split into $l=5$ sub-domains $[(j-1)l^{-1}, jl^{-1}]$, $j=1,\ldots l$ of the lengths l^{-1} and the sensor locations were set to

$$\phi_j(r)=\begin{cases}1, & \text{if }\ (j-1)l^{-1}\le r\le jl^{-1},\\ 0, & \text{otherwise.}\end{cases} \tag{8.136}$$

For the simulations, the model parameters of (8.79) were specified to $\kappa=0.12$, $F_0=25$. For the case of energy pumping, parameter A in (8.134) was set to $A=10^{-3}$ (in this case, the initial energy was less than the desired one). An alternative case with $A=0.02$, where the initial energy was greater than the desired one, was also investigated in the simulations. The *output feedback* synthesis was investigated where both the system state variables and the energy were estimated by means of the proposed observer (8.109)–(8.111). The value of the observer gains μ_j, $j=1,\ldots,5$ were taken identical $\mu_j=\mu$ and the value $\mu=10$ was chosen to optimize the observer error decay rate. The velocity sensor gains $\nu_k=0$ were formally set to

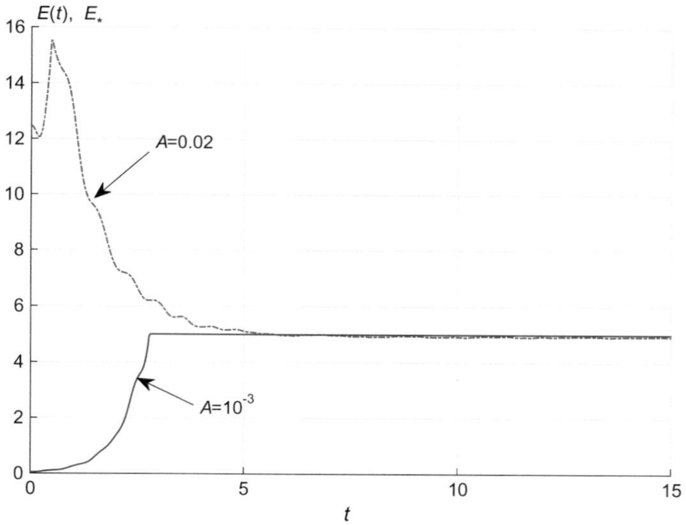

Fig. 8.17 The output feedback performance in terms of the energy evolution $E(t)$: solid and dashed lines are for initial conditions, required energy pumping ($A = 10^{-3}$) and dissipation ($A = 0.02$), respectively. © 2019 $IEEE$. Reprinted, with permission, from Orlov et al. (2019)

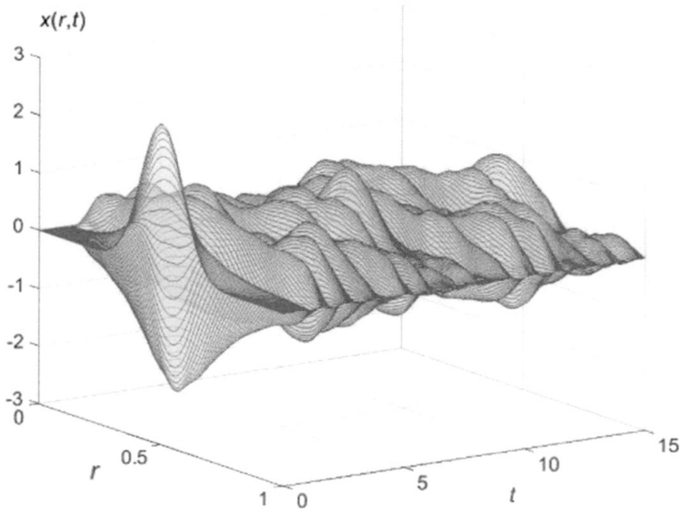

Fig. 8.18 Energy dissipation ($A = 0.02$): spatial-temporal graph of $x(r, t)$. © 2019 $IEEE$. Reprinted, with permission, from Orlov et al. (2019)

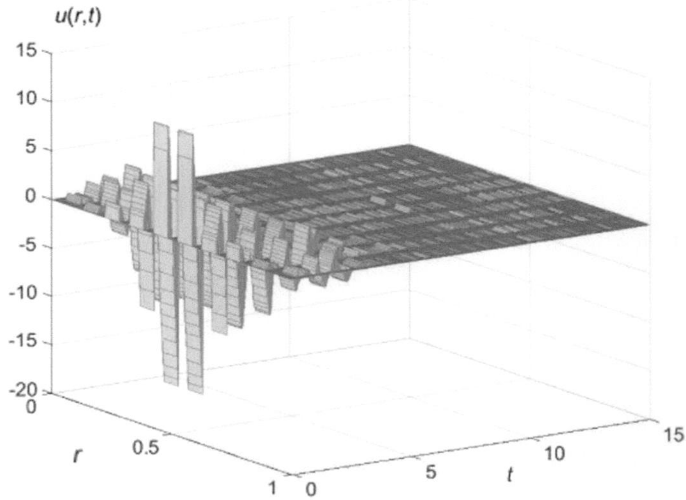

Fig. 8.19 Energy dissipation ($A = 0.02$): spatial-temporal graph of control action $u(r, t)$. © 2019 $IEEE$. Reprinted, with permission, from Orlov et al. (2019)

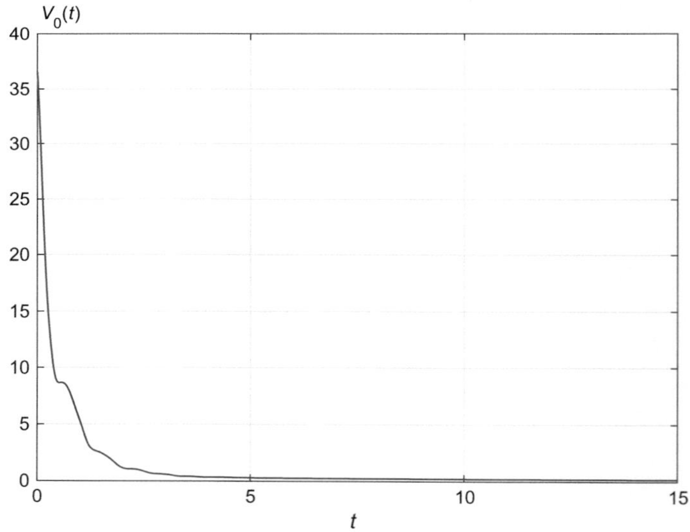

Fig. 8.20 Energy dissipation ($A = 0.02$): time history of the goal observer functional (8.121). © 2019 $IEEE$. Reprinted, with permission, from Orlov et al. (2019)

zero for all k to reflect the absence of the velocity sensors. In the simulations, the trivial observer initial conditions $\xi(0, r) = 0$, $\zeta(0, r) = 0$ were chosen.

The plots on Fig. 8.17 demonstrate the energy evolution $E(t)$ for various initial conditions and for the desired energy level $E_* = 5$. Spatial-temporal graphs of Figs. 8.18 and 8.19 reflect complex oscillatory modes of the closed-loop system to persist even after achieving the energy goal. The time history of the observer goal functional (8.121) $V_0(t)$, depicted in Fig. 8.20, illustrates the state estimation error decay. Good performance of the output feedback synthesis is concluded from Figs. 8.17, 8.18, 8.19, and 8.20.

8.3 Concluding Remarks

Modern hybrid and PDE control applications are involved to support capabilities of the revised Lyapunov approach to actually be available for a broad class of dynamic systems. Chosen to add practical value, these applications illustrate actual challenges in periodic tracking of bipedal gait with ground constraints and in energy control of nonlinear sine-Gordon PDE models. Synthesis of hybrid mechanical systems, which are not fully actuated, and energy control of more general DPS with non-collocated sensing and actuation are among open problems to hopefully be tackled within the Lyapunov approach, advanced properly.

Appendix A: Proof of Theorem 8.4

Proof A unique solution of the closed-loop system is first shown to locally exist beyond the discontinuity manifold (8.98). For proving this, the BVP (8.81), (8.83) is embedded into the Hilbert space $H^1 \times H^0$ where it takes the form

$$x_{tt} + A_x x = -F_0 \sin x + \sum_{i=1}^{m} B_i u_i(t). \tag{8.137}$$

The above Hilbert space-valued equation is specified with the bounded operators

$$B_i = b_i(r) : H^0 \to H^0, \ i = 1, \dots, m$$

of multiplication by $b_i(r) \in H^2$ and with the strictly positive definite self-adjoint operator

$$A_x = -\frac{d^2}{dr^2}, \tag{8.138}$$

which is defined on the domain

$$\mathscr{D}(A_x) = \{x \in H^2 : x(0) = 0, \ x(1) = 0\} \tag{8.139}$$

and which possesses a compact inverse A_x^{-1}. The eigenfunctions

$$p_i(r) = \frac{1}{2} \sin \pi i r, \ i = 1, 2, \ldots \tag{8.140}$$

of the operator A_x solve the Sturm–Liouville problem

$$\begin{aligned} p_{rr}(r) + \lambda p(r) = 0 \\ p(0) = 0, \ p(1) = 0 \end{aligned} \tag{8.141}$$

with the associated eigenvalues

$$\lambda_i = (\pi i)^2. \, i = 1, 2, \ldots, \tag{8.142}$$

These eigenfunctions constitute the complete orthonormal system such that the Fourier expansion

$$A_x^\sigma \hat{x} = \sum_{i=1}^{\infty} \lambda_i^\sigma \int_0^1 \hat{x}(r) p_i(r) \, dr \ p_i \tag{8.143}$$

is correctly introduced (see Krasnoselskii et al. 1976, p. 463) for an arbitrary $\hat{x} \in \mathscr{D}(A_x)$ and for the operator A_x when $\sigma = 1$, for its inverse A_x^{-1} when $\sigma = -1$, and for its arbitrary fractional degree A_x^σ when $\sigma \in (-1, 1)$.

Now, let us differentiate (8.81), (8.83), (8.102) in the spatial variable r to derive auxiliary relations

$$\omega_{tt} = \kappa \omega_{rr} - F_0 \cos x \cdot \omega + \sum_{i=1}^{m} b_i'(r) u_i(t), \tag{8.144}$$

$$\omega_r(0, t) = 0, \ \omega_r(1, t) = 0 \tag{8.145}$$

$$\omega(r, 0) = x_r^0(r) \in H^2, \ \omega_t(r, 0) = x_r^1(r) \in H^1, \tag{8.146}$$

governing the spatial state derivative $\omega(r, t) = x_r(r, t), \ t \geq 0, \ 0 \leq r \leq 1$. Note that the boundary condition (8.145) on the auxiliary variable $\omega(r, t) = x_r(r, t)$ is formally obtained from the PDE (8.81). Indeed, by employing the boundary condition (8.83), it follows from (8.81) that $x_{rr}(0, t) = x_{rr}(1, t) = 0$, thereby ensuring (8.145). The auxiliary BVP (8.144), (8.145), thus derived, is represented in the form

$$\omega_{tt} + A_\omega \omega = -F_0 \cos x \cdot \omega + \sum_{i=1}^{m} B_i'(r) u_i(t), \tag{8.147}$$

where the operators B_i', $i = 1, \ldots, m$ in (8.144) stand for the multiplication by $b_i'(r,t) \in H^1$ and the double differentiation operator $A_\omega = -\frac{d^2}{dr^2}$ is now defined on the domain

$$\mathscr{D}(A_\omega) = \{\omega \in H^2 : \omega_r(0) = 0,\ \omega_r(1) = 0\}. \tag{8.148}$$

Then the corresponding Sturm–Liouville problem

$$q_{rr}(r) + \lambda q(r) = 0, \quad q_r(0) = 0,\ q_r(1) = 0 \tag{8.149}$$

governs the orthonormal eigenfunctions

$$q_0(r) = 1,\ q_j(r) = 2\cos \pi j r,\ i = 1, 2, \ldots \tag{8.150}$$

of the operator A_ω and its eigenvalues

$$\lambda_0 = 0, \lambda_i = (\pi i)^2,\ i = 1, 2, \ldots . \tag{8.151}$$

Similar to (8.143), fractional degrees of the operator A_ω, being applied to an arbitrary $\hat{\omega}(\cdot) \in \mathscr{D}(A_\omega)$, admit the following Fourier representation

$$A_\omega^\sigma \hat{\omega} = \sum_{i=1}^{\infty} \lambda_i^\sigma \int_0^1 \hat{\omega}(r) q_i(r)\, dr\, q_i \tag{8.152}$$

for nonzero $\sigma \in [-1, 1]$.

Next, let us consider the BVPs (8.81), (8.83) and (8.144), (8.145), coupled together by means of the control input (8.96). Since by construction, the spatial derivative x_r of the output x of (8.81), (8.83) is mimicked by the output ω of (8.144), (8.145), properly initialized with (8.146), it is further assumed without loss of generality that the auxiliary state variable ω is substituted into the control law (8.96) for x_r so that the control input $u_\gamma(x, x_t, x_r)$ takes the state feedback form $u_\gamma(x, x_t, \omega)$. Thus interpreted, the coupled system (8.81), (8.83), (8.144), (8.145) comes with the nonlinear terms $(F_0 + u_\gamma(x, x_t, \omega)) \sin x$ and $(F_0 + u_\gamma(x, x_t, \omega))\omega \cos x$, which are Lipschitz continuous in the state variables x, x_t, ω.

Taking into account that the infinitesimal operators of the Hilbert space-valued Eqs. (8.137) and (8.147) are well-known (Curtain and Zwart 1995) to generate (strongly continuous) C_0-semigroups on $H^1 \times H^0$, the result of Krasnoselskii et al. (1976, Theorem 23.2) is applicable to the closed-loop system (8.81)–(8.83), (8.144)–(8.145) whose nonlinear terms are Lipschitz continuous. Thus by Krasnoselskii et al. (1976, Theorem 23.2), there locally exist unique solutions $x(t), \omega(t) \in H^2$ of the corresponding integral equations

$$x(t) = \cos(A_x^{\frac{1}{2}} t)x^0 + A_x^{-\frac{1}{2}} \sin(A_x^{\frac{1}{2}} t)x^1 -$$
$$A_x^{-\frac{1}{2}} \int_0^t \sin(A_x^{\frac{1}{2}} t - A_x^{\frac{1}{2}} \tau)(F_0 + u_\gamma(\tau)) \sin x(\tau)\, d\tau \tag{8.153}$$

$$\omega(t) = \cos(A_\omega^{\frac{1}{2}} t)x_r^0 + A_\omega^{-\frac{1}{2}} \sin(A_\omega^{\frac{1}{2}} t)x_r^1 -$$
$$A_\omega^{-\frac{1}{2}} \int_0^t \sin(A_\omega^{\frac{1}{2}} t - A_\omega^{\frac{1}{2}} \tau)(F_0 + u_\gamma(\tau))\omega(\tau) \cos x(\tau)\, d\tau \tag{8.154}$$

where

$$\cos(A_x^\sigma t)x = \sum_{i=1}^{\infty} \cos(\lambda^\sigma t) \int_0^1 x(r)p_i(r)\, dr\, p_i$$
$$\cos(A_\omega^\sigma t)\omega = \sum_{i=1}^{\infty} \cos(\lambda^\sigma t) \int_0^1 \omega(r)q_i(r)\, dr\, q_i$$
$$\sin(A_x^\sigma t)x = \sum_{i=1}^{\infty} \sin(\lambda^\sigma t) \int_0^1 x(r)p_i(r)\, dr\, p_i$$
$$\sin(A_\omega^\sigma t)\omega = \sum_{i=1}^{\infty} \sin(\lambda^\sigma t) \int_0^1 \omega(r)q_i(r)\, dr\, q_i, \quad \sigma = \pm\frac{1}{2}$$

and the initial values x^0, x^1, x_r^0, x_r^1 respect (8.146).

Moreover, by applying (Krasnoselskii et al. 1976, Theorem 22.4) the Fourier approximations

$$P_n x = \sum_{i=1}^{n} \int_0^1 x(r)p_i(r)\, dr\, p_i, \; P_n\omega = \sum_{i=1}^{n} \int_0^1 \omega(r)q_i(r)\, dr\, q_i \tag{8.155}$$

of the solutions (8.153) and (8.154) are established to converge to the corresponding solutions of (8.137) and (8.147) as $n \to \infty$, thereby proving the existence of the classical solutions of the coupled system (8.81)–(8.83), (8.144)–(8.146). Since these solutions are well-recognized (Krasnoselskii et al. 1976) to satisfy the integral Eqs. (8.153) and (8.154), their uniqueness is straightforwardly concluded from the earlier established uniqueness of the solutions of (8.153) and (8.154).

To complete the proof of Theorem 8.4 it remains to note that along the discontinuity manifold (8.98), the solution has already been established in the previous subsection to evolve in the sliding mode, governed by the open-loop boundary-value problem (8.81)–(8.83). ■

Appendix B: Proof of Theorem 8.6

Proof The line of reasoning used in the proof of Theorem 8.4 is applied here as well. Indeed, by taking the estimate errors (8.112), (8.113) into account, the underlying PDE of the boundary-value problem (8.111), (8.114) can be represented in the form

$$\xi_{tt} = \kappa\xi_{rr} - F_0 \sin\xi + \hat{u}(r,t) - \sum_{j=1}^{l} \mu_j \phi_j(r) \int_0^1 \phi_j(r)\dot{\xi}(r,t)dr$$
$$- \sum_{k=1}^{n} \nu_k \psi_k(r) \int_0^1 \psi_k(r)\zeta(r,t)dr \tag{8.156}$$

where

$$\hat{u}(r,t) = u(r,t) + \sum_{j=1}^{l} \dot{y}_j(t) + \sum_{k=1}^{n} z(t). \tag{8.157}$$

The external term (8.157) in (8.156) is composed of the state measurements (8.84), (8.85) and the input function (8.96), which is the same as in Theorem 8.4. Thus, the present boundary-value problem (8.111), (8.156) is of the same structure as that of the closed-loop system (8.81), (8.83), (8.96) of Theorem 8.4 in a particular case of $F_0 = 0$, and it can therefore be addressed in the same manner of expanding its solutions in the Fourier series as in the proof of Theorem 8.4. The rest of the proof of Theorem 8.6 is similar to that of Theorem 8.4, and it is therefore omitted. ■

References

Andrievsky B, Guzenko P (2014) Energy speed-gradient control of nonlinear satellite oscillations. Cybern Phys 3(1):9–15

Åström KJ, Furuta K (2000) Swinging up a pendulum by energy control. Automatica 36:287–295

Bamieh B, Paganini F, Dahleh M (2002) Distributed control of spatially-invariant systems. IEEE Trans Auto Control 47:1091–1107

Biemond J, Van de Wouw N, Heemels W, Nijmeijer H (2013) Tracking control for hybrid systems with state-triggered jumps. IEEE Trans Auto Control 58:876–890

Bonnard B, Caillau JB, Cots O (2011) Energy minimization in two-level dissipative quantum control: the integrable case. Discret Contin Dyn Syst 31:198–208

Chevallereau C, Abba G, Aoustin Y, Plestan F, Westervelt E, Canudas De Wit C, Grizzle J (2003) Rabbit: a testbed for advanced control theory. IEEE Control Syst Mag 23(5):57–79

Cuevas-Maraver J, Kevrekidis P, Williams F (eds) (2014) The sine-Gordon model and its applications: from Pendula and Josephson junctions to gravity and high-energy physics. Springer, Heidelberg

Curtain R, Zwart H (1995) An introduction to infinite-dimensional linear systems. Springer, New York

Dolgopolik M, Fradkov A, Andrievsky B (2019) Observer-based boundary control of the sine-Gordon model energy. Automatica (to be published)

Dormand JR, Prince PJ (1980) A family of embedded Runge–Kutta formulae. J Comput Appl Math 6:19–26

Fantoni I, Lozano R, Spong M (2000) Energy based control of the pendubot. IEEE Trans Auto Control 45:725–729

Formalskii AM (2009) Ballistic walking design via impulsive control. J Aerosp Eng 23(2):129–138

Fradkov AL (1979) Speed-gradient scheme and its applications in adaptive control. Autom Remote Control 40(9):13331342

Fradkov AL (1996) Swinging control of nonlinear oscillations. Int J Control 64:1189–1202

Fradkov AL (2007) Cybernetical physics: from control of chaos to quantum control. Springer, Berlin

Goebel R, Sanfelice R, Teel A (2009) Hybrid dynamical systems. IEEE Control Syst Mag 29:28–93

Garofalo G, Ott C (2017) Energy based limit cycle control of elastically actuated robots. IEEE Trans Auto Control 62:2490–2497

Haq A, Aoustin Y, Chevallereau C (2012) Effects of knee locking and passive joint stiffness on energy consumption of a seven-link planar biped. In: Proceedings of the IEEE international conference on robotics and automation (ICRA), pp 870–876

Hamed K, Grizzle J (2013) Robust event-based stabilization of periodic orbits for hybrid systems: application to an underactuated 3D bipedal robot. In: American control conference, pp 6206 –6212

Isidori A, Astolfi A (1992) Disturbance attenuation and $\mathscr{H}_\infty$-control via measurement feedback in nonlinear systems. IEEE Trans Auto Control 37:1283–1293

Kajita S, Yamaura T, Kobayashi A (1992) Dynamic walking control of a biped robot along a potential energy conserving orbit. IEEE Trans Robot Autom 8:431–438

Kobayashi T (2003) Boundary feedback stabilization of the sine-Gordon equation without velocity feedback. J Sound Vib 266:775–784

Kobayashi T (2004) Adaptive stabilization of the sine-Gordon equation by boundary control. Math Methods Appl Sci 27:957–970

Krasnoselskii MA, Zabreyko PP, Pustylnik EI, Sobolevski PE (1976) Integral operators in spaces of summable functions. Noordhoff, Leyden

Krstic M, Smyshlyaev A (2008) Boundary control of PDEs: a course on backstepping designs. SIAM, Philadeplhia

Lasiecka I, Triggiani R (1992) Uniform stabilization of the wave equation with dirichlet or neumann feedback control without geometric conditions. Appl Math Optim 25:189–224

Lasiecka I, Triggiani R (2000) Control theory for partial differential equations: continuous and approximation theories. Cambridge University Press, Cambridge

Leong AS, Dey S, Quevedo DE (2018) Transmission scheduling for remote state estimation and control with an energy harvesting sensor. Automatica 91:54–60

Mantile A (2008) Point interaction controls for the energy transfer in 3-D quantum systems. Int J Control 81:752–763

Montano O, Orlov Y, Aoustin Y (2014) Nonlinear state feedback $\mathscr{H}_\infty$-control of mechanical systems under unilateral constraints. Proc 19th IFAC World Congress 3833–3838

Montano O, Orlov Y, Aoustin Y (2016) Nonlinear $\mathscr{H}_\infty$-control under unilateral constraints. Int J Control 89:2549–2571

Montano O, Orlov Y, Aoustin Y, Chevallereau C (2016) Orbital stabilization of an underactuated bipedal gait via nonlinear $\mathscr{H}_\infty$-control using measurement feedback. Auton Robot 41:1277–1295

Naldi R, Sanfelice RG (2013) Passivity-based control for hybrid systems with applications to mechanical systems exhibiting impacts. Automatica 49:1104–1116

Nesic D, Teel AR, Valmorbida G, Zaccarian L (2013) Finite-gain stability for hybrid dynamical systems. Automatica 49:23842396

Nikpoorparizi P, Deodhar N, Vermillion C (2018) Modeling, control design, and combined plant/controller optimization for an energy-harvesting tethered wing. IEEE Trans Control Syst Technol 26(4):1157–1169

Orlov Y (2000) Sliding mode observer-based synthesis of state derivative-free model reference adaptive control of distributed parameter systems. J Dyn Syst Meas Control 122:725–731

Orlov Y, Aguilar L (2014) Advanced $\mathscr{H}_\infty$ control - towards nonsmooth theory and applications. Birkhäuser, New York

Orlov Y, Fradkov A, Andrievsky B (2017a) Energy control of distributed parameter systems via speed-gradient method: case study of string and sine-Gordon benchmark models. Int J Control 90:2554–2566

Orlov Y, Fradkov A, Andrievsky B (2017b) Sliding mode-based speed-gradient control of the string energy. IFAC PapersOnLine 50(1):8484–8489

Orlov Y, Fradkov A, Andrievsky B (2019) Output feedback energy control of the sine-Gordon PDE model using collocated spatially sampled sensing and actuation. IEEE Trans Auto Ctrl. https://doi.org/10.1109/TAC.2019.2921620

Petcu M, Temam R (2004) Control for the sine-Gordon equation. ESAIM Control Optim Calc Var 10:553–573

Rengifo C, Aoustin Y, Plestan F, Chevallereau C (2011) Contact forces computation in a 3d bipedal robot using constrained-based and penalty-based approaches. In: Proceedings of the multibody dynamics

Shiriaev A, Fradkov AL (2000) Stabilization of invariant sets for nonaffine nonlinear systems. Automatica 36:1709–1715

Shiriaev A, Fradkov AL (2001) Stabilization of invariant sets for nonlinear systems with applications to control of oscillations. Int J Robust Nonlinear Control 11:215–240

Siang J, Lim M, and Liong M (2018) Review of vibration-based energy harvesting technology: Mechanism and architectural approach. Int J Energy Res 42

Spong M (1995) The swing up control problem for the acrobot. IEEE Control Syst Mag 15:49–55

Tang X, Zuo L (2012) Simultaneous energy harvesting and vibration control of structures with tuned mass dampers. J Intell Mater Syst Struct 23:2117–2127

Tlalolini D, Aoustin Y, Chevallereau C (2010) Design of a walking cyclic gait with single support phases and impacts for the locomotor system of a thirteen-link $3D$ biped using the parametric optimization. Multibody Syst Dyn 23(1):33–56

Wang Y, Cheng D, Li C, Ge Y (2003) Dissipative Hamiltonian realization and energy-based-disturbance attenuation control of multimachine power systems. IEEE Trans Auto Control 48:1428–1433

Xin X, Kaneda M (2005) Analysis of the energy-based control for swinging up two pendulums. IEEE Trans Auto Control 50:679–685

Index

A
Aizerman–Pyatnitskii solution, 22

B
Barbalat's lemma, 17, 18, 97, 108, 118, 235, 271, 278
Boundary-layer regularization, 23, 24

C
Complementarity constraint, 28
Constrained Lagrange system, 28

D
Derivative
- contingent hyper-/epiderivative, 69
- Dini, 69, 73, 102, 129
- proximal super-/subderivative, 69, 71, 73
- sliding mode, 5

Dynamic system, 28, 29, 45, 74–79, 81, 83, 84, 89, 95, 96, 98–100, 104–110, 123, 133, 135, 137, 145, 205–207, 219, 232, 242, 243, 319, 325, 331

E
Equivalent control method, 17, 20–22, 27, 31, 153, 192, 231, 232

F
Filippov solution, 19
First-Order Sliding Mode (FOSM), 4, 5
Fuller phenomenon, 6, 7, 12

G
Generalized Form (GF), 115, 116, 119, 121, 129–134
Generalized solution, 26

H
Hamilton–Jacobi–Isaacs PDI, 292
Hamilton–Jacobi PDE, 149, 283
Higher-Order Sliding Mode (HOSM), 4, 14
Hilbert space-valued system, 8, 25
Homogeneity, 74, 83–89, 102, 110, 121, 124, 127, 133

I
Implicit Euler integration, 17, 28, 29
Inequality
- Agmon, 47, 278
- Cauchy–Schwartz, 89
- Poincaré, 46, 47, 89, 198, 269, 270
- Wirtinger, 47, 89, 270

Infinite-dimensional sliding mode, 24

L
LaSalle–Krasovskii invariance principle, 105, 106, 223, 248, 266, 271
$\mathcal{L}_2$-gain, 45, 169
Limit cycle, 12–14, 17, 33, 34, 36–43, 286, 298, 299, 303–306

Y. Orlov, *Nonsmooth Lyapunov Analysis in Finite and Infinite Dimensions*,
Communications and Control Engineering,
https://doi.org/10.1007/978-3-030-37625-3

Lyapunov equation, 113, 114, 129
Lyapunov function
- multiple, 95, 98, 113, 135
- non-strict, 95, 96, 104–107, 109, 116, 117
- semi-global, 95, 99, 116
- strict, 95, 96, 98–102, 104, 108, 109, 115, 120, 142, 145
Lyapunov–Krasovskii functional, 208, 220, 222, 248, 269, 270, 273, 276, 278
Lyapunov–Razumikhin function, 207, 220, 235

M
Mild solution, 81

N
Newton restitution rule, 10, 11

P
Poincaré–Andronov–Hopf bifurcation, 12, 33
Poincaré–Bendixon criterion, 36
Proximal super-/sub-gradient, 69, 73

S
Second-Order Sliding Mode (SOSM), 6
Sliding Mode (SM), 9, 26, 80
- First-Order (FOSM), 3, 4
- Higher-Order (HOSM), 3, 4
- Second-Order (SOSM), 6
- infinite-dimensional SM, 8, 24, 27, 28, 81
Sliding mode equation, 19
Solution
- Aizerman–Pyatnitskii, 22, 24
- Clarke, 71, 73
- Filippov, 17, 19, 20, 23, 24, 73, 75, 88, 192
- generalized, 26, 28, 80, 81, 191–193, 195, 196, 200
- mild, 65, 66, 75, 82, 89, 189
- proximal, 71, 73, 140–142, 145, 283
- strong, 25, 26, 28, 51, 70, 75, 79–81, 138
- Utkin, 21–24
- viscosity, 68–71, 89
Space
- Banach, 76, 83, 104
- Hilbert, 8, 25–27, 46, 53, 74, 76, 78–80, 82, 83, 89, 95, 105, 107–109, 137, 138, 208, 230, 255, 319, 331
- Sobolev, 45, 46, 89, 189, 317
Speed gradient method, 113, 312
Stability
- ISS, 95, 189
- finite time, 34, 74, 76, 77, 83, 89, 100, 102, 122–124, 127, 128, 134, 170, 171
Strong solution, 25
Supertwisting algorithm, 29, 115, 121, 162, 166

T
Twisting algorithm, 115, 116, 121, 163, 184

U
Utkin solution, 21

V
Van der Pol oscillator, 12–14, 32–38, 43, 286, 298, 299, 305, 306
Variable Structure System (VSS), 71
Viscosity solution, 68

Z
Zeno mode, 11, 14
Zhuravlev–Ivanov transformation, 3, 10

MIX
Papier aus verantwortungsvollen Quellen
Paper from responsible sources
FSC® C105338

If you have any concerns about our products,
you can contact us on
ProductSafety@springernature.com

In case Publisher is established outside the EU,
the EU authorized representative is:
Springer Nature Customer Service Center GmbH
Europaplatz 3, 69115 Heidelberg, Germany

Printed by Libri Plureos GmbH
in Hamburg, Germany